FUNDAMENTALS OF STRUCTURAL STABILITY

FUNDAMENTALS OF STRUCTURAL STABILITY

GEORGE J. SIMITSES

Professor Emeritus
Daniel Guggenheim School of Aerospace Engineering
Georgia Institute of Technology

DEWEY H. HODGES

Professor
Daniel Guggenheim School of Aerospace Engineering
Georgia Institute of Technology

AMSTERDAM • BOSTON • HEIDELBERG • LONDON
NEW YORK • OXFORD • PARIS • SAN DIEGO
SAN FRANCISCO • SINGAPORE • SYDNEY • TOKYO

Butterworth-Heinemann is an imprint of Elsevier

ELSEVIER

Butterworth–Heinemann is an imprint of Elsevier
30 Corporate Drive, Suite 400, Burlington, MA 01803, USA
Linacre House, Jordan Hill, Oxford OX2 8DP, UK

∞ Recognizing the importance of preserving what has been written, Elsevier prints its books on
acid-free paper whenever possible.

Library of Congress Cataloging-in-Publication Data
Simitses, George J., 1932-
  Fundamentals of structural stability / George J. Simitses, Dewey H. Hodges.
    p. cm.
  Includes bibliographical references and index.
  ISBN 0-7506-7875-5 (alk. paper)
  1. Structural stability—Textbooks. 2. Buckling (Mechanics)—Textbooks. I. Hodges, Dewey H. II. Title.
  TA656.S547 2005
  624.1′71—dc22   2005029831

British Library Cataloguing-in-Publication Data
A catalogue record for this book is available from the British Library.

ISBN 13: 978-0-7506-7875-9
ISBN 10: 0-7506-7875-5

For information on all Elsevier Butterworth–Heinemann publications
visit our Web site at www.books.elsevier.com

Printed in the United States of America
05 06 07 08 09 10   10 9 8 7 6 5 4 3 2 1

£60.95

624.171 SIM

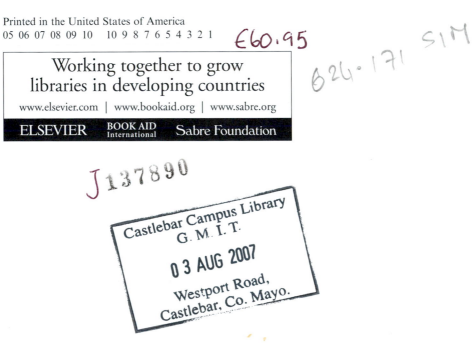

To our wives, Nena and Margaret

Contents

PREFACE

Knowledge of structural stability theory is of paramount importance to the practicing structural engineer. In many instances, buckling is the primary consideration in the design of various structural configurations. Because of this, formal courses in this important branch of mechanics are available to students in Aerospace Engineering, Civil Engineering, Engineering Science and Mechanics, and Mechanical Engineering at many institutions of higher learning. This book is intended to serve as a text in such courses. The emphasis of the book is on the fundamental concepts and on the methodology developed through the years to solve structural stability problems.

The material contained in this text is ideally suited for a two-semester Master's level course, although with judicious deletion of topics, the text may be adopted for a one-semester course.

The first chapter introduces the basic concepts of elastic stability and the approaches used in solving stability problems. It also discusses the different buckling phenomena that have been observed in nature. In Chapter 2, the basic concepts and methodology are applied to some simple mechanical models with finite degrees of freedom. This is done to help the student understand the fundamentals without getting involved with lengthy and complicated mathematical operations, which is usually the case when dealing with the continuum (infinitely many degrees of freedom). In Chapter 3, a complete treatment of the elastic stability of columns is presented, including effects of elastic restraints. New to this edition are treatments of the elastica theory of beams and of the buckling of thin-walled beam-columns. This new material facilitates the solutions of several problems in later chapters. Some simple frame problems are discussed in Chapter 4. Moreover, a nonlinear analysis of frames is presented, which clearly shows that in some cases, buckling occurs through limit-point instability. This chapter is of special importance to the Civil Engineering student. Since energy-based methods have been successfully used in structural mechanics, Chapter 5 presents a comprehensive treatment of the energy criterion for stability and contains many energy-related methods. The study of this chapter requires some knowledge of work- and energy-related principles and theorems.

These topics are presented in the Appendix for the benefit of the student who never had a formal course in this area. Columns on elastic foundations are discussed in Chapter 6. Chapter 7 presents a comprehensive treatment of the buckling of thin rings and high and low arches. In this chapter, a complete analysis is given for a shallow, pinned sinusoidal arch on an elastic foundation subject to a sinusoidal transverse loading. This is an interesting model for stability studies because, depending upon the values of the different parameters involved, it exhibits all types of buckling that have been observed in different structural systems: top-of-the-knee buckling, stable bifurcation (Euler-type), and unstable bifurcation. The use of elastica theory augments the more traditional treatment illustrating how a buckling analysis can be carried out with very few restrictive assumptions. Chapter 8 treats the buckling of shafts, making use of both the elastica theory and energy methods. This chapter is important for Mechanical and Aerospace Engineering students, showing that torques which differ by infinitesimal amounts can have buckling loads that radically differ, and that compressive forces and spin can affect stability as well. Chapter 9 is devoted to lateral-torsional buckling of deep beams, emphasizing the role of certain secondary effects such as the Vlasov phenomenon, initial curvature, the offset of the load, the way torque is applied, etc. In Chapter 10 we examine various instabilities of rotating rods and beams. Chapter 11 is devoted to the stability of nonconservative systems undergoing follower forces. An extended version of the elastica theory is shown to facilitate analysis of such systems, which must be analyzed according to kinetic theory. Chapter 12 classifies the various "dynamic instability" phenomena by taking into consideration the nature of the cause, the character of the response and the history of the problem. Moreover, the various concepts and methodologies, as developed and used by different investigators, are fully described. Finally, the concepts and criteria for dynamic stability are demonstrated through simple mechanical models. The emphasis here is on suddenly applied loads of constant magnitude and infinite duration or extremely small duration (ideal pulse).

The authors are indebted to the late Profs. J. N. Goodier and N. J. Hoff and to Prof. George Herrmann for introducing many topics and for valuable suggestions. Special thanks are due to Professor M. E. Raville for providing tangible and intangible support, for reading large sections of the manuscript for earlier editions, and for making many corrections. Numerous discussions with Profs. W. W. King, G. M. Rentzepis, C. V. Smith Jr., David A. Peters, M. Stallybrass, A. N. Kounadis and Izhak Sheinman are gratefully acknowledged. Thanks are also due to several former students of the first author: C. M. Blackmon, V. Ungbhakorn, J. Giri, A. S. Vlahinos, D. Shaw and J. G. Simitses; and of the second author: A. R. Atilgan, R. R. Bless, and V. V. Volovoi.

George J. Simitses
Dewey H. Hodges
Georgia Institute of Technology

FUNDAMENTALS OF STRUCTURAL STABILITY

1

INTRODUCTION AND FUNDAMENTALS

1.1 MOTIVATION

Many problems are associated with the design of modern structural systems. Economic factors, availability and properties of materials, interaction between the external loads (e.g. aerodynamic) and the response of the structure, dynamic and temperature effects, performance, cost, and ease of maintenance of the system are all problems which are closely associated with the synthesis of these large and complicated structures. Synthesis is the branch of engineering which deals with the design of a system for a given mission. Synthesis requires the most efficient manner of designing a system (i.e., most economical, most reliable, lightest, best, and most easily maintained system), and this leads to *optimization*. An important part of system optimization is structural optimization, which is based on the assumption that certain parameters affecting the system optimization are given (i.e., overall size and shape, performance, nonstructural weight, etc.). It can only be achieved through good theoretical analyses supported by well-planned and well-executed experimental investigations.

Structural analysis is that branch of structural mechanics which associates the behavior of a structure or structural elements with the action of external causes. Two important questions are usually asked in analyzing a structure: (1) What is the response of the structure when subjected to external causes (loads and temperature changes)? In other words, if the external causes are known, can we find the deformation patterns and the internal load distribution? (2) What is the character of the response? Here we are interested in knowing if the equilibrium is stable or if the motion is limited (in the case of dynamic causes). For example, if a load is periodically applied, will the structure oscillate within certain bounds or will it tend to move without bounds?

If the dynamic effects are negligibly small, in which case the loads are said to be applied quasistatically, then the study falls in the domain of structural *statics*. On the other hand, if the dynamic effects are not negligible, we are dealing with structural *dynamics*.

The branch of structural statics that deals with the character of the response is called stability or instability of structures. The interest here lies in the fact that stability criteria are often associated directly with the load-carrying capability of the structure. For example, in some cases instability is not directly associated with the failure of the overall system, i.e., if the skin wrinkles, this does not mean that the entire fuselage or wing will fail. In other cases though, if the portion of the fuselage between two adjacent rings becomes unstable, the entire fuselage will fail catastrophically. Thus, stability of structures or structural elements is an important phase of structural analysis, and consequently it affects structural synthesis and optimization.

1.2   STABILITY OR INSTABILITY OF STRUCTURES

There are many ways a structure or a structural element can become unstable, depending on the structural geometry and the load characteristics. The spatial geometry, the material along with its distribution and properties, the character of the connections (riveted joints, welded, etc.), and the supports comprise the structural geometry. By load characteristics we mean spatial distribution of the load, load behavior (whether or not the load is affected by the deformation of the structure, e.g., if a ring is subjected to uniform radial pressure, does the load remain parallel to its initial direction, does it remain normal to the deformed ring, or does it remain directed towards the initial center of curvature?), and/or whether the force system is conservative.

1.2.1   CONSERVATIVE FORCE FIELD

A mechanical system is conservative if subjected to conservative forces. If the mechanical system is rigid, there are only external forces; if the system is deformable, the forces may be both external and internal. Regardless of the composition, a system is conservative if all the forces are conservative. A force acting on a mass particle is said to be conservative if the work done by the force in displacing the particle from position 1 to position 2 is independent of the path. In such a case, the force may be derived from a potential. A rigorous mathematical treatment is given below for the interested student.

The work done by a force $\mathbf{F}$ acting on a mass particle in moving the particle from position P_0 (at time t_0) to position P_1 (at time t_1) is given by

$$W = \oint_C \int_{\mathbf{r}_0}^{\mathbf{r}_1} \mathbf{F} \cdot d\mathbf{r} \tag{1}$$

Thus the integral, W (a scalar), depends on the initial position, $\mathbf{r}_0$, the final position, $\mathbf{r}_1$, and the path C. If a knowledge of the path C is not needed and the work is a function of the initial and final positions only, then

$$W = W(\mathbf{r}_0, \mathbf{r}_1, \mathbf{F}) \tag{2}$$

and the force field is called *conservative*. (Lanczos 1960; Langhaar 1962; Whittaker 1944).

Parenthesis.   If S denotes some surface in the space and C some space curve, then by *Stokes' theorem*

$$\oint_C \mathbf{U} \cdot d\ell = \iint_S \text{curl } \mathbf{U} \cdot \mathbf{n} ds \tag{3}$$

where $\mathbf{n}$ is a unit vector normal to the surface S (see Fig. 1.1), and $\mathbf{U}$ some vector quantity.

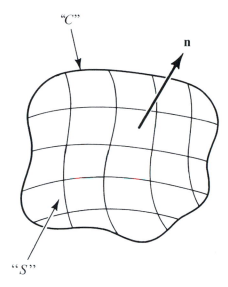

FIGURE 1.1 "S"

If $\oint_c \mathbf{U} \cdot d\ell = 0$, then

$$\iint_S \text{curl}\, \mathbf{U} \cdot \mathbf{n}\, ds = 0 \qquad (4)$$

for all surfaces S and spanning curves C. If this is so, then the curl of $\mathbf{U}$ (some vector quantity) must be identically equal to zero, or

$$\text{curl}\, \mathbf{U} \equiv 0 \qquad (5)$$

Next, if we apply this result to a conservative force field where $\mathbf{U}$ is replaced by $\mathbf{F}$, then according to the previous result

$$\text{curl}\, \mathbf{F} \equiv 0$$

It is well known from *vector analysis* that the curl of the gradient of any scalar function vanishes identically. Therefore, for a conservative field we may write

$$\mathbf{F} = -\nabla V \qquad (6)$$

where:

1. The negative sign is arbitrary,
2. V is some scalar function, and
3. ∇ is the vector operator

$$\frac{\partial}{\partial x}\mathbf{i} + \frac{\partial}{\partial y}\mathbf{j} + \frac{\partial}{\partial z}\mathbf{k}$$

where $\mathbf{i}$, $\mathbf{j}$, $\mathbf{k}$ form an orthogonal unit vector triad along x, y, z, respectively. This implies that the force can be derived from a potential.

Note that in this case the work done by the force in a conservative force field is given by

$$W = \oint_{r_0}^{r_1} \mathbf{F} \cdot d\mathbf{r} = -\oint_{r_0}^{r_1} \mathbf{\nabla} V \cdot d\mathbf{r} = -\oint_{r_0}^{r_1} (\mathbf{\nabla} \cdot d\mathbf{r}) V$$

and since

$$\mathbf{\nabla} = \frac{\partial}{\partial x} \mathbf{i} + \frac{\partial}{\partial y} \mathbf{j} + \frac{\partial}{\partial z} \mathbf{k} \text{ and } d\mathbf{r} = (dx) \mathbf{i} + (dy) \mathbf{j} + (dz) \mathbf{k}$$

then

$$(\mathbf{\nabla} \cdot d\mathbf{r}) V = \frac{\partial V}{\partial x} dx + \frac{\partial V}{\partial y} dy + \frac{\partial V}{\partial z} dz = dV$$

or

$$W = -\int_{V_0}^{V_1} dV = V_0 - V_1 = -\delta(V) \tag{7}$$

where δ denotes a change in the potential of the conservative force $\mathbf{F}$ from position $\mathbf{r}_0$ to position $\mathbf{r}_1$.

Thus a system is conservative if the work done by the forces in displacing the system from deformation state 1 to deformation state 2 is independent of the path. If this is the case, the force can be derived from a potential.

There are many instances where systems are subjected to loads which cannot be derived from a potential. For instance, consider a column clamped at one end and subjected to an axial load at the other, the direction of which is tangential to the free end at all times (follower force). Such a system is nonconservative and can easily be deduced if we consider two or more possible paths that the load can follow in order to reach a final position. In each case the work done will be different. Systems subject to time-dependent loads are also nonconservative. Nonconservative systems have been given special consideration (Bolotin, 1963; Hermann, 1967), and the emphasis in this text will be placed on conservative systems Ziegler (1968) has a detailed description of forces and systems.

1.2.2   THE CONCEPT OF STABILITY

As the external causes are applied quasistatically, the elastic structure deforms and static equilibrium is maintained. If now at any level of the external causes, "small" external disturbances are applied, and the structure reacts by simply performing oscillations about the deformed equilibrium state, the equilibrium is said to be *stable*. The disturbances can be in the form of deformations or velocities, and by "small" we mean as small as desired. As a result of this latter definition, it would be more appropriate to say that the equilibrium is stable in the small. In addition, when the disturbances are applied, the level of the external causes is kept constant. On the other hand, if the elastic structure either tends to and does remain in the disturbed position or tends to and/or diverges from the deformed equilibrium state, the equilibrium is said to be *unstable*. Some authors prefer to distinguish these two conditions and call the equilibrium *neutral* for the former case and *unstable* for the latter. When either of these two cases occurs, the level of the external causes is called *critical*.

This can best be demonstrated by the system shown in Fig. 1.2. This system consists of a ball of weight W resting at different points on a surface with zero curvature normal to the plane of the figure. Points of zero slope on the surface denote positions of static equilibrium (points A, B, and C). Furthermore, the character of

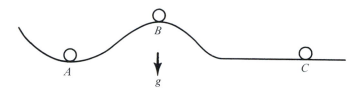

FIGURE 1.2   Character of static equilibrium positions.

equilibrium at these points is substantially different. At *A*, if the system is disturbed through infinitesimal disturbances (small displacements or small velocities), it will simply oscillate about the static equilibrium position *A*. Such equilibrium position is called *stable* in the small. At point *B*, if the system is disturbed, it will tend to move away from the static equilibrium position *B*. Such an equilibrium position is called *unstable* in the small. Finally, at point *C*, if the system is disturbed, it will tend to remain in the disturbed position. Such an equilibrium position is called *neutrally stable* or *indifferent* in the small. The expression "in the small" is used because the definition depends on the *small* size of the perturbations. If the disturbances are allowed to be of finite magnitude, then it is possible for a system to be unstable in the small but stable in the large (point *B*, Fig. 1.3a) or stable in the small but unstable in the large (point *A*, Fig. 1.3b).

In most structures or structural elements, loss of stability is associated with the tendency of the configuration to pass from one deformation pattern to another. For instance, a long, slender column loaded axially, at the critical condition, passes from the straight configurations (pure compression) to the combined compression and bending state. Similarly, a perfect, complete, thin, spherical shell under external hydrostatic pressure, at the critical condition, passes from a pure membrane state (uniform radial displacement only; shell stretching) to a combined stretching and bending state (nonuniform radial displacements). This characteristic has been recognized for many years and it was first used to solve stability problems of elastic structures. It allows the analyst to reduce the problem to an eigenvalue problem, and many names have been given to this approach: the classical method, the bifurcation method, the equilibrium method, and the static method.

1.2.3   CRITICAL LOADS VERSUS BUCKLING LOAD

At this point nomenclature merits some attention. There is a definite difference in principle between the buckling load observed in a loading process where the loads

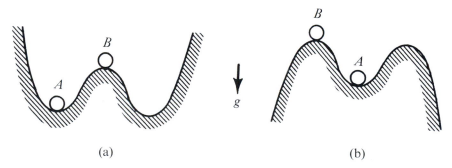

(a)   (b)

FIGURE 1.3   Character of static equilibrium positions in the large.

keep changing (observed physical phenomenon) and the buckling load calculated from some mathematical model, which always refers to a system with prescribed loads. Since the latter is based on theory and is usually obtained as the characteristic or eigenvalue of some eigen-boundary-value problem, it is properly called the critical load.

In the process of buckling in the testing machine, in the static or dynamic testing of a structural configuration, and in the failure of the structure in actual use, we are confronted with the physical aspects of buckling. The load at which a structure buckles should preferably be designated as the buckling load.

The compound term *critical buckling load* is unnecessary and should be avoided. It may have originated from the observation that theory (for the ideal column, for instance) predicts several critical loads (eigenvalues) corresponding to different deflection patterns (eigenfunctions). In an experiment, however, only one buckling pattern is observed, namely, the one that corresponds to the lowest eigenvalue. This lowest eigenvalue is no more critical than any of the higher ones, but it is the one that corresponds to the observed buckling load. If it is desired to give it a special designation, it should be called the lowest critical load, rather than the critical buckling load.

1.2.4   BASIC APPROACHES OF STABILITY ANALYSIS

A number of approaches have been successfully used in determining critical conditions for elastic structures which are subject to instability. The oldest approach, while is applicable to many problems, is concerned with the answer to the following question. If an external cause is applied quasistatically to an elastic structure, is there a level of the external cause at which two or more different but infinitesimally close equilibrium states can exist? By different equilibrium states we mean that the response of the structure is such that equilibrium can be maintained with different deformation patterns. An example of this is the long perfect column loaded axially in compression. As the load increases quasistatically from zero, the column is compressed but remains straight. At some value of the load though, a bent position of infinitesimal amplitude also represents an equilibrium position. Since at this value of the load there are two different equilibrium states infinitesimally close, a bifurcation point exists (adjacent equilibrium positions). Mathematically, in this approach, the problem is reduced to an eigen-boundary-value problem, and the critical conditions are denoted by the eigenvalues. This approach is usually referred to as the *classical* approach, *equilibrium* approach, or *bifurcation* approach. Many examples will be discussed in the chapters to follow.

Another approach is to write the equations governing small free vibrations of the elastic structures at some level of the external causes (treated as a constant) and try to find out for what level of the external cause the motion ceases to be bounded in the small. In writing the governing equations, one must allow all possible modes of deformation. The form of equilibrium is said to be stable if a slight disturbance (in the form of displacement or velocity) causes a small deviation of the system from the considered equilibrium configuration, but by decreasing the magnitude of the disturbance, the deviation can be made as small as required. On the other hand, a critical condition is reached if a disturbance, however small, causes a finite deviation of the system from the considered form of equilibrium. This approach is known as the *kinetic* or *dynamic* approach, and it is a direct application of the stability concept demonstrated in Fig. 1.2.

Next, if a system is conservative, the forces can be derived from a potential, and the total potential of the system can be expressed in terms of the generalized coordinates and the external forces. The generalized coordinates are the parameters needed to express the deflectional shapes which the elastic structure could possibly assume. In this case, the equilibrium is stable in the small if the total potential is a relative minimum. This approach is completely equivalent to the kinetic approach (a proof is given in Whittaker, 1944) for conservative systems, and it is known as the *potential energy* method or simply the *energy* method. This definition of stability requires special attention, and it will be fully justified in the next section.

Finally, there is a fourth approach in dealing with stability problems of elastic structures. This method is usually called the *imperfection* method. The question in this case is: "What is the value of the load (level of external causes) for which the deflections of an imperfect system increase beyond any limit?" It should be pointed out that certain systems, when subjected to certain external causes, are imperfection sensitive. This means that the critical conditions of the perfect system are different from those of the imperfect one. Imperfection sensitivity has served to explain the discrepancy between theory and experiment for such systems. It will also be demonstrated that there are systems for which the perfect and imperfect systems have the same critical conditions according to the approaches defined above. It is the opinion of the authors that the imperfection approach should not be associated with the stability of the perfect system, but simply characterize the response of the imperfect system. In short, the stability of a system, whether perfect or imperfect, should be investigated by the first three methods (whichever is applicable).

1.2.5 THE ENERGY METHOD

This method is based on the kinetic criterion of stability, and it is an association of this criterion with characteristics of the total potential (relative minimum) surface at a position of static equilibrium. Since it requires the existence of a total potential surface, this method is applicable only to conservative systems.

Before the energy criterion is justified, let us describe in analytical form the kinetic criterion of stability. This concept was first introduced by Lagrange (1788) for a system with a finite number of degrees of freedom.

A more strict definition of stability of equilibrium was given by Lyapunov (see Chetayev, 1961; Krasoskii, 1963; Langhaar, 1962; LaSalle and Lefschetz, 1961; Liapunor, 1952) as a particular case of motion. Let us assume that the position of a system depends on n generalized coordinates q_i $(i = 1, 2, \ldots, n)$ and that a static equilibrium state is characterized by $q_i = 0$. Let the system be at this static equilibrium position, and at time $t = 0$ we allow small bounded disturbances $|q_i^0| < \delta$ and $|\dot{q}_i^0| < \delta$. The response of the system at any instant $t > 0$ is characterized by $q_i(t)$ and $\dot{q}_i(t)$. If the response is also bounded

$$|q_i(t)| < \varepsilon \quad \text{and} \quad |\dot{q}_i(t)| < \varepsilon \tag{8}$$

then we say that the static equilibrium position $q_i = 0$ is stable. In other words, in the case of stable static equilibrium (in the small) positions, we can always select such small initial conditions that the generalized coordinates and velocities are bounded.

The energy criterion is based on the Lagrange-Dirichlet theorem, which states: If the total potential has a relative minimum at an equilibrium position (stationary value), then the equilibrium position is stable. This theorem can easily be proven if we simply employ the principle of conservation of energy for conservative systems, which states that the sum of the kinetic energy and the total potential is a constant $(T + U = c)$. Now if we define the equilibrium position by $q_i = 0$ and let $U(0) = 0$, then, if $U(0)$ is a minimum, $U(q_i)$ must have a positive lower bound $\bar{c}$ on the boundary of any sufficiently close neighborhood of $q_i = 0$. It is now always possible to select q_i^0 and $\dot{q}_i^0$ such that $T + U = c$ and $c < \bar{c}$. In other words, since the sum of the total potential and the nonnegative kinetic energy is a constant c, if $c < \bar{c}$ the boundary of the neighborhood of $q_i = 0$ can never be reached, and the equilibrium position $q_i = 0$ is stable (bounded motion). Unfortunately, it is very difficult to prove the converse of the Lagrange-Dirichlet theorem. A statement of this converse theorem is as follows: If the equilibrium is stable at an equilibrium position characterized by $q_i = 0$, then $U(0)$ is a relative minimum. Proof of this theorem under certain restrictive assumptions has been given by Chetayev (1930). Although there is no general proof of this converse theorem, its validity has been accepted and the energy criterion has been used as both a necessary and sufficient condition for stability. This criterion for stability can be generalized for systems with infinitely many degrees of freedom (cohesive, continuous, deformable configurations).

The energy criterion can be used to arrive at critical conditions by simply seeking load conditions at which the response of the system ceases to be in stable equilibrium. This implies that we are interested in knowing explicitly the conditions under which the change in the total potential is positive definite. If the total potential is expressed as a Taylor series about the static equilibrium point characterized by $q_i = 0$, then

$$U(q_1, q_2, \ldots, q_N) = U(0, 0, \ldots 0) + \sum_{i=1}^{N} \frac{\partial U}{\partial q_i}\Big|_0 q_i$$
$$+ \frac{1}{2} \sum_{i=1}^{N} \sum_{j=1}^{N} \frac{\partial^2 U}{\partial q_i \partial q_j}\Big|_0 q_i q_j + \ldots \tag{9}$$

Since $q_i = 0$ characterizes a position of static equilibrium, then

$$\frac{\partial U}{\partial q_i}\Big|_0 = 0 \tag{10}$$

and

$$U(q_1, q_2, \ldots, q_N) - U(0, 0, \ldots, 0) = \Delta U = \frac{1}{2} \sum_{i=1}^{N} \sum_{j=1}^{N} c_{ij} q_i q_j \tag{11}$$

where

$$c_{ij} = \frac{\partial^2 U}{\partial q_i \, \partial q_j}\Big|_0$$

The energy criterion requires that the homogeneous quadratic form given by Eq. (11) be positive definite.

THEOREM   The homogeneous quadratic form

$$U(q_1, q_2, \ldots, q_N) = \frac{1}{2} \sum_{i=1}^{N} \sum_{j=1}^{N} c_{ij} q_i q_j \tag{12}$$

is positive definite if and only if the determinant D of its coefficients, c_{ij}, and its principal minors, D_i, are all positive.

$$
\begin{array}{cccc}
D_1 & D_2 & D_3 & \cdots
\end{array}
$$

$$
\left|
\begin{array}{cccccc}
c_{11} & c_{12} & c_{13} & c_{14} & \cdots & c_{1N} \\
c_{21} & c_{22} & c_{23} & c_{24} & \cdots & c_{2N} \\
c_{31} & c_{32} & c_{33} & c_{34} & \cdots & c_{3N} \\
\vdots & \vdots & \vdots & \vdots & & \vdots \\
c_{N1} & c_{N2} & c_{N3} & c_{N4} & \cdots & c_{NN}
\end{array}
\right| > 0
\tag{13}
$$

Proof: The proof will be given in a number of steps.

1. If U is positive for any set of coordinates $[q_i] \neq [0]$ (not all zero), then

$$
U(q_1, 0, 0, 0, \ldots, 0) = \frac{1}{2} c_{11} q_1^2 > 0
\tag{14}
$$

which requires that $c_{11} > 0$. Note that if c_{11} is positive, then $U(q_1, 0, 0, \ldots, 0) > 0$.

2. Assuming that $c_{11} \neq 0$, we can make the following transformation:

$$
\begin{aligned}
q_1^* &= q_1 + \frac{c_{12}}{c_{11}} q_2 + \frac{c_{13}}{c_{11}} q_3 + \cdots + \frac{c_{1N}}{c_{11}} q_N \\
&= q_1 + \sum_{i=2}^{N} \frac{c_{1i}}{c_{11}} q_i
\end{aligned}
\tag{15}
$$

With this transformation we note that

$$
\begin{aligned}
\frac{1}{2} c_{11} \left(q_1^*\right)^2 &= \frac{1}{2} c_{11} \left(q_1 + \sum_{i=2}^{N} \frac{c_{1i}}{c_{11}} q_i\right)^2 \\
&= \frac{1}{2} c_{11} q_1^2 + q_1 \sum_{i=2}^{N} c_{1i} q_i + \frac{1}{2} \sum_{i=2}^{N} \sum_{j=2}^{N} \frac{c_{1i} c_{1j}}{c_{11}} q_i q_j
\end{aligned}
\tag{16}
$$

From Eq. (16)

$$
\frac{1}{2} c_{11} q_1^2 = \frac{1}{2} c_{11} \left(q_1^*\right)^2 - \left(q_1 \sum_{i=2}^{N} c_{1i} q_i + \frac{1}{2} \sum_{i=2}^{N} \sum_{j=2}^{N} \frac{c_{1i} c_{1j}}{c_{11}} q_i q_j\right)
\tag{17}
$$

Next we rewrite Eq. (12) in the following form:

$$
U(q_1, q_2, \ldots, q_N) = \frac{1}{2} c_{11} q_1^2 + q_1 \sum_{i=2}^{N} c_{1i} q_i + \frac{1}{2} \sum_{i=2}^{N} \sum_{j=2}^{N} c_{ij} q_i q_j
\tag{18}
$$

Substitution of Eq. (17) into Eq. (18) yields

$$
U\left(q_1^*, q_2, \ldots, q_N\right) = \frac{1}{2} c_{11} q_1^{*2} + \frac{1}{2} \sum_{i=2}^{N} \sum_{j=2}^{N} \left(c_{ij} - \frac{c_{1i} c_{1j}}{c_{11}}\right) q_i q_j
\tag{19}
$$

If we let

$$
c_{ij} - \frac{c_{1i} c_{1j}}{c_{11}} = \alpha_{ij}
\tag{20}
$$

then Eq. (19) becomes

$$U(q_1^*, q_2, \ldots, q_N) = \frac{1}{2}c_{11}q_1^{*2} + \frac{1}{2}\sum_{i=2}^{N}\sum_{j=2}^{N}\alpha_{ij}q_iq_j \tag{21}$$

3. If U is positive for $q_1^* \neq 0$ and $q_i = 0$ ($i = 2, 3, \ldots, N$), then $c_{11} > 0$. If U is positive for $q_1^* = 0$, $q_2 \neq 0$, and $q_i = 0 (i = 3, \ldots, N)$, then $\alpha_{22} > 0$. Note that the converse is also true for the same condition, i.e., if c_{11} is positive, U is positive, and if α_{22} is positive, U is positive.

These conditions for positive U can be written solely in terms of c_{ij} by use of Eq. (20), or

$$c_{11} > 0 \text{ and } c_{11}c_{22} - c_{12}^2 > 0 \tag{22}$$

Note that the second inequality is equivalent to the requirement $D_2 > 0$ if $c_{12} = c_{21}$. This requirement is by no means restrictive since Eq. (12) represents a homogeneous quadratic form.

4. Next step 2 is repeated with $c_{22} \neq 0$ and the following transformation:

$$q_2^* = q_2 + \sum_{i=3}^{N}\frac{\alpha_{2i}}{\alpha_{22}}q_i \tag{23}$$

This transformation leads to the following expression for U:

$$U(q_1^*, q_2^*, q_3, \ldots, q_N) = \frac{1}{2}c_{11}q_1^{*2} + \frac{1}{2}\alpha_{22}q_2^{*2} + \frac{1}{2}\sum_{i=3}^{N}\sum_{j=3}^{N}\beta_{ij}q_iq_j \tag{24}$$

where

$$\beta_{ij} = \alpha_{ij} - \frac{\alpha_{2i}\alpha_{2j}}{\alpha_{22}} \tag{25}$$

As in step 3

$$U(q_1^*, 0, 0, \ldots, 0) > 0 \quad \text{if and only if} \quad c_{11} > 0$$
$$U(0, q_2^*, 0, 0, \ldots, 0) > 0 \quad \text{if and only if} \quad \alpha_{22} > 0$$

and

$$U(0, 0, q_3, 0, \ldots, 0) > 0 \quad \text{if and only if} \quad \beta_{33} > 0$$

This requirement implies that

$$\alpha_{22}\alpha_{33} - \alpha_{23}^2 > 0 \tag{26}$$

By Eq. (20)

$$\left(c_{22} - \frac{c_{12}^2}{c_{11}}\right)\left(c_{33} - \frac{c_{13}^2}{c_{11}}\right) - \left(c_{23} - \frac{c_{12}c_{13}}{c_{11}}\right)^2 > 0 \tag{27}$$

This last requirement is equivalent to $D_3 > 0$ provided $c_{ij} = c_{ji}$.

5. The continuation of this procedure eventually leads to the representation of the homogeneous quadratic form as a linear combination of squares:

$$U = \frac{1}{2}c_{11}q_1^{*2} + \frac{1}{2}\alpha_{22}q_2^{*2} + \frac{1}{2}\beta_{33}q_3^{*2} + \cdots \tag{28}$$

From this form, it is clearly seen that U is positive definite if and only if

$$c_{11} > 0, \quad \alpha_{22} > 0, \quad \beta_{33} > 0 \quad \text{QED} \tag{29}$$

Use of this theorem in the energy criterion implies that a position of static equilibrium is stable if and only if

$$D_N = \begin{vmatrix} \dfrac{\partial^2 U_T}{\partial q_1^2} & \dfrac{\partial^2 U_T}{\partial q_1 \, \partial q_2} & \dfrac{\partial^2 U_T}{\partial q_1 \, \partial q_3} & \cdots & \dfrac{\partial^2 U_T}{\partial q_1 \, \partial q_N} \\[2mm] \dfrac{\partial^2 U_T}{\partial q_2 \, \partial q_1} & \dfrac{\partial^2 U_T}{\partial q_2^2} & \dfrac{\partial^2 U_T}{\partial q_2 \, \partial q_3} & \cdots & \dfrac{\partial^2 U_T}{\partial q_2 \, \partial q_N} \\[2mm] \dfrac{\partial^2 U_T}{\partial q_3 \, \partial q_1} & \dfrac{\partial^2 U_T}{\partial q_3 \, \partial q_2} & \dfrac{\partial^2 U_T}{\partial q_3^2} & \cdots & \dfrac{\partial^2 U_T}{\partial q_3 \, \partial q_N} \\[2mm] \vdots & \vdots & \vdots & & \vdots \\[2mm] \dfrac{\partial^2 U_T}{\partial q_N \, \partial q_1} & \dfrac{\partial^2 U_T}{\partial q_N \, \partial q_2} & \dfrac{\partial^2 U_T}{\partial q_N \, \partial q_3} & \cdots & \dfrac{\partial^2 U_T}{\partial q_N^2} \end{vmatrix} > 0 \tag{30}$$

and all its principal minors $D_1 > 0$, $D_2 > 0$, etc.

In all problems in mechanics, dealing with the stability of elastic systems under external causes, the total potential of the system depends not only on the generalized coordinates (variables defining the position of the system) but also on certain parameters that characterize the external cause or causes.

The general theory of equilibrium positions of such systems with various values of the parameters was established by Poincaré (1885; see also Chetayev, 1930). Among the findings of Poincaré are the following (simplified in this text for the sake of understanding):

1. The requirements

$$\frac{\partial U_T}{\partial q_i} = 0 \text{ and } D_N = 0$$

define a point of bifurcation (intersection of static equilibrium branches at the same value of the external cause parameter). See for example Figs. 1.4 and 1.5 (points A and A').
2. Changes in stability along the primary path (from stable to unstable equilibrium positions) do occur at points of bifurcation. Consider, for example, branch OAB of Fig. 1.4. If the part of this branch characterized by OA denotes stable static equilibrium positions, the part characterized by AB must denote unstable static equilibrium positions.

These findings support the classical approach to stability problems which only seeks bifurcation points. The external cause condition at such a point is called a *critical condition*.

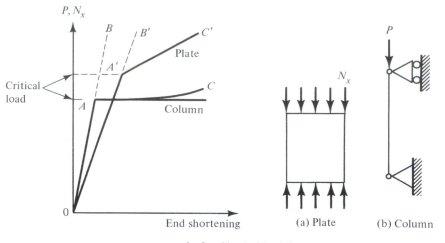

FIGURE 1.4   Classical buckling.

1.2.6   TYPES OF BUCKLING

When the external causes are applied quasistatically and the level at which instability occurs is reached, the elastic structure assumes an equilibrium configuration which is distinctly different from the ones assumed during the quasistatic application of the causes. When this occurs, we say that the elastic structure has *buckled*. Since there are different ways by which the new equilibrium configuration may be reached, buckling can be classified by the use of proper adjectives.

The type of buckling that was first studied and has been given the most attention is the so-called *classical* or *bifurcation buckling*. This type of buckling is characterized by the fact that, as the load passes through its critical stage, the structure passes from its unbuckled equilibrium configuration to an infinitesimally close buckled equilibrium configuration. As will be demonstrated in later chapters, buckling of long straight columns loaded axially, buckling of thin plates loaded by inplane loads, and buckling of rings are classical examples of this kind of buckling (see Fig. 1.4).

Another type of buckling is what Libove (Flügge, 1962) calls *finite-disturbance buckling*. For some structures, the loss of stiffness after buckling is so great that the buckled equilibrium configuration can only be maintained by returning to an earlier level of loading. Classical examples of this type are buckling of thin cylindrical shells under axial compression and buckling of complete, spherical, thin shells under uniform external pressure (see Fig. 1.5). In Fig. 1.5a, N_x denotes the applied axial load per unit length. In Fig. 1.5b, q denotes the uniform external pressure, V_0 the initial volume of the sphere, and ΔV the change in the volume during loading. The reason for the name is that in such structures a finite disturbance during the quasistatic application of the load can force the structure to pass from an unbuckled equilibrium configuration to a nonadjacent buckled equilibrium configuration before the classical buckling load, P_{cr}, is reached. A third type of buckling is known as *snapthrough* buckling or oil-canning (Durchschlag). This phenomenon is characterized by a visible and sudden jump from one equilibrium configuration to another equilibrium configuration for which displacements are larger than in the first (nonadjacent equilibrium states). Classical examples of this type are snapping of a low

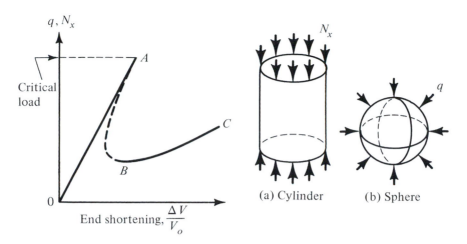

FIGURE 1.5   Finite-disturbance buckling.

pinned arch under lateral loads (see Fig. 1.6) and snapping of clamped shallow spherical caps under uniform lateral pressure.

The above discussion shows that there is some similarity between finite-disturbance buckling and snapthrough buckling. It should also be mentioned that, for many systems, nonlinear theory must be used to either evaluate critical conditions and/or explain the buckling phenomena.

It will become evident in subsequent chapters that there are two different viewpoints as far as types of buckling are concerned and two classifications within each viewpoint. The first viewpoint is based on the existence of a bifurcation point. For the examples shown in Figs. 1.4 and 1.5, there is a bifurcation point (A or A'). For the example shown in Fig. 1.6, there is no bifurcation point at A.

The second viewpoint is based on the expected response of the system under deadweight loading. For the examples in Fig. 1.4, the branches AC and $A'C'$ correspond to stable static equilibrium positions, and under deadweight loading there exists the possibility for the system to pass from one deformation configuration (the straight for the column) to another deformation configuration (the bent or buckled) with no appreciable dynamic effects (time-independent response). For the example shown in Fig. 1.5, since the branch AB is unstable when the system reaches point A (under deadweight loading), it will tend to snap through toward a far stable equilibrium position with a time-dependent response. This is very much the same situation for the system of Fig. 1.6. When point A is reached, the system will snap through toward a far stable equilibrium position.

In a later chapter, a model is considered which exhibits all types of buckling, top-of-the-knee (Fig. 1.6), stable bifurcation (Fig. 1.4), and unstable bifurcation (Fig. 1.5). This model is a low half-sine arch simply supported at both ends under quasistatic application of a half-sine transverse loading resting on an elastic foundation.

In investigating stability problems, one should always consider the effect of load behavior. In the case of a circular ring loaded uniformly by a radial pressure, different critical conditions are obtained depending on the behavior of the applied load. If the load behaves as hydrostatic pressure does (remains normal to the deflected shape), the critical condition is different from the case for which the load

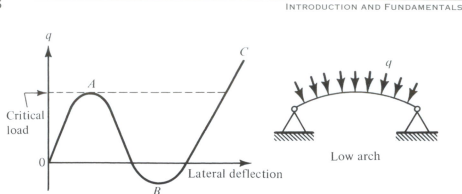

FIGURE 1.6   Snapthrough buckling.

remains directed toward the center of the ring at all times (point sink). On the other hand, the effect of load behavior for certain structures is negligible.

For some systems there are certain constraints on the loading mechanism. This case can also come under the classifications of load behavior problems. For instance, suppose that the axial load on a long, straight, elastic bar is applied through a rigid bar. At buckling, the loading member (rigid bar) may tilt, and then the load behavior is such that it aggravates the situation. Such problems are known as tilt-buckling problems, and they will be discussed in later chapters also.

So far, the different types of buckling of conservative systems under quasistatic application of the external causes have been discussed. We will also discuss buckling of a limited number of nonconservative systems (such as the "follower-force" problem) under quasistatic application of the load. Finally, buckling can occur under dynamic application of the external causes. In general, if the load depends explicitly on time, the system is nonconservative. A large class of such problems, where the load varies sinusoidally with time, have been discussed by Bolotin (1964). This type of dynamic loading is called by Bolotin *parametric*, and the associated phenomenon of loss of stability, *parametric excitation*. Bolotin has shown that all systems which are subject to loss of stability under quasistatically applied loads are also subject to loss of stability under parametric loads.

If the dynamic load does not depend explicitly on time, the system can be conservative (see Ziegler, 1968). Typical examples of such loads are (1) loads suddenly applied with constant magnitude and infinite duration and (2) ideal impulsive loads. When such loads are applied to elastic structures we may ask: "Is buckling possible under such loads, and if so, what are the critical conditions?" We may note that such loads are obvious idealizations of two extreme cases of blast loading: blasts of low decay rates and large decay times and blasts of large decay rates and short decay times, respectively (see Chapter 12).

1.3   CONTINUOUS DEFORMABLE ELASTIC BODIES

A continuous body is called *deformable* if the relative distance between any two material points changes when the system is experiencing changes in the externally applied causes. The changes in the deformations and their gradients are related to the changes in the load intensities and their rates through the constitutive relations.

If the loading path, characterized by the constitutive relations, is the same as the unloading path, the continuous deformable body is called *elastic*. If, in addition, these paths are characterized by linear relations in the absence of dynamic effects (generalized Hooke's law equations), the continuous deformable body is called *linearly elastic*. Furthermore, if the properties (modulus of elasticity, Poisson's ratio, etc.) of such a body do not depend on the position of the material point, the body is termed *homogeneous*. If these properties at a material point are independent of direction, the body is called *isotropic*. If this is not so, the body is called *anisotropic*. A particular case of anisotropic elastic bodies is orthotropic elastic bodies. An elastic body is called *orthotropic* if some or all of the properties of the elastic body differ in mutually orthogonal directions.

The branch of mechanics that deals with the behavior of elastic bodies is called theory of elasticity. Since only a relatively small number of problems can be solved by means of the exact field equations of the theory of elasticity, the structural engineer is forced to make a number of simplifying assumptions in dealing with structural problems. These simplifying assumptions depend heavily on the relative dimensions of the structural element in three-dimensional space. Depending on these assumptions, all of the structural elements fall in one of the following four categories:

1. All three dimensions are of the same magnitude (spheres, short or moderate length cylinders, etc.).
2. One of the dimensions is much larger than the other two, which are of the same order of magnitude (columns, thin beams, shafts, rings, etc.).
3. One of the dimensions is much smaller than the other two, which are of the same order of magnitude (thin plates, thin shells).
4. All three dimensions are of different order of magnitude (thin-walled, open-section beams).

Structural elements of the first category are not subject to instability. All other elements are. Typical stability problems associated with elements of the second category will be discussed in subsequent chapters.

1.4   BRIEF HISTORICAL SKETCH

Structural elements that are subject to instability have been used for many centuries. Although their use is ancient, the first theoretical analysis of one such structural element (long column) was performed only a little over two hundred years ago. This first theoretical analysis is due to Leonhard Euler. Other men of the 18th and 19th centuries who are associated with theoretical and experimental investigations of stability problems are Bresse, G.H. Bryan, Considére, Fr. Engesser, W. Fairbairn, A.G. Greenhill, F. Jasinski, Lagrange, and M. Lévy. An excellent historical review is given by Hoff (1954) and Timoshenko (1953).

Bryan's work (1888, 1891) merits special attention because of his mathematical rigor and the novelty of the problems treated.

Among the existing texts on the subject, in addition to those cited so far, we should mention the books of Biezeno and Grammel (1956), Bleich (1952), Hoff (1956), Leipholz (1970), and Timoshenko and Gere (1961).

Significant contributions to the understanding of the concept of stability are those of Koiter (1945 and 1963), Pearson (1956), Thompson (1963), and Trefftz (1933).

REFERENCES

Biezeno, C. B. and Grammel, R. (1956). *Engineering Dynamics.* Vol. II, Part IV, Blackie and Son Ltd., London.

Bleich, F. (1952). *Buckling Strength of Metal Structures.* McGraw-Hill Book Co., New York.

Bolotin, V. V. (1963). *Nonconservative Problems of the Theory of Elastic Stability*, edited by G. Herrmann (translated from the Russian). The Macmillan Co., New York.

Bolotin, V. V. (1964). *The Dynamic Stability of Elastic Systems.* Holden-Day, Inc., San Francisco.

Bryan, G. H. (1888). On the stability of elastic systems, *Proc. Cambridge Phil. Soc.* Vol. 6, p. 199.

Bryan, G. H. (1891). Buckling of plates, *Proc. the London Math. Soc.* Vol. 22, p. 54.

Chetayev, N. G. (1930). Sur la réciproque du théorème de Lagrange, *Comptes Rendues.*

Chetayev, N. G. (1961). *Stability of Motion*, Translated from the second Russian edition by M. Nadler. Pergamon Press, London.

Flügge, W. (1962). *Handbook of Engineering Mechanics.* McGraw-Hill Book Company, New York, chaps. 44 and 45.

Herrmann, G. (1967). Stability of the equilibrium of elastic systems subjected to nonconservative forces, *Applied Mechanics Review*, 20.

Hoff, N. J. (1954). Buckling and stability, *J. Royal Aero. Soc.* Vol. 58, January.

Hoff, N. J. (1956). *The Analysis of Structures.* John Wiley & Sons, Inc., New York.

Koiter, W. T. (1945). "The Stability of Elastic Equilibrium," Thesis, Delft. (English translation NASA TT-F-10833, 1967).

Koiter, W. T. (1963). Elastic stability and post-buckling behavior, *Nonlinear Problems*, edited by R. E. Langer, University of Wisconsin Press, Madison, pp. 257–275.

Krasovskii, N. N. (1963). *Stability of Motion*, Translated from the Russian by J. L. Brenner. Stanford University Press, Stanford, California.

Lagrange, J. L. (1788). *Mecanique Analytique.* Paris.

Lanczos, C. (1960). *The Variational Principles of Mechanics.* University of Toronto Press, Toronto.

Langhaar, H. L. (1962). *Energy Methods in Applied Mechanics.* John Wiley & Sons, Inc., New York.

LaSalle, J. P. and Lefschetz, S. (1961). *Stability of Liapunov's Direct Method with Applications.* Academic Press, New York.

Leipholz, H. (1970). *Stability Theory.* Academic Press, New York.

Liapunov, A. (1952). "Problémé générale de la Stabilité du Mouvement," Traduit du Russe par E. Davaux. Annales de la Faculté des Sciences de Toulouse, Zéme serie, **9**, 1907. Reprinted by Princeton University Press, Princeton.

Pearson, C. E. (1956). General theory of elastic stability, *Quarterly of Applied Mathematics*, Vol. 14, p. 133.

Poincaré, H. (1885). "Sur le Equilibre d'une Masse Fluide Animée d'un Mouvement de Rotation," *Acta Mathematica*, Vol. 7, pp. 259–380, Stockholm.

Thompson, J. M. T. (1963). Basic principles in the general theory of elastic stability, *J. of the Mechanics and Physics of Solids*, Vol. 11, p. 13.

Timoshenko, S. P. (1953). *History of Strength of Materials.* McGraw-Hill Book Co., New York.

Timoshenko, S. P. and Gere, J. M. (1961). *Theory of Elastic Stability.* McGraw-Hill Book Co., New York.

Trefftz, E. (1933). "Zur Theorie der Stabilität des Elastischen Gleichgewichts." *Zeitschr. f. angew. Math. u. Mech*, Bd. 13, p. 160.

Whittaker, E. T. (1944). *Analytical Dynamics.* Dover Publications, New York.

Ziegler, H. (1968). *Principles of Structural Stability.* Blaisdell Publishing Co., Waltham, Massachusetts.

2

MECHANICAL STABILITY MODELS

Before undertaking the study of stability of elastic structures, the different methods available for understanding and obtaining critical conditions will be demonstrated through the use of simple mechanical models. The discussion will be limited to conservative systems. It is also intended to demonstrate the effect of geometric imperfections and load eccentricity on the response of the system. For a number of models, both the small-deflection (linear) and large-deflection (nonlinear) theories will be used for the sake of comparison. Finally, a comprehensive discussion of the different types of behaviors will be given to enhance understanding of buckling, critical conditions, and advantages or disadvantages of the approaches used.

2.1   MODEL *A*; A ONE-DEGREE-OF-FREEDOM MODEL

Consider a rigid bar of length l, hinged at one end, free at the other, and supported through a frictionless ring connected to a spring that can move only horizontally (see Fig. 2.1). The free end is loaded with a force P in the direction of the bar. It is assumed that the direction of the force remains unchanged. We may now ask: "Will the rigid bar remain in the upright position under the quasistatically applied load P?"

In trying to answer this question, we must consider all possible deflectional modes and study the stability of the system equilibrium. One possible deflectional mode allows rotation θ about the hinged end. In writing equilibrium conditions for some θ positions, we could be interested in small θ as well as large θ values.

2.1.1   SMALL-θ ANALYSIS

In the casual small-θ analysis, we make the usual assumption that θ is so small that $\theta \approx \sin\theta \approx \tan\theta$. With this restriction, we can only investigate the stability of the equilibrium configuration corresponding to $\theta = 0$. This type of investigation is sufficient to answer the posed question. The three approaches will be used separately.

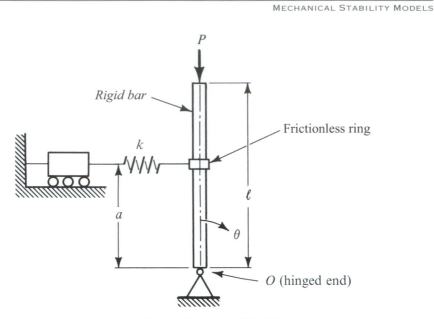

FIGURE 2.1   Geometry of Model *A*.

1. The Classical or Equilibrium Method.   The equilibrium equation corresponding to a deflected position is written under the assumption of small θ's (see Fig. 2.2). The expression for the moment about O is given by

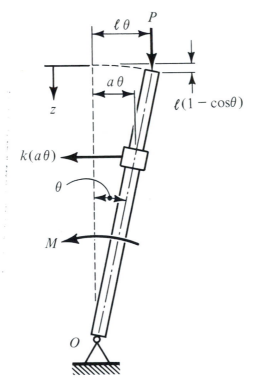

FIGURE 2.2   Small θ deflected position (Model *A*).

$$M = -P\ell\theta + (ka\theta)a$$

Since the bar is hinged at O, then

$$M = 0 \quad \text{or} \quad (P\ell - ka^2)\theta = 0 \tag{1}$$

Thus a nontrivial solution exists if

$$P\ell = ka^2$$

and the bifurcation point is located by

$$P = \frac{ka^2}{\ell} \quad \text{or} \quad P_{cr} = \frac{ka^2}{\ell}$$

Fig. 2.3 shows a plot of the load parameter $p = P\ell/ka^2$ versus θ. Note that the bifurcation point is located at $p = 1$, and the $\theta \neq 0$ equilibrium positions are limited by the assumption of small θ.

2. Kinetic or Dynamic Approach.   In this approach, we are interested in the character of the motion for small disturbances about the $\theta = 0$ position and at a constant P value. The equation of motion is given by

$$I\ddot{\theta} + M = 0$$

or   (2)

$$I\ddot{\theta} - (P\ell - ka^2)\theta = 0$$

where dots above θ denote differentiation with respect to time, and I is the moment of inertia of the rigid bar about the hinged end O. It is easily seen from the differential equation that if

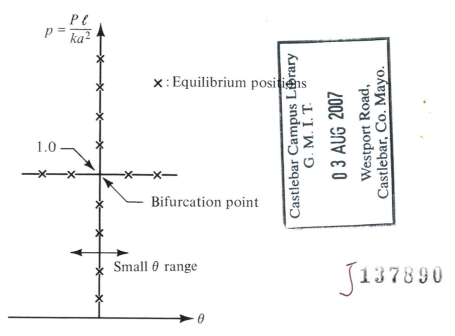

FIGURE 2.3   Load-deflection curve (Model *A*; small θ analysis).

$$P\ell - ka^2 < 0$$

the motion is oscillatory and the equilibrium is stable. If

$$P\ell - ka^2 > 0$$

the motion is diverging and the equilibrium is unstable. If

$$P\ell - ka^2 = 0$$

the motion can still be considered diverging (constant or linear with respect to time) and the equilibrium is unstable (neutrally stable).

Note that the frequency f is given by

$$\frac{1}{2\pi}\left(\frac{ka^2 - P\ell}{I}\right)^{\frac{1}{2}} \tag{3}$$

and at the critical condition $f = 0$, or

$$P_{cr} = \frac{ka^2}{\ell}$$

3. Energy Approach. Since the system is conservative, the externally applied force P can be derived from a potential. Thus (see Fig. 2.2)

$$U_p = P(z_0 - z)$$

and letting $z_0 = 0$, then $U_p = -Pz$ and

$$P = -\frac{dU_p}{dz} \tag{4}$$

On the other hand, the energy, U_i, stored in the system is given by

$$U_i = \frac{1}{2}k(a\theta)^2 \tag{5}$$

Thus the total potential U_T is given by

$$U_T = U_i + U_P$$
$$= -Pz + \frac{1}{2}k(a\theta)^2 \tag{6}$$

But since $z = \ell(1 - \cos\theta)$, then

$$U_T = -P\ell(1 - \cos\theta) + \frac{1}{2}ka^2\theta^2 \tag{7}$$

For static equilibrium, the total potential must have a stationary value. Thus

$$\frac{dU_T}{d\theta} = 0$$

or

$$-P\ell\sin\theta + ka^2\theta = 0$$

and since $\sin\theta \approx \theta$, then

$$(-P\ell + ka^2)\theta = 0 \tag{8}$$

This is the same equilibrium equation we derived previously. Furthermore, if the second variation is positive definite, the static equilibrium is stable. If the second variation is negative definite, the static equilibrium is unstable; if it is zero, no conclusion can be drawn.

It is seen in this case that

$$\frac{d^2 U_T}{d\theta^2} = ka^2 - P\ell \tag{9}$$

Therefore, for $P < ka^2/\ell$ the static equilibrium positions ($\theta = 0$) are stable, while for $P > ka^2/\ell$ they are unstable. Thus, as before,

$$P_{cr} = \frac{ka^2}{\ell}$$

2.1.2 LARGE-Θ ANALYSIS

In this particular approach, the only limitation on θ is dictated from geometrical considerations. Note from Fig. 2.4 that $-\cos^{-1} a/\ell < \theta < \cos^{-1} a/\ell$. For θ values outside this range, the ring will fly off the rigid bar. As before, the three approaches shall be treated separately.

1. The Classical or Bifurcation Method.   Since the ring is frictionless, the force R, normal to the rigid bar, is related to the spring force through the following expression (see Fig. 2.4):

$$k(a \tan \theta) = R \cos \theta$$

Then the moment about pin O is given by

$$M = -P\ell \sin \theta + \frac{ka^2 \sin \theta}{\cos^3 \theta} \tag{10}$$

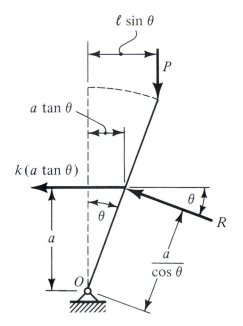

FIGURE 2.4   Geometry for large θ analysis (Model *A*).

For static equilibrium we require that $M = 0$. Thus, the equilibrium positions are characterized by the equation

$$\left(\frac{ka^2}{\cos^3 \theta} - P\ell\right) \sin \theta = 0 \tag{11}$$

which implies that either

$$\theta = 0 \tag{12a}$$

or

$$\frac{P\ell}{ka^2} = \sec^3 \theta \tag{12b}$$

It is clearly seen that a nontrivial solution ($\theta \neq 0$) can exist for $P\ell/ka^2 > 1$ and a bifurcation point exists at $P\ell/ka^2 = 1$ (see Fig. 2.5).

The answer to the original question is yes, and $P_{cr} = ka^2/\ell$.

2. Kinetic or Dynamic Approach. In this approach, as before, we are interested in the character of the motion for small disturbances about the static equilibrium positions, keeping P constant. The equation of motion is given by

$$I\ddot{\theta} + M = 0$$

But

$$M = -P\ell \sin \theta + \frac{ka^2 \sin \theta}{\cos^3 \theta} \tag{13}$$

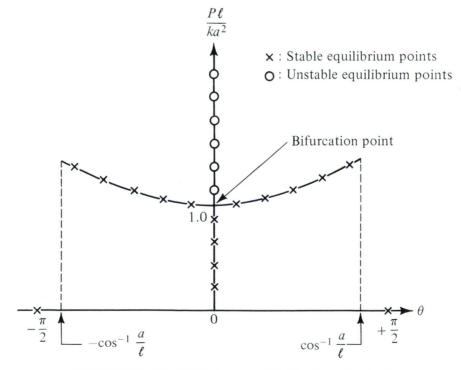

FIGURE 2.5 Load-deflection curve (Model *A*; large θ analysis).

If we denote the equilibrium positions by θ_0 and the disturbed positions by $\theta = (\theta_0 + \varphi)$, the Taylor-series expansion for the moment is given by

$$M(\theta_0 + \varphi) = M(\theta_0) + \varphi\left(\frac{dM}{d\theta}\right)_{\theta=\theta_0} + \frac{\varphi^2}{2!}\left(\frac{d^2M}{d\theta^2}\right)_{\theta=\theta_0} + \cdots \tag{14}$$

At the static equilibrium positions, $M(\theta_0) = 0$. Through differentiation, we may write

$$\left(\frac{dM}{d\theta}\right)_{\theta_0} = \left(\frac{ka^2}{\cos^3\theta_0} - P\ell\right)\cos\theta_0 + \sin\theta_0\left(3ka^2\frac{\sin\theta_0}{\cos^4\theta_0}\right) \tag{15}$$

The equilibrium positions (see Fig. 2.4) are denoted by Eqs. (12). Thus, the equation of motion under the assumption of small disturbances for the equilibrium positions corresponding to $\theta_0 \neq 0$ is given by

$$I\ddot{\varphi} + 3ka^2\frac{\sin^2\theta_0}{\cos^4\theta_0}\varphi = 0 \tag{16a}$$

and since

$$3ka^2\frac{\sin^2\theta_0}{\cos^4\theta_0} > 0$$

these equilibrium positions are stable. Note that $\theta_0 \neq 0$.

The equation of motion for the positions corresponding to $\theta_0 = 0$ is given by

$$I\ddot{\varphi} + \left(ka^2 - P\ell\right)\varphi = 0 \tag{16b}$$

If $P\ell < ka^2$, the equilibrium is stable, while if $P\ell > ka^2$, the equilibrium is unstable.

The equation of motion for the particular position corresponding to $\theta_0 = 0$ and $P\ell = ka^2$ is given below. We obtain this equation by taking more terms in the series expansion for M, Eq. (14).

$$I\ddot{\varphi} + \frac{3}{2}ka^2\varphi^3 = 0 \tag{16c}$$

A study of this differential equation (see the following Section, *Parenthesis*) indicates that the motion is stable. Although the equilibrium position $\theta_0 = 0$, $P\ell = ka^2$ is stable, the answer to the original question is that the bar will not remain in the upright position, and the critical value of the load is given by

$$P_{cr} = \frac{ka^2}{\ell}$$

Parenthesis.   If the equation of motion of a nonlinear system is given by

$$\ddot{x} + k^2x^{2n+1} = 0 \tag{17}$$

where the dots above x denote differentiation with respect to time, k^2 is a positive number, and n is a positive integer, then the system is conservative and the position $x = 0$ is stable (see Krasovskii, 1963; LaSalle, 1961; Malkin, 1958; and Stoker, 1950). What this implies physically is that, depending on the initial conditions, the total energy of the system is constant (sum of kinetic and potential energies is constant, thus the system is conservative), and the system performs nonlinear oscillations about the null position $x = 0$ within a bounded region enclosing the position $x = 0$.

The following computations will further clarify the above statements. Since

$$\ddot{x} = \frac{d\dot{x}}{dt} = \frac{d\dot{x}}{dx} \cdot \frac{dx}{dt} = \dot{x} \frac{d\dot{x}}{dx}$$

then Eq. (17) may now be written as

$$\dot{x} \frac{d\dot{x}}{dx} = -k^2 x^{2n+1}$$

or

$$\dot{x} d\dot{x} = -k^2 x^{2n+1} dx$$

If the initial conditions are denoted by $\dot{x}_0$ and x_0, then integration of this last equation yields

$$\frac{1}{2} \left[\dot{x}^2 - \dot{x}_0^2 \right] = \frac{k^2}{2(n+1)} \left[x_0^{2(n+1)} - x^{2(n+1)} \right]$$

This equation expresses the law of conservation of energy. The left side denotes the change in kinetic energy, and the right side denotes the change in potential energy. With reference to Eq. (16), x and $\dot{x}$ denote the size of the response to initial disturbances $\dot{x}_0$ and/or x_0. If we let the disturbance be x_0 only ($\dot{x}_0 \equiv 0$), then

$$\dot{x}^2 = \frac{k^2}{(n+1)} \left[x_0^{2(n+1)} - x^{2(n+1)} \right]$$

which implies that the response is bounded.

3. *Energy Approach.* The total potential of the system is given by

$$U_T = -P\ell(1 - \cos\theta) + \frac{1}{2} k a^2 \tan^2\theta \tag{18}$$

The static equilibrium positions are characterized by the equation

$$\frac{dU_T}{d\theta} = 0$$

or

$$\left(-P\ell + \frac{ka^2}{\cos^3\theta} \right) \sin\theta = 0$$

Furthermore, the second variation is given by

$$\frac{d^2 U_T}{d\theta^2} = \left(\frac{ka^2}{\cos^3\theta} - P\ell \right) \cos\theta + 3ka^2 \frac{\sin^2\theta}{\cos^4\theta} \tag{19}$$

It is easily concluded that the static equilibrium positions characterized by $\theta \neq 0$ are stable. Similarly, the positions $\theta = 0$ for $P\ell > ka^2$ are unstable. It can also be concluded, by considering higher variations, that the position denoted by $\theta_0 = 0$ and $P\ell = ka^2$ is stable (see Chapter 1). The answer to the original question, though, still remains the same and

$$P_{cr} = \frac{ka^2}{\ell}$$

2.2   MODEL B; A ONE-DEGREE-OF-FREEDOM MODEL

Consider two rigid links pinned together and supported by hinges on rollers at the free ends (Fig. 2.6a). The system is supported at the middle hinge by a vertical linear spring and is acted upon by two collinear horizontal forces of equal intensity. The two links are initially horizontal. Can the system buckle? What is the critical load? To answer these questions, we may use small-deflection theory. The classical method and the energy method shall be used in this case.

1. The Classical or Bifurcation Method.   Using casual small-deflection theory, we put the system into a deflected position (Fig. 2.6b) and write the equilibrium equations.

Since the system is symmetric, the vertical reactions at the hinges are $k\delta/2$. Furthermore, the moment about the middle hinge must vanish. This requirement leads to the equilibrium equation

$$\left(P - \frac{k\ell}{2}\right)\delta = 0 \tag{20}$$

Thus, the equilibrium positions are defined by either $\delta = 0$ (trivial solution) or $P = k\ell/2$. In plotting $2P/k\ell$ versus δ, we notice that a bifurcation point exists at $2P/k\ell = 1$ (Fig. 2.6c) and

$$P_{cr} = \frac{k\ell}{2} \tag{21}$$

2. Energy Method.   The total potential is the sum of the energy stored in the spring and the potential of the external forces. Thus

$$U_T = \frac{k\delta^2}{2} - 2P\left(\ell - \sqrt{\ell^2 - \delta^2}\right)$$

For static equilibrium $dU_T/d\delta = 0$ and

$$k\delta - 2P \cdot \frac{1}{2}\frac{2\delta}{\left(\ell^2 - \delta^2\right)^{\frac{1}{2}}} = 0 \tag{22}$$

which, under the assumption of $\delta^2 \ll \ell^2$, is identical to Eq. (20).

For the static equilibrium positions to be stable, the second variation must be positive definite, or

$$\frac{d^2 U_T}{d\delta^2} = k - \frac{2P}{\left(\ell^2 - \delta^2\right)^{\frac{1}{2}}} + \frac{2P\delta^2}{\left(\ell^2 - \delta^2\right)^{\frac{3}{2}}} \approx k - \frac{2P}{\ell} \tag{23}$$

Thus the equilibrium positions denoted by $\delta = 0$ and $P < k\ell/2$ are stable, and the critical load is given by Eq. (21).

2.3   MODEL C; A TWO-DEGREE-OF-FREEDOM MODEL

Consider the system shown in Fig. 2.7a, composed of three rigid bars of equal length hinged together as shown. The linear springs are of equal intensity. This is a two-degree-of-freedom system and it is acted upon by a horizontal force, P, applied

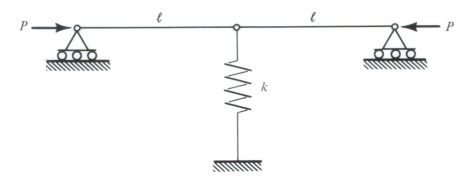

(a) Geometry; Model *B*

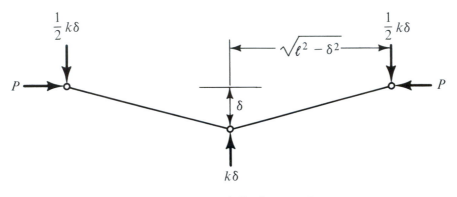

(b) Forces and displacements

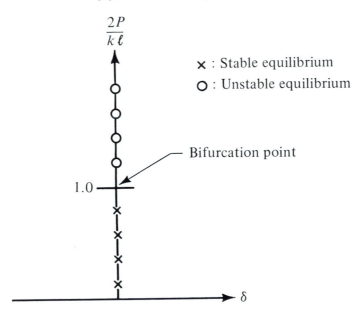

(c) Load-deflection curve

FIGURE 2.6   Model *B*.

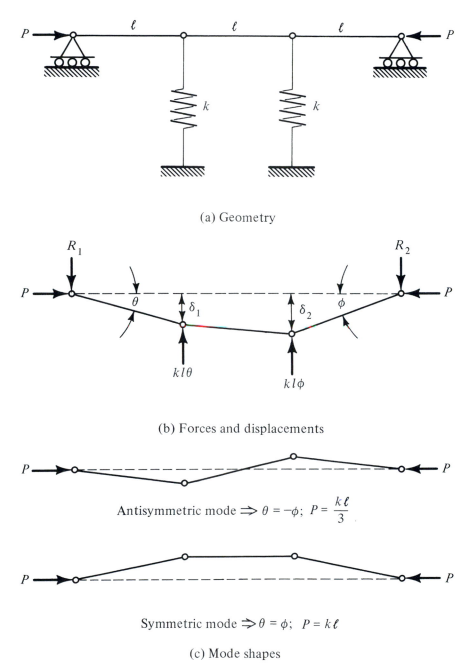

(a) Geometry

(b) Forces and displacements

Antisymmetric mode $\Rightarrow \theta = -\phi$; $P = \dfrac{k\ell}{3}$

Symmetric mode $\Rightarrow \theta = \phi$; $P = k\ell$

(c) Mode shapes

FIGURE 2.7   Model *C*.

quasistatically. We must determine whether or not the system will buckle and the critical value of the applied load. The load is assumed to remain horizontal.

1. The Classical or Bifurcation Method.   In solving this problem, we will first use the classical method under the assumption of small deflections.

Denoting by θ and φ the rotations about the support pins (see Fig. 2.7b) and by R_1 and R_2 the vertical reactions at the pins, we may write the following equilibrium equations for the deflected system:

$$\left.\begin{array}{c} 3\ell R_1 = 2k\ell^2\theta + k\ell^2\varphi \\ 3\ell R_2 = k\ell^2\theta + 2k\ell^2\varphi \\ R_1\ell = P\ell\theta \\ R_2\ell = P\ell\varphi \end{array}\right\} \tag{24}$$

Elimination of R_1 and R_2 yields the following system of linear homogeneous algebraic equations:

$$\left.\begin{array}{c} \left(P - \dfrac{2k\ell}{3}\right)\theta - \dfrac{k\ell\varphi}{3} = 0 \\ \dfrac{k\ell\theta}{3} - \left(P - \dfrac{2k\ell}{3}\right)\varphi = 0 \end{array}\right\} \tag{25}$$

The critical condition is derived if we require the existence of a nontrivial solution. This leads to the characteristic equation

$$\begin{vmatrix} \left(P - \dfrac{2k\ell}{3}\right) & -\dfrac{k\ell}{3} \\ \dfrac{k\ell}{3} & -\left(P - \dfrac{2k\ell}{3}\right) \end{vmatrix} = 0$$

from which

$$P = \left\{\begin{array}{c} \dfrac{k\ell}{3} \\ k\ell \end{array}\right\} \tag{26}$$

Thus, there are two solutions (eigenvalues) corresponding to two modes of deformation (Fig. 2.7c):

$$P = \frac{k\ell}{3} \text{ and } \varphi = -\theta$$
$$P = k\ell \text{ and } \varphi = \theta$$

This shows that the smallest load corresponds to the antisymmetric mode.

2. The Energy Method. In Fig. 2.7b, the total potential for the system, which consists of the energy stored in the springs and the potential of the external forces, is given by

$$U_T = U_i + U_p = \frac{1}{2}k\ell^2\theta^2$$
$$+ \frac{1}{2}k\ell^2\varphi^2 - P\ell[(1 - \cos\theta) + (1 - \cos\varphi) + 1 - \cos(\varphi - \theta)] \tag{27}$$

By assuming that the angles φ and θ can be made as small as desired, we may rewrite Eq. (27) as

$$U_T = \frac{1}{2}k\ell^2\theta^2 + \frac{1}{2}k\ell^2\varphi^2 - P\ell(\theta^2 + \varphi^2 - \varphi\theta) \tag{28}$$

For static equilibrium, the total potential must be stationary; therefore

$$\frac{\partial U_T}{\partial \theta} = \frac{\partial U_T}{\partial \varphi} = 0 \qquad (29)$$

which leads to the following equilibrium equations:

$$\left. \begin{array}{l} (k\ell^2 - 2P\ell)\theta + P\ell\varphi = 0 \\ P\ell\theta + (k\ell^2 - 2P\ell)\varphi = 0 \end{array} \right\} \qquad (30)$$

The nontrivial solution is the same as the one obtained by the classical approach.

$$P = \frac{k\ell}{3}, \quad P = k\ell \qquad (31)$$

Study of the stability of the equilibrium positions characterized by $\theta = \varphi = 0$ for the entire range of values of P requires knowledge of the second variations

$$\frac{\partial^2 U_T}{\partial \theta^2} = k\ell^2 - 2P\ell \qquad (32)$$

$$\frac{\partial^2 U_T}{\partial \varphi^2} = k\ell^2 - 2P\ell \qquad (33)$$

$$\frac{\partial^2 U_T}{\partial \theta \partial \varphi} = P\ell \qquad (34)$$

The equilibrium positions are stable if and only if (see Chapter 1) both of the following inequalities are satisfied.

$$\frac{\partial^2 U}{\partial \theta^2} > 0$$
$$\frac{\partial^2 U_T}{\partial \theta^2} \cdot \frac{\partial^2 U_T}{\partial \varphi^2} > \left(\frac{\partial^2 U_T}{\partial \theta \partial \varphi} \right)^2 \qquad (35)$$

In terms of the applied load and the structural geometry, these inequalities are

$$k\ell > 2P$$
$$(k\ell - P)\left(\frac{k\ell}{3} - P \right) > 0 \qquad (36)$$

From these expressions, we see that equilibrium positions for which $P < k\ell/3$ are stable, while all equilibrium positions for which $P > k\ell/3$ are unstable. Therefore

$$P_{cr} = \frac{k\ell}{3}$$

2.4   MODEL D; A SNAPTHROUGH MODEL

In the analysis of this model, we will demonstrate the type of buckling known as snapthrough or oil-canning.

Consider two rigid bars of length ℓ pinned together, with one end of the system pinned to an immovable support, and the other pinned to a linear horizontal spring (see Fig. 2.8a). The rigid bars make an angle α with the horizontal when the spring is unstretched and the system is loaded laterally through a force P applied quasistatically at the connection of the two rigid bars. As the load is increased quasistatically from zero, the spring will be compressed and the two bars will make an angle θ with the horizontal ($\theta < \alpha$). The question then arises whether it is possible for the system to snap through toward the other side at some value of the applied load. In seeking the answer to this question, we will first use the equilibrium approach and then analyze the system by considering the character of the equilibrium positions. The latter will be accomplished through the energy approach.

1. The Equilibrium Approach. Let the horizontal reaction of the spring be F. This force is equal to k times the compression in the spring (Fig. 2.8b), or

$$F = 2k\ell(\cos\theta - \cos\alpha) \tag{37}$$

Furthermore, from symmetry the vertical reactions at the ends are $P/2$. Since no moment can be transferred through the middle joint, the equilibrium states are characterized by the following equation

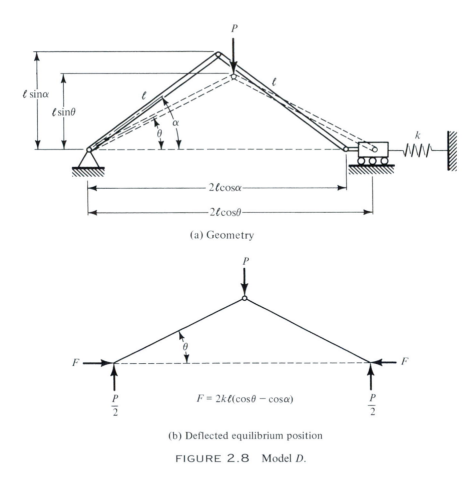

(a) Geometry

(b) Deflected equilibrium position

FIGURE 2.8   Model *D*.

$$\frac{P\ell}{2}\cos\theta = F\ell\sin\theta \tag{38}$$

Use of Eq. (37) yields

$$\frac{P}{4k\ell} = \sin\theta - \cos\alpha\tan\theta \tag{39}$$

Note that $-\pi/2 < \theta < \alpha < \pi/2$.

The equilibrium states, Eq. (39), are plotted in Fig. 2.9b. Note that loading starts at point A and it is increased quasistatically. When point B is reached, we see that with no appreciable change in the load the system will tend to snap through toward the CD portion of the curve. The load corresponding to position B is a critical one, and its magnitude may be obtained from the fact that

$$\frac{dP}{d\theta} = 0 \tag{40}$$

Note that the right side of Eq. (39) is a continuous function of θ with continuous first derivatives.

If we denote by θ_B the angles corresponding to positions B and B', then

$$\theta_B = \pm\cos^{-1}(\cos\alpha)^{\frac{1}{3}} \tag{41}$$

and

$$\left|\frac{P}{4k\ell}\right|_{cr} = |\sin\theta_B - \cos\alpha\tan\theta_B| \tag{42}$$

2. Energy Approach. The total potential, U_T, for the system, which is equal to the potential of the external force and the energy stored in the spring, is given by

$$U_T = 2k\ell^2(\cos\theta - \cos\alpha)^2 - P\ell(\sin\alpha - \sin\theta) \tag{43}$$

Static equilibrium positions are characterized by the vanishing of the first variation of the total potential, or

$$\frac{dU_T}{d\theta} = 4k\ell^2(\cos\theta - \cos\alpha)(-\sin\theta) + P\ell\cos\theta = 0$$

This leads to the equilibrium equation

$$\frac{P}{4k\ell} = \sin\theta - \cos\alpha\tan\theta \tag{39}$$

The character of the equilibrium positions is governed by the second variation, or

$$\frac{d^2U_T}{d\theta^2} = 4k\ell^2(\cos\theta - \cos\alpha)(-\cos\theta) + 4k\ell^2\sin^2\theta - P\ell\sin\theta \tag{44}$$

Making use of the equilibrium condition, Eq. (39), we may write

$$\frac{d^2U_T}{d\theta^2} = 4k\ell^2\left(\frac{\cos\alpha}{\cos\theta} - \cos^2\theta\right) \tag{45}$$

Thus in the region

$$-\cos^{-1}(\cos\alpha)^{\frac{1}{3}} < \theta < +\cos^{-1}(\cos\alpha)^{\frac{1}{3}}$$

the second derivative is negative, and the equilibrium positions are unstable. Outside this region, the second derivative is positive, and the equilibrium positions are stable. Thus, P_{cr} is given by Eq. (42). Note that points between B and B' represent "hills" on the total potential curve, while points outside this region represent "valleys" (see Fig. 2.9a).

A critical condition is reached when the load is such that the near equilibrium point coincides with the unstable point.

Note from Fig. 2.9 that the stationary $(dU_T/d\theta = 0)$ points on the total potential curve corresponding to different values of the applied load make up the load-deflection curve (equilibrium states).

2.5   MODELS OF IMPERFECT GEOMETRIES

In many cases it is possible to predict critical conditions for a system of perfect geometry by studying the behavior of the system under the same load conditions but with slight geometric imperfections.

Consider, for instance, model B with a small imperfection δ_0 (Fig. 2.10a) when the spring is unstretched. The problem is to find the behavior of the imperfect system under the quasistatic application of the horizontal forces. Once this behavior has been established, the question arises whether or not we can predict the critical condition for the system of perfect geometry.

From the conditions of symmetry, the vertical reactions at the end pins are equal to $\frac{1}{2} k\delta$. The equilibrium condition is obtained if we require the moment about the middle hinge to vanish.

$$\begin{aligned}
P(\delta + \delta_0) &= \frac{k\delta}{2}\sqrt{\ell^2 - (\delta + \delta_0)^2} \\
&\approx \frac{k\delta\ell}{2}
\end{aligned} \tag{46}$$

This equation can be written in the form

$$\left(P - \frac{k\ell}{2}\right)(\delta + \delta_0) = -\frac{k\ell}{2}\delta_0 \tag{47}$$

If we divide both sides by $(k\ell/2)\delta_0$, the equilibrium equation becomes

$$\left(\frac{2P}{k\ell} - 1\right)\left(1 + \frac{\delta}{\delta_0}\right) = -1 \tag{48}$$

This represents a hyperbola in the coordinate system of $(2P/k\ell - 1)$ and $(1 + \delta/\delta_0)$ (see Fig. 2.10b). When a translation of axes is used, it appears that the load-deflection curve, in the coordinate system of $2P/k\ell$ and δ/δ_0, approaches the line $2P/k\ell = 1$ asymptotically. Furthermore, when $2P/k\ell$ is plotted versus δ, the single curve of Fig. 2.10b becomes a family of curves dependent on the value of the imperfection δ_0. (see Fig. 2.10c). We see from this last illustration that as $\delta_0 \to 0$, the behavior of the system is such that δ remains zero until $2P/k\ell$ becomes equal to unity. Thus, for

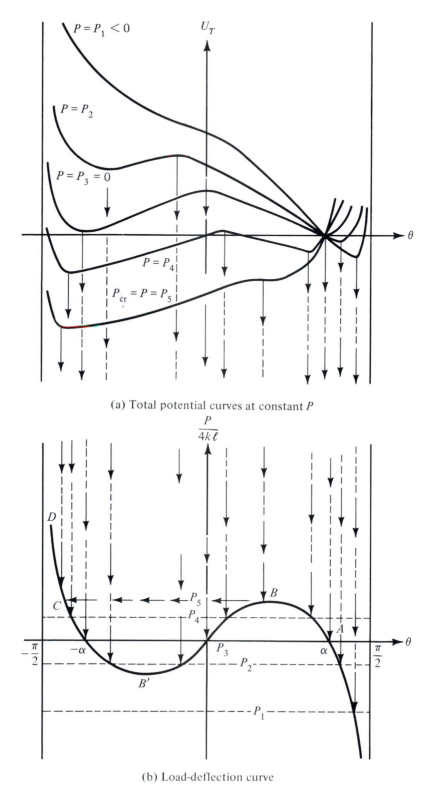

(a) Total potential curves at constant P

(b) Load-deflection curve

FIGURE 2.9   Critical conditions for Model D.

$\delta_0 = 0$, $P_{cr} = k\ell/2$. This conclusion is the same as that reached when the system of perfect geometry was analyzed.

As a second example, consider the imperfect model shown in Fig. 2.11. Note that as the load eccentricity approaches zero, we have the corresponding perfect geometry model given in Problem 1 at the end of this chapter. For this particular problem, we want to find the effect of the eccentricity, e, on the critical load, P_{cr}. Once this effect is established, we can predict P_{cr} for the perfect configuration by letting the eccentricity approach zero. We will use the energy approach to solve the problem.

The total potential is given by

$$U_T = \frac{1}{2}ka^2 \sin^2\theta - P\ell\left(1 - \cos\theta + \frac{e}{\ell}\sin\theta\right) \tag{49}$$

For equilibrium

$$\frac{\partial U_T}{\partial\theta} = 0 = ka^2 \sin\theta\cos\theta - P\ell\left(\sin\theta + \frac{e}{\ell}\cos\theta\right) \tag{50}$$

From this equation we obtain the load-deflection curve for a given load eccentricity e:

$$p = \frac{P\ell}{ka^2} = \frac{\sin\theta}{\tan\theta + (e/\ell)} \tag{51}$$

Note that if e is replaced by $-e$ and θ by $-\theta$, we have the same load-deflection relation.

If we restrict the range of θ values to $0 < \theta < \pi/2$, we may study the second variation.

$$\frac{2}{ka^2} \cdot \frac{\partial^2 U_T}{\partial\theta^2} = \cos 2\theta - p\left(\cos\theta - \frac{e}{\ell}\sin\theta\right) \tag{52}$$

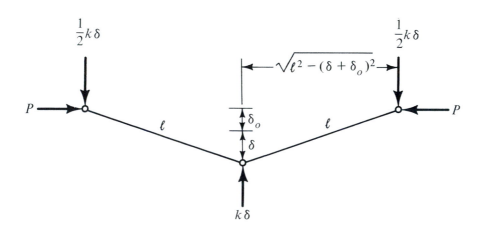

(a) Geometry

FIGURE 2.10   Model B with initial imperfection.

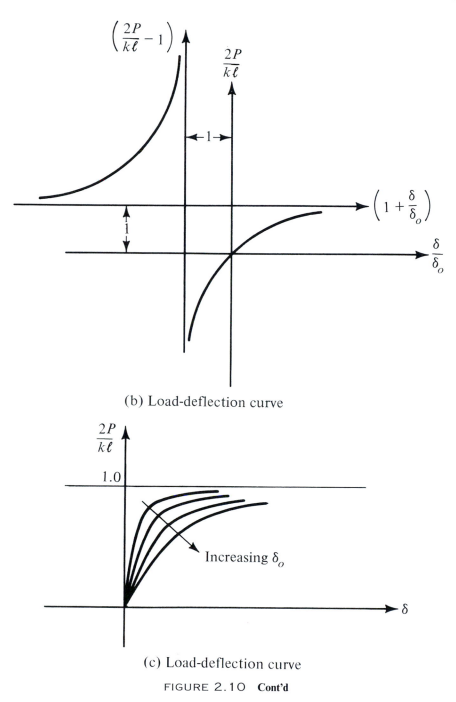

(b) Load-deflection curve

(c) Load-deflection curve

FIGURE 2.10 **Cont'd**

If we eliminate p, through Eq. (51), and use some well-known trigonometric identities, we finally obtain

$$\frac{2}{ka^2} \cdot \frac{\partial^2 U_T}{\partial \theta^2} = \frac{\cos^2 \theta}{\tan \theta + (e/\ell)} \left(-\tan^3 \theta + \frac{e}{\ell} \right)$$

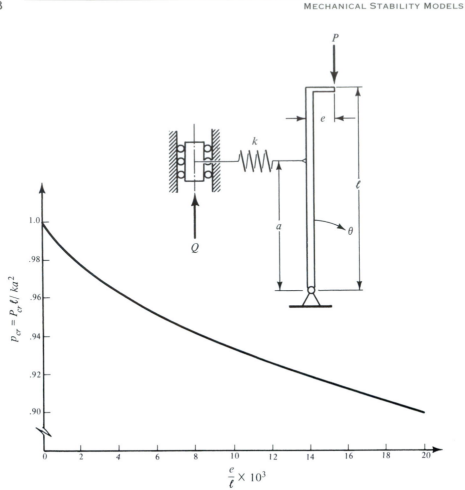

FIGURE 2.11   Effect of imperfection on the critical load.

Clearly, if $\tan^3 \theta < e/\ell$, the equilibrium positions are stable, and if $\tan^3 \theta > e/\ell$, the equilibrium positions are unstable. When $\tan^3 \theta = e/\ell$, $p = p_{cr}$, substitution of this expression for θ into Eq. (51) yields

$$p_{cr} = \left[1 + \left(\frac{e}{\ell} \right)^{\frac{2}{3}} \right]^{-\frac{3}{2}} \qquad (53)$$

A plot of p_{cr} versus e/ℓ is shown in Fig. 2-11.

A (qualitative) plot of p versus θ for this model is given in Fig. 2.12b (imperfect geometry), and this model exhibits snapthrough buckling.

Finally, if we let the eccentricity approach zero, $P_{cr} = 1$ and

$$P_{cr} = \frac{ka^2}{\ell}$$

2.6   DISCUSSION OF THE METHODS

After having considered these few mechanical models, certain observations must be discussed in order to enhance the understanding of the question of critical loads as well as the question of stability of elastic systems in general. In particular, attention is given to the relationship between the classical approach and the energy approach, which is completely equivalent to the dynamic approach for conservative systems (a proof of this is found in Whittaker 1944), and to the need for using large-deflection theories in certain problems.

We first noticed that, whenever the model exhibited a bifurcation point (Models A and B), regardless of the approach used, the same result is obtained. On the other hand, when there is no bifurcation point (Model D), the classical approach could only lead us to a load-deflection curve, and the criticality of the load at point B (see Fig. 2.9b) was explained as follows: If one wishes to increase the load any further, the system will visibly snap through toward a far equilibrium position. This argument, of course, implies deadweight-type of loading (prescription of the load rather than deflection), and it seems rather arbitrary. When the energy approach is used, it is very clear that the equilibrium positions between B and B' (Fig. 2.9b) are unstable, and therefore the load at B is critical because the slightest possible disturbance at this equilibrium position will make the system snap toward a far equilibrium position. In the absence of damping and assuming that the spring remains elastic, if the load at B is maintained, the system will simply oscillate (nonlinearly) between θ_B and some angle past $-\alpha$ (see Fig. 2.9a).

The second observation deals with the question of using large-deflection theories for predicting instability of perfect geometries (Model A). It is clear that, when dealing with systems characterized by model D, large-deflection theory cannot be avoided. Therefore this question is directed to systems that exhibit bifurcational buckling (adjacent equilibrium position). From the examples considered, we may suspect that small-deflection theory suffices to predict critical loads. Since the analysis (models A, B and C) is based on the assumption that there are no imperfections in the geometry of the system, large-deflection theories are needed because they clearly indicate through the load-deflection curves (equilibrium positions) whether geometrical imperfections are likely to have a significant effect on the buckling of the real structure. Consider, for example, model A (Fig. 2.5). Small geometric imperfections have little effect on this system. This can be verified by the introduction of a small imperfection θ_0 and the use of a large-θ analysis on the imperfect system. The result is qualitatively shown in Fig. 2.12a. However, it can be demonstrated that small geometric imperfections can cause a dramatic reduction in the buckling load when the load-deflection curve is characterized by either Fig. 2.12b or Fig. 2.12c. Note that in all three cases (Fig. 2.12) the small-deflection theory can only predict the bifurcation load.

The stability of structures immediately after buckling (bifurcation) was first investigated systematically by Koiter (1967), and alternative formulations of the general theory have subsequently been given by Sewell (1966) and by Thompson (1963 and 1964). Pope (1968) and Thompson (1964) show that the derivative $dp/d\theta$ at the bifurcation can be calculated exactly, for complicated elastic systems, by a finite-deformation analysis of the Rayleigh-Ritz type. Some remarks on Koiter's theory are presented in Chapter 5.

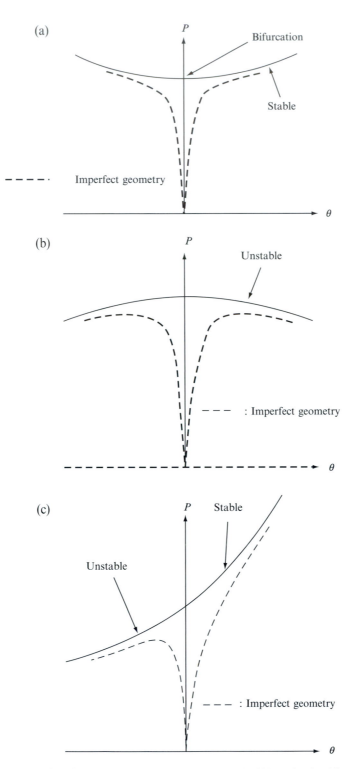

FIGURE 2.12   Possible load-deflection curves for bifurcation buckling.

PROBLEMS

1. Analyze the system shown using large-deflection theory. Give the load-deflection curve and the critical load.
    (a) Use the classical approach.
    (b) Use the kinetic approach.
    (c) Use the energy approach.
2. A uniform disc can rotate freely about O, except that it is restrained by a rotational spring giving a restoring couple $\alpha\theta$ for angular displacement θ. A weight W is attached at radius a and vertically above O.
    (a) Show that a stable tilted position, θ_0, of equilibrium is possible when $W > \alpha/a$.

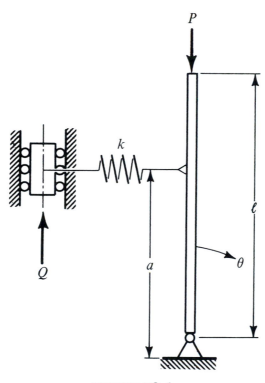

FIGURE P2.1

(b) Show that when $W > \alpha/a$, the frequency of small oscillations about the position of stable equilibrium is

$$\frac{1}{2\pi} \sqrt{\frac{\alpha - Wa\cos\theta_0}{I}}$$

where I is the moment of inertia (including Wa^2/g).

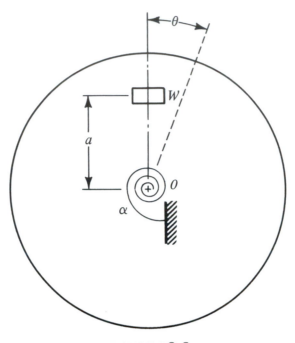

(c) Show that when $W = \alpha/a$, the differential equation for small oscillations is

$$I\frac{d^2\theta}{dt^2} + \frac{1}{6}\alpha\theta^3 = 0$$

3. In the mechanism shown, a light stiff rod is pinned at O. There is no friction. P
   remains vertical if the bar tilts.
   (a) By using any method, find the P-θ relation for equilibrium positions, and plot
       the curve.
   (b) Discuss the stability or instability of all the equilibrium positions in the entire
       practical range of θ values.
4. In the coplanar system shown, the initially vertical rod is rigid. The block to which
   the spring is attached slides in the inclined guide and is controlled so that the
   spring is always horizontal. All parts have negligible mass except the weight W.
   (a) Show that tilted equilibrium positions are characterized by

$$W = \frac{ka^2}{\ell}\left(1 - \tan\beta\tan\frac{\theta}{2}\right)\cos\theta$$

   (b) Sketch the curve for the two cases $\tan\beta$ small (e.g., 1/20) and $\tan\beta$ large (e.g.,
       10). What conclusions can you draw as to the stability of the tilted position?
       Give reasons.
   (c) Show that the vertical position is stable with respect to sufficiently small
       disturbances so long as $W\ell < ka^2$, and find a formula for the frequency of
       small oscillations.

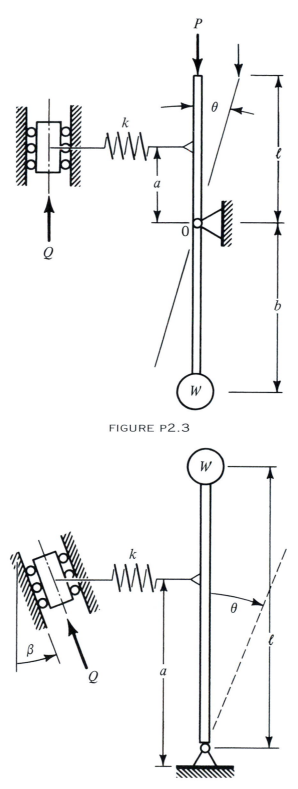

FIGURE P2.3

FIGURE P2.4

(d) Show that when $W\ell = ka^2$, the beginning of the motion from $\theta = 0$ follow-ing a slight disturbance will be governed by the equation

$$\frac{W\ell^2}{g}\ddot{\theta} - \frac{1}{2}ka^2 \tan\beta \cdot \theta^2 = 0$$

5. Analyze model C by assuming that the lengths of the rigid bars are unequal. Let these lengths be ℓ_1, ℓ_2, and ℓ_3 starting from the left. Let the spring constants be k for both.
   (a) Use the classical approach.
   (b) Use the energy approach.
6. Consider the rigid bar shown with an initial rotation θ_0 and initial stretch c of the spring. Use small-deflection theory, and through a complete analysis of the behavior of the imperfect system, predict critical conditions for the perfect system ($\theta_0 = c = 0$). c is the initial stretch of the spring.
7. Repeat Problem 6 assuming that the initial stretch, c, is zero.

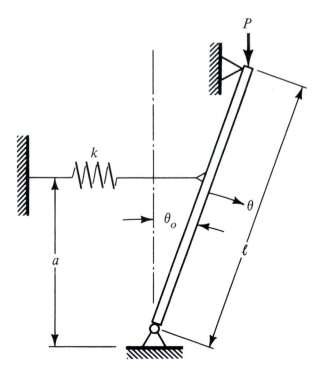

FIGURE P2.6 AND 7

8. Two rigid bars are connected by rotational springs to each other and to the support at C. Find P_{cr}, assuming that the load remains vertical.
9. Find the critical condition for model D through the kinetic approach.
   Hint: Consider the left leg as a free body and study its motion about the immovable support.
10. Consider the model shown loaded by a vertical force, P, applied quasistatically. Establish critical conditions for the system (for $C = 0$).
    (a) Use the equilibrium approach.
    (b) Use the energy approach.

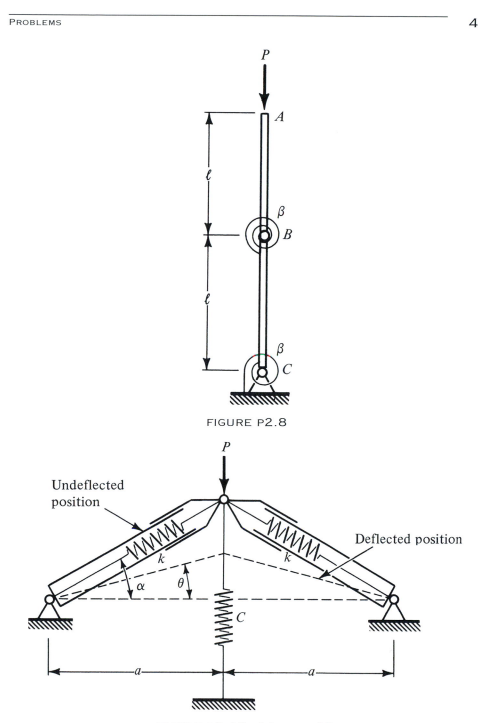

FIGURE P2.8

FIGURE P2.10, 11, AND 12

11. Repeat Problem 10 assuming C is constant.
12. Repeat Problem 10 assuming $C = A + B \sin \theta + D \sin^2 \theta$ (nonlinear spring). Note that the numerical work involved is complicated and a computer program is needed as well as knowledge of the values of the different parameters.

REFERENCES

Hayashi, C. (1964). *Nonlinear Oscillations in Physical Systems*. McGraw-Hill Book Co., New York.

Koiter, W. T. (1945). "The Stability of Elastic Equilibrium," Thesis, Delft. (English translation NASA TT-F-10833, 1967).

Krasovskii, N. N. (1963). *Stability of Motion*, Translated from the Russian by J. L. Brenner. Stanford University Press, Stanford, California.

LaSalle, J. P. and Lefschetz, S. (1961). *Stability of Liapunov's Direct Method With Applications*. Academic Press, New York.

Malkin, I. G. (1958). *Theory of Stability of Motion*. A.E.C. Translation 3352, Dept. of Commerce, U.S.A.

Pope, G. G. (1968). On the bifurcational buckling of elastic beams, plates and shallow shells, *The Aeronautical Quarterly*, p. 20, Feb.

Sewell, M. J. (1966). On the connexion between stability and the shape of the equilibrium surface, *Journal of the Mechanics and Physics of Solids*, Vol. 14, p. 203.

Stoker, J. J. (1950). *Nonlinear Vibrations*. Interscience Publishers, Inc., New York.

Thompson, J. M. T. (1963). Basic principles in the general theory of elastic stability, *Journal of the Mechanics and Physics of Solids*, Vol. 11, p. 13.

Thompson, J. M. T. (1964). Eigenvalue branching configurations and the Rayleigh-Ritz procedure, *Quarterly of Appl. Math*, Vol. 22, p. 244.

Whittaker, E. T. (1944). *Analytical Dynamics*. Dover Publications, New York.

3

ELASTIC BUCKLING OF COLUMNS

In this chapter, the problem of elastic buckling of bars will be studied using the approach discussed in Chapter 1 and demonstrated in Chapter 2. To accomplish this, we will derive the equations governing equilibrium for structural elements of class 2 in Chapter 1 (Section 1.3) along with the proper boundary conditions. This derivation is based on the Euler-Bernoulli assumptions, listed below, and principle of the stationary value of the total potential (Appendix).

In analyzing slender rods and beams, we make the following basic engineering assumptions:

1. The material of the element is homogeneous and isotropic.
2. Plane sections remain plane after bending.
3. The stress-strain curve is identical in tension and compression.
4. No local type of instability will occur.
5. The effect of transverse shear is negligible.
6. No appreciable initial curvature exists.
7. The loads and the bending moments act in a plane passing through a principal axis of inertia of the cross section.
8. Hooke's law holds.
9. The deflections are small as compared to the cross-sectional dimensions.

In addition, the loads are assumed to be coplanar, applied quasistatically, and are either axial or transverse. The transverse loads include distributed loads, $q(x)$, concentrated loads, P_i, and concentrated couples, C_j. Finally, the ends of the structures are supported in such a way that primary degrees of freedom (translation and rotation as a rigid body) are constrained. Before proceeding with the derivation of the equilibrium equations and boundary conditions, it is desirable to define and discuss the properties of some special functions.

3.1   SPECIAL FUNCTIONS

The following special functions and their properties will be used in the development of the theory of slender rods and beams. See Fig. 3.1 for their graphical representation.

1. Macauley's Bracket

$$[x - x_i] = \begin{cases} 0 & \text{for } x < x_i \\ x - x_i & \text{for } x > x_i \end{cases} \tag{1}$$

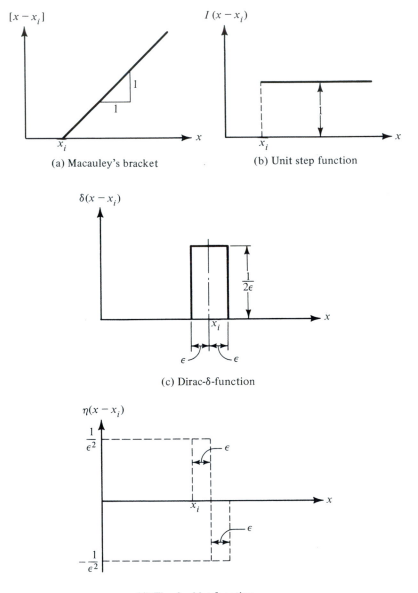

(a) Macauley's bracket   (b) Unit step function

(c) Dirac-δ-function

(d) The doublet function

FIGURE 3.1   Special functions.

2. Unit Step Function

$$I(x - x_i) = \begin{cases} 0 & \text{for } x < x_i \\ 1 & \text{for } x \geq x_i \end{cases} \tag{2}$$

Note that

$$[x - x_i] = (x - x_i)I(x - x_i)$$

and similarly

$$[x - x_i]^2 = (x - x_i)^2 I(x - x_i)$$

3. Dirac δ-Function (Carslaw, 1947). The Dirac δ-function in all applications is considered as a result of a limiting process which involves a function $\delta(x, \varepsilon)$ subject to the following conditions:

$$\delta(x, \varepsilon) \geq 0 \text{ for } -\infty \leq x \leq \infty \quad \text{and} \quad 0 < \varepsilon < \infty$$

$$\int_{-\infty}^{\infty} \delta(x, \varepsilon)dx = 1 \quad \text{for} \quad 0 < \varepsilon < \infty$$

An example of such a function is the following:

$$\delta(x - x_i) = \begin{cases} 0 & \text{if } x < x_i - \varepsilon \\ \dfrac{1}{2}\varepsilon & \text{if } x_i - \varepsilon \leq x \leq x_i + \varepsilon \\ 0 & \text{if } x > x_i + \varepsilon \end{cases} \tag{3}$$

Note that

$$\int_{-\infty}^{\infty} f(x)\delta(x - x_i)dx = f(x_i)$$

4. The Doublet Function. Let this function be denoted by $\eta(x - x_i)$. This is a special function such that

$$\frac{d\delta(x - x_i)}{dx} = \eta(x - x_i)$$

Another property of this function (see Shames 1964 for detailed discussion) is that

$$\int_{-\infty}^{+\infty} f(x)\eta(x - x_i)dx = -\frac{df}{dx}(x_i)$$

A particular function that has the foregoing properties is defined as

$$\eta(x - x_i) = \begin{cases} 0 & x < x_i \\ \dfrac{1}{\varepsilon^2} & x_i < x < x_i + \varepsilon \\ 0 & x = x_i + \varepsilon \\ -\dfrac{1}{\varepsilon^2} & x_i + \varepsilon < x < x_i + 2\varepsilon \\ 0 & x > x_i + 2\varepsilon \end{cases} \tag{4}$$

3.2   BEAM THEORY

The equilibrium equations and proper boundary conditions for an initially straight beam under transverse and axial loads (beam-column) are derived using the principle of the stationary value of the total potential. (See Part II of Hoff 1956 and Appendix A.)

Consider the beam of length L, shown in Fig. 3.2, under the action of a distributed local $q(x)$, n concentrated forces, P_i, m concentrated couples, C_j, and boundary forces and couples as shown. If u and w denote the displacement components of the reference surface (actually, here we deal with a two-dimensional problem, and the reference plane is the locus of the centroids), the extensional strain of any material point, z units from the reference surface, is given by

$$\varepsilon_{xx} = \varepsilon_{xx}^0 + z k_{xx} \tag{5}$$

where ε_{xx}^0 is the extensional strain on the reference plane (average strain) and κ_{xx} is the change in curvature of the reference plane.

The first-order nonlinear strain-displacement relation is given by

$$\varepsilon_{xx}^0 = u_{,x} + \frac{1}{2} w_{,x}^2 \tag{6}$$

where the comma denotes differentiation with respect to coordinate x, u, and w are displacement components of the reference plane.

The curvature for the reference plane is approximated by

$$k_{xx} = -w_{,xx} \tag{7}$$

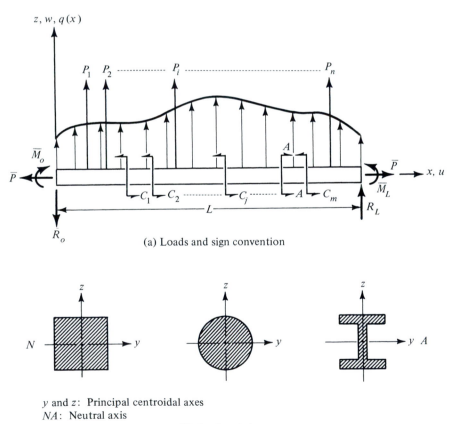

(a) Loads and sign convention

y and z: Principal centroidal axes
NA: Neutral axis

(b) Section A-A

FIGURE 3.2   Beam geometry and sign convention.

In addition, the mathematical expression of Hooke's law is

$$\sigma_{xx} = E\varepsilon_{xx} \tag{8}$$

If U_i and U_p denote the strain energy and potential of external forces, respectively, and U_T, the total potential of the system, then

$$U_i + U_p = U_T$$

Use of the principle of the stationary value of the total potential leads to

$$\delta_\varepsilon U_T = \delta_\varepsilon U_i + \delta_\varepsilon U_p = 0 \tag{9}$$

where ε denotes variations with respect to strains and displacements.

Since the variation of the strain energy (see Part II of Hoff, 1956) is given by

$$
\begin{aligned}
\delta_\varepsilon U_i &= \int_V \sigma_{xx}\delta\varepsilon_{xx}dV \\
&= \int_0^L \int_A E\left(\varepsilon_{xx}^0 + zk_{xx}\right)\left(\delta\varepsilon_{xx}^0 + z\delta k_{xx}\right)dA\ dx
\end{aligned} \tag{10}
$$

and since ε_{xx}^0, κ_{xx}, and their variations are only functions of x (note that x is a centroidal axis), then

$$\int_A z\ dA = 0$$

and

$$\delta_\varepsilon U_i = \int_0^L \left(P\delta\varepsilon_{xx}^0 + EIk_{xx}\delta k_{xx}\right)dx$$

where

$$P = E\varepsilon_{xx}^0 \int_A dA = EA\varepsilon_{xx}^0$$

and

$$I = \int_A z^2 dA$$

Next, replacing the change in curvature and the variations in strain and change in curvature by the displacement components and their variations, we obtain

$$\delta_\varepsilon U_i = \int_0^L \left[P\left(\delta u_{,x} + w_{,x}\delta w_{,x}\right) + EIw_{,xx}\delta w_{,xx}\right]dx \tag{11}$$

Note that

$$\delta_\varepsilon\left(\frac{1}{2}w_{,x}^2\right) = \frac{1}{2}\left[\left(w_{,x} + \delta w_{,x}\right)^2 - w_{,x}^2\right] = \frac{1}{2}\left[2w_{,x}\delta w_{,x} + \left(\delta w_{,x}\right)^2\right]$$

and neglecting higher-order terms (assuming small changes),

$$\delta_\varepsilon\left(\frac{1}{2}w_{,x}^2\right) = w_{,x}\delta w_{,x}$$

Integration by parts of Eq. (11) yields

$$\delta_\varepsilon U_i = P\delta u\Big|_0^L + Pw_{,x}\delta w\Big|_0^L + EIw_{,xx}\delta w_{,x}\Big|_0^L - (EIw_{,xx})_{,x}\delta w\Big|_0^L$$
$$+ \int_0^L \left[-P_{,x}\delta u - (Pw_{,x})_{,x}\delta w + (EIw_{,xx})_{,xx}\delta w\right]dx \tag{12}$$

Similarly, the variation in the potential of the external forces is given by

$$\delta_\varepsilon U_p = -\int_0^L q\delta w \, dx - \sum_{i=1}^n P_i\delta w(x_i) - \sum_{j=1}^m C_j\delta w_{,x}(x_j)$$
$$- \overline{P}u(L) + \overline{P}u(0) + M_0\delta w_{,x}(0) - M_L\delta w_{,x}(L) + R_0\delta w(0) - R_L\delta w(L) \tag{13}$$

If we introduce the special functions $\delta(x - x_i)$ and $\eta(x - x_i)$, the expression for the variation of the potential of the external forces becomes

$$\delta_\varepsilon U_p = -\int_0^L \left[q + \sum_{i=1}^n P_i\delta(x - x_j) - \sum_{j=1}^m C_j\eta(x - x_j)\right]\delta w \, dx$$
$$- (\overline{P}\delta u)\Big|_0^L - (\overline{M}\delta w_{,x})\Big|_0^L - (R\delta w)\Big|_0^L \tag{14}$$

Substitution of Eqs. (12) and (14) into Eq. (9) finally yields

$$\delta_\varepsilon U_T = \int_0^L \left[-P_{,x}\delta u + \left\{(EIw_{,xx})_{,xx} - (Pw_{,x})_{,x} - q - \sum_{i=1}^n P_i\delta(x - x_i)\right.\right.$$
$$\left.\left. + \sum_{j=1}^m C_j\eta(x - x_j)\right\}\delta w\right]dx + (P - \overline{P})\delta u\Big|_0^L$$
$$+ \left[Pw_{,x} - (EIw_{,xx})_{,x} - R\right]\delta w\Big|_0^L + \left[EIw_{,xx} - \overline{M}\right]\delta w_{,x}\Big|_0^L = 0 \tag{15}$$

The identical satisfaction of Eq. (15) (since δu and δw are arbitrary displacement functions) leads to the governing differential equations and the proper boundary conditions.
The differential equations are

$$P_{,x} = 0$$
$$(EIw_{,xx})_{,xx} - Pw_{,xx} = q + \sum_{i=1}^n P_i\delta(x - x_i) - \sum_{j=1}^m C_j\eta(x - x_j) \tag{16}$$

The proper boundary conditions are given by

Either	Or
$P = \overline{P}$	$\delta u = 0 \Rightarrow u = \overline{u}$
$\overline{P}w_{,x} - (EIw_{,xx})_{,x} = R$	$\delta w = 0 \Rightarrow w = 0$
$EIw_{,xx} = \overline{M}$	$\delta w_{,x} = 0 \Rightarrow w_{,x} = 0$

It is clearly shown above that, at the boundaries ($x = 0, L$), we may prescribe either the forces and moments or the displacements and rotations, but not both.

Examples:
A free edge with no moment or shear force applied is characterized by

$$\overline{P}w_{,x} - (EIw_{,xx})_{,x} = 0, \quad EIw_{,xx} = 0, \quad \text{and} \quad P = \overline{P} \quad \text{or} \quad u = \overline{u} \tag{17a}$$

A simply supported edge is characterized by

$$w = 0, \quad EIw_{,xx} = 0, \quad \text{and} \quad P = \overline{P} \quad \text{or} \quad u = \overline{u} \tag{17b}$$

Finally, a clamped edge is characterized by

$$w = 0, \quad w_{,x} = 0, \quad \text{and} \quad P = \overline{P} \quad \text{or} \quad u = \overline{u} \tag{17c}$$

Note that the first of Eqs. (16) implies that $P = \text{constant}$, and from the boundary condition, this constant is equal to $\overline{P}$. In the case where the end shortening is prescribed ($\overline{u}$), there is a $P = \text{constant}$ corresponding to each value of $\overline{u}$.

3.3 BUCKLING OF COLUMNS

When a bar is initially straight and of perfect geometry and it is subjected to the action of a compressive force without eccentricity, then it is called an *ideal column*. When the load is applied quasistatically, the column is simply compressed but remains straight. We then need to know if the column will remain straight no matter what the level of the applied force is. To determine this, we seek nontrivial solutions ($w \neq 0$) for the equations governing the bending (see Eqs. 16 with $q = 0$, $P_i = 0$, and $C_j = 0$) of this column under an axial compressive load ($-\overline{P}$) and subject to the particular set of boundary conditions. Note that in deriving the governing differential equations, it was assumed that the applied compressive load remained parallel to its original direction and there was no eccentricity in either the geometry or the applied load. Thus, the problem has been reduced to an eigen-boundary-value problem.

3.3.1 SOLUTION

In this case the solution of the problem will be discussed for a number of boundary conditions. It will be shown that the manner in which the column is supported at the two ends affects the critical load considerably. This approach to the problem is known as the classical, equilibrium, or bifurcation approach. In addition, the other approaches (dynamic and energy) will be demonstrated.

1. Simply Supported Ideal Column. The mathematical formulation of this problem is given below.

$$\text{D.E.} \quad \left(EIw_{,xx}\right)_{,xx} + \overline{P}w_{,xx} = 0 \tag{18a}$$

$$\text{B.C.'s} \quad \begin{aligned} w(0) = w(L) = 0 \\ w_{,xx}(0) = w_{,xx}(L) = 0 \end{aligned} \tag{18b}$$

Assuming that the bending stiffness (EI) of the column is constant and introducing the parameter $k^2 = \overline{P}/EI$ allows us to write the governing differential equation in the following form

$$w_{,xxxx} + k^2 w_{,xx} = 0 \tag{18c}$$

The general solution of this equation is given by

$$w = A_1 \sin kx + A_2 \cos kx + A_3 x + A_4 \tag{19}$$

This solution must satisfy the prescribed boundary conditions. This requirement leads to four linear homogeneous algebraic equations in the four constants A_i. A nontrivial solution then exists if all four constants are not identically equal to zero. This can happen only if the determinant of the coefficients of the A_i's vanishes or

$$\begin{vmatrix} 0 & 1 & 0 & 1 \\ \sin kL & \cos kL & L & 1 \\ 0 & -k^2 & 0 & 0 \\ -k^2 \sin kL & -k^2 \cos kL & 0 & 0 \end{vmatrix} = 0 \tag{20}$$

The expansion of this determinant leads to

$$\sin kL = 0$$

The solution of this equation is

$$kL = n\pi \quad n = 1, 2, \ldots$$

or

$$\bar{P} = \frac{n^2 \pi^2 EI}{L^2}$$

and the smallest of these corresponds to $n = 1$. Thus

$$P_{cr} = \frac{\pi^2 EI}{L^2}$$

This is known as the Euler equation because the problem was first solved by Leonhard Euler (see Timoshenko, 1953).

Note that if A denotes the cross-sectional area of the column and ρ is the radius of gyration of the cross-sectional area, the critical stress is given by

$$\sigma_{cr} = \frac{\pi^2 EI}{AL^2} = \frac{\pi^2 \rho^2 E}{L^2} = \frac{\pi^2 E}{(L/\rho)^2} \tag{21a}$$

and the corresponding strain is

$$\varepsilon_{cr} = \left(\frac{\pi\rho}{L}\right)^2 \tag{21b}$$

The displacement function corresponding to $n = 1$ is $w = A_1 \sin \pi x / L$.

2. *Clamped Ideal Column.* For this particular problem, the mathematical formulation is given below

$$\text{D.E.} \quad w_{,xxxx} + k^2 w_{,xx} = 0$$
$$\text{B.C.'s} \quad w(0) = w(L) = 0 \tag{22}$$
$$w_{,x}(0) = w_{,x}(L) = 0$$

The solution is given by Eq. (19) and it must satisfy the prescribed boundary conditions. The characteristic equation for this case is given by

$$\begin{vmatrix} 0 & 1 & 0 & 1 \\ \sin kL & \cos kL & L & 1 \\ k & 0 & 1 & 0 \\ k \cos kL & -k \sin kL & 1 & 0 \end{vmatrix} = 0 \tag{23}$$

Expansion of this determinant yields the following equation

$$2(\cos kL - 1) + kL \sin kL = 0 \tag{24}$$

Since

$$\cos kL - 1 = -2 \sin^2 \frac{kL}{2}$$

and

$$\sin kL = 2 \sin \frac{kL}{2} \cos \frac{kL}{2}$$

then Eq. (24) becomes

$$\sin \frac{kL}{2} \left(\frac{kL}{2} \cos \frac{kL}{2} - \sin \frac{kL}{2} \right) = 0$$

Then either

$$\sin \frac{kL}{2} = 0$$

or

$$\frac{kL}{2} \cos \frac{kL}{2} = \sin \frac{kL}{2}$$

The first of these solutions leads to

$$\overline{P} = \frac{4n^2 \pi^2 EI}{L^2} \quad n = 1, 2, \dots$$

from which

$$P_{cr} = \frac{4\pi^2 EI}{L^2} \tag{25a}$$

and

$$\sigma_{cr} = \frac{4\pi^2 E}{(L/\rho)^2}$$

$$\varepsilon_{cr} = 4 \left(\frac{\pi \rho}{L} \right)^2 \tag{25b}$$

The second of the solutions leads to $P_{cr} > 4\pi^2 EI/L^2$. The displacement function corresponding to $n = 1$ is

$$w = A_2 \left(\cos \frac{2\pi x}{L} - 1 \right)$$

3. *Ideal Column with One End Clamped and the Other Free.* For this case the boundary conditions are (assuming that the fixed end is at $x = 0$)

$$\left. \begin{array}{r} w(0) = w_{,x}(0) = 0 \\ w_{,xx}(L) = 0 \\ k^2 w_{,x}(L) + w_{,xxx}(L) = 0 \end{array} \right\} \tag{26}$$

The solution is still given by Eq. (19) and the characteristic equation is

$$
\begin{vmatrix}
0 & 1 & 0 & 1 \\
k & 0 & 1 & 0 \\
-k^2 \sin kL & -k^2 \cos kL & 0 & 0 \\
0 & 0 & k^2 & 0
\end{vmatrix} = 0 \tag{27}
$$

which expanded results in the following equation

$$
\left.\begin{aligned}
k^5 \cos kL &= 0 \\[1em]
\cos kL &= 0
\end{aligned}\right\} \tag{28}
$$

or

This equation leads to the following result

$$
\overline{P} = \left(\frac{2m-1}{2}\right)^2 \frac{\pi^2 EI}{L^2} \quad m = 1, 2, \ldots
$$

and

$$
\left.\begin{aligned}
P_{cr} &= \frac{\pi^2 EI}{4L^2} \\[0.8em]
\sigma_{cr} &= \frac{\pi^2 E}{4(L/\rho)^2} \\[0.8em]
\varepsilon_{cr} &= \frac{1}{4}\left(\frac{\pi\rho}{L}\right)^2
\end{aligned}\right\} \tag{29}
$$

The displacement function for this case is given by

$$
w = A_2\left(\cos\frac{\pi x}{2L} - 1\right)
$$

3.3.2 REDUCTION OF THE ORDER OF THE DIFFERENTIAL EQUATION

If the moment and shear are prescribed at $x = 0$, then it is possible to reduce the order of the governing differential equation from four to two.

Starting with Eq. (18a) under the assumption of constant flexural stiffness, two consecutive integrations yield the following equations

$$
EIw_{,xxx} + \overline{P}w_{,x} = B_1
$$

and

$$
EIw_{,xx} + \overline{P}w = B_1 x + B_2
$$

If, in addition, the normal displacement w is measured from the left end, then $w(0) = 0$.

Then the constants B_1 and B_2 can be evaluated from the known moment and shear and (see Fig. 3.2; R_1 opposite to positive w).

$$
B_1 = -R_0
$$
$$
B_2 = +\overline{M}_0
$$

Thus the governing differential equation reduces to

$$EIw_{,xx} + \overline{P}w = -R_0x + \overline{M}_0 \tag{30}$$

Note that, for the case of simply supported ideal columns, $\overline{R}_0 = \overline{M}_0 = 0$ and the equation becomes

$$EIw_{,xx} + \overline{P}w = 0 \tag{31}$$

3.3.3   EFFECTIVE SLENDERNESS RATIO

We have seen from the previous sections that the critical load of a compressed ideal column is affected by the boundary conditions. For all possible boundary conditions, the critical load may be expressed by

$$P_{cr} = C\frac{\pi^2 EI}{L^2} \tag{32}$$

where C is a constant that depends on the boundary conditions and is called the *end fixity factor*. Thus for both ends simply supported, $C = 1$; for one end fixed and the other free, $C = {}^1\!/_4$; and so on (see Table 3.1).

Similarly, the corresponding critical stress is given by

$$\sigma_{cr} = C\frac{\pi^2 E}{(L/\rho)^2} \tag{33}$$

where the parameter L/ρ is known as the slenderness ratio of the ideal column. For the case of simply supported ends, the end fixity factor is equal to unity. By defining a new parameter L' through

$$L' = \frac{L}{\sqrt{C}} \tag{34}$$

we may use the simply-supported-ends equation, for all cases of boundary conditions, through the equivalent slenderness ratio L'/ρ. Thus

$$\sigma_{cr} = \frac{\pi^2 E}{(L'/\rho)^2} \tag{35}$$

Note that a single curve of σ_{cr} plotted versus L'/ρ represents critical loads for all possible geometries and boundary conditions of ideal columns (see Fig. 3.3).

3.3.4   IMPERFECT COLUMNS

So far, we have concentrated on ideal columns. In practice, though, no column is of perfect geometry, and the applied load does not necessarily pass through the

TABLE 3.1   End fixity factors

Boundary Conditions	C	L'
Both ends simply supported	1	L
One end fixed, the other free	$\dfrac{1}{4}$	$2L$
Both ends fixed	4	$\dfrac{L}{2}$
One end fixed, the other simply supported	$\left(\dfrac{4.493}{\pi}\right)^2$	$0.699L$

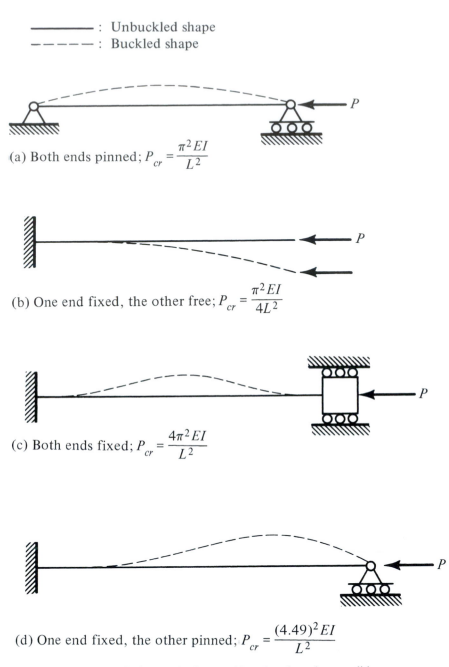

——————— : Unbuckled shape

— — — — — : Buckled shape

(a) Both ends pinned; $P_{cr} = \dfrac{\pi^2 EI}{L^2}$

(b) One end fixed, the other free; $P_{cr} = \dfrac{\pi^2 EI}{4L^2}$

(c) Both ends fixed; $P_{cr} = \dfrac{4\pi^2 EI}{L^2}$

(d) One end fixed, the other pinned; $P_{cr} = \dfrac{(4.49)^2 EI}{L^2}$

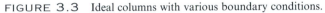

FIGURE 3.3   Ideal columns with various boundary conditions.

centroid of the column cross section. It is therefore necessary to study the behavior of columns of imperfect geometries and of columns for which the load is applied eccentrically. Another reason for studying the behavior of columns of imperfect geometries is that by letting the imperfection disappear (limiting process), we can predict the behavior of the perfect system.

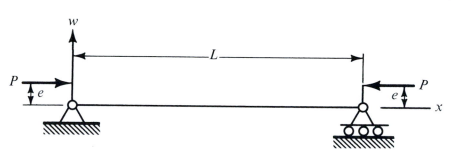

FIGURE 3.4   Eccentrically loaded column.

1. Eccentrically Loaded Columns.   Consider first the case of a simply supported column (see Fig. 3.4) loaded eccentrically with the same eccentricity e at both ends. The equilibrium equation for this case is

$$w_{,xxxx} + k^2 w_{,xx} = 0 \qquad (36a)$$

The boundary conditions are given by

$$\left.\begin{aligned} w(0) &= 0 \\ w(L) &= 0 \\ EIw_{,xx}(0) &= Pe \Rightarrow w_{,xx}(0) = k^2 e \\ EIw_{,xx}(L) &= Pe \Rightarrow w_{,xx}(L) = k^2 e \end{aligned}\right\} \qquad (36b)$$

The solution for this equation is

$$w = A_1 \sin kx + A_2 \cos kx + A_3 k + A_4 \qquad (19)$$

Use of the boundary conditions, Eqs. (36b), yields

$$w(x) = e\left(1 - \cos kx - \tan\frac{kL}{2}\sin kx\right) \qquad (37)$$

Note that as the load P increases from zero, the value of k, and consequently $\tan kL/2$, increases. Therefore the displacement function w becomes unbounded as $\tan kL/2$ approaches infinity and the corresponding load P approaches $\pi^2 EI/L^2$.

If Eq. (37) is evaluated at some characteristic point (say $x = L/2$), it may serve as a load-displacement relation. Denoting by δ the displacement at the midpoint, we may write

$$\delta = -e\left(1 - \cos\frac{kL}{2} - \tan\frac{kL}{2}\sin\frac{kL}{2}\right)$$

or

$$\delta = e\left(\sec\sqrt{\frac{P}{EI}}\frac{L}{2} - 1\right) \qquad (38)$$

This result is plotted in Fig. 3.5, which shows that as $e \to 0$, the plot represents the behavior of the perfect system.

Finally, if the eccentricities are equal in magnitude but opposite in direction, the last two of the boundary conditions, Eqs. (36b), are given by

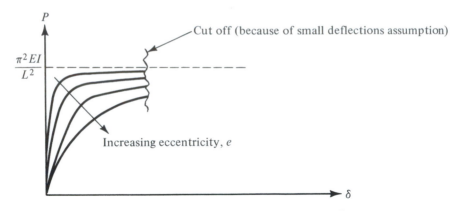

FIGURE 3.5   P-δ diagram with eccentricity effect.

$$\left.\begin{array}{l} w_{,xx}(0) = -k^2 e \\ w_{,xx}(L) = k^2 e \end{array}\right\} \tag{39}$$

and the solution is

$$w(x) = e\left(-1 + \frac{2x}{L} + \cos kx - \cot\frac{kL}{2}\sin kx\right) \tag{40}$$

For this case the displacement becomes unbounded when the load P approaches $4\pi^2 EI/L^2$, and

$$w\left(\frac{L}{2}\right) = 0$$

and

$$w\left(\frac{L}{4}\right) = \frac{e}{2}\left(\sec\sqrt{\frac{P}{EI}}\frac{L}{4} - 1\right) \tag{41}$$

2. Columns with Geometric Imperfections.   Next consider a simply supported column with an initial geometric imperfection, $w_0(x)$, loaded along the axis joining the end points (see Fig. 3.6).

The governing differential equation for this case becomes

$$EI\left(w_{,xx} - w_{0,xx}\right) + Pw = 0 \tag{42}$$

Note that the change in curvature for this case is $\left(-w_{,xx} + w_{0,xx}\right)$.

The boundary conditions for this problem are $w(0) = w(L) = 0$. Note that $w_0(0) = w_0(L) = 0$.

Since the initial shape is continuous and has a finite number of maxima and minima in the range $0 < x < L$, then the shape can be represented by a sine series (there is no initial moment at the boundaries).

$$w_0(x) = \sum_{n=1}^{\infty} a_n \sin\frac{n\pi x}{L}$$

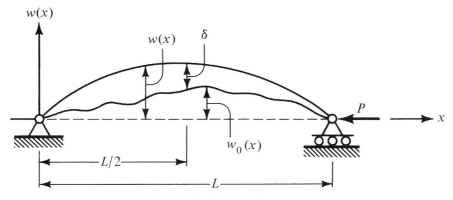

FIGURE 3.6   Imperfect column.

Then the differential equation, Eq. (42), assumes the following form

$$EIw_{,xx} + Pw = EI \sum_{n=1}^{\infty} \left(\frac{n\pi}{L}\right)^2 a_n \sin\frac{n\pi x}{L} \tag{43}$$

A series solution may be assumed, each term of which satisfies the boundary conditions

$$w(x) = \sum_{j=1}^{\infty} A_j \sin\frac{j\pi x}{L} \tag{44}$$

Substitution into Eq. (43) leads to the following equation:

$$-\left[EI\left(\frac{n\pi}{L}\right)^2 - P\right]A_n = EI\left(\frac{n\pi}{L}\right)^2 a_n \tag{45}$$

Thus the solution becomes

$$w(x) = \sum_{n=1}^{\infty} \frac{P_n}{P_n - P} a_n \sin\frac{n\pi x}{L} \tag{46}$$

and

$$w - w_0 = \sum_{n=1}^{\infty} \frac{P}{P_n - P} a_n \sin\frac{n\pi x}{L} \tag{47}$$

where

$$P_n = \frac{n^2 \pi^2 EI}{L^2}$$

Southwell (1936) considered this problem of initial geometric imperfections; and he concluded that, as long as the imperfection is such that a_1 exists, then a critical condition arises as $P \to \pi^2 EI/L^2$. Furthermore, since in a test one can measure P and the net deflection of the midpoint $[w(L/2) - w_0(L/2) = \delta]$, he devised a plot from which $P_1 = P_E$ can be determined through use of the experimental data. This plot is known as the Southwell plot.

Assume that $a_1 \neq 0$ and all $a_n = 0$ for $n = 2, 3, 4, \dots$. Then

$$\delta = w\left(\frac{L}{2}\right) - w_0\left(\frac{L}{2}\right) = \frac{Pa_1}{P_1 - P} \tag{48}$$

From this we obtain

$$P_1 \left(\frac{\delta}{P} \right) - a_1 = \delta \tag{49}$$

And thus δ/P is a linear function of δ.

Since, as $P \to P_1$, the first term of the series is the predominant one, Eq. (47), then for large values of P (but $P < P_1$) it can safely be assumed that Eq. (49) holds and δ/P varies linearly with δ. Thus, if for a test we plot δ/P vs δ at the higher values of δ, the relation is linear and the intercepts give a_1 and a_1/P_1 (see Fig. 3.7).

3.3.5   TILTING OF FORCES

In many practical applications where stability of columns is a basic design criterion, the force does not remain fixed in direction, but it passes through a fixed point. This is a case of load-behavior (during the buckling process) problems where the system may still be considered as conservative. Note that the follower-force problem does not fall in this category. Examples of these types of problems are shown in Figs. 3.8 and 3.9.

Consider an elastic column, as shown in Fig. 3.10, with the applied load P passing through a fixed point C. Using the free end as the origin, for displacements and the x-coordinate, the governing equation (equilibrium) is

$$M = EIw_{,xx} = -(P \cos \varphi)w - (P \sin \varphi)x$$

If φ is a small constant (this implies that P is initially slightly tilted), the equation becomes

$$w_{,xx} + k^2 w = -k^2 \varphi x \tag{50}$$

The associated boundary conditions are

$$w(0) = 0, \quad w_{,x}(L) = 0$$

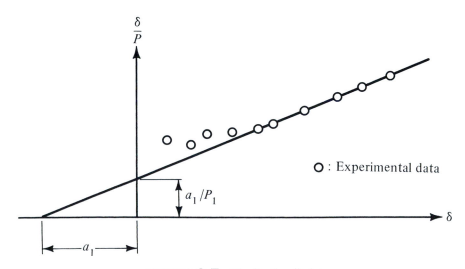

FIGURE 3.7   The Southwell plot.

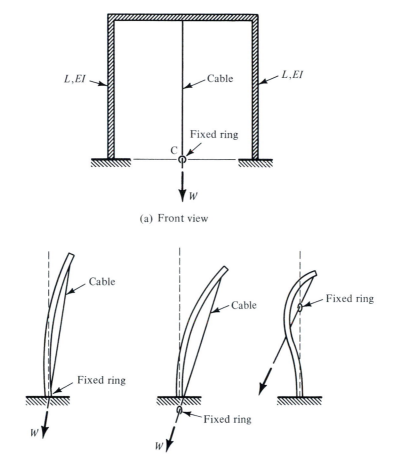

(a) Front view

(b) Side views

FIGURE 3.8 Loaded frame.

and the auxiliary condition

$$w(L) = \delta \approx a\varphi$$

The general solution of Eq. (50) is

$$w = -\varphi x + C_1 \sin kx + C_2 \cos kx \qquad (51)$$

Use of the above conditions yields

$$a\varphi = -\varphi L \left(1 - \frac{\tan kL}{kL} \right) \qquad (52)$$

Since we are only interested in the existence of a nontrivial solution, $\varphi \neq 0$, the characteristic equation is

$$\tan kL = kL \left(1 + \frac{a}{L} \right) \qquad (53)$$

This is a transcendental equation and it may be solved either numerically or graphically (see Fig. 3.11).

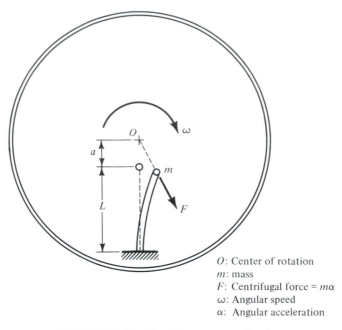

O: Center of rotation
m: mass
F: Centrifugal force = $m\alpha$
ω: Angular speed
α: Angular acceleration

FIGURE 3.9   Elastic bar on a rotating disc.

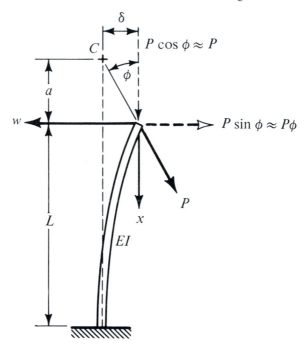

FIGURE 3.10   Tilt-buckled column.

It is seen from Fig. 3.11 that the critical value for (kL) lies between zero and 4.493. Note the following special cases:

1. If $a = -L$, $(kL)_{cr} = \pi$ and $P_{cr} = \pi^2 EI/L^2$.
2. If $a = \pm\infty$, $(kL)_{cr} = \pi/2$ and $P_{cr} = \pi^2 EI/4L^2$.

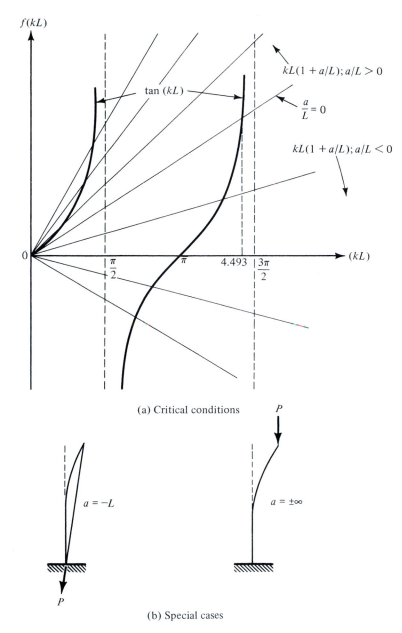

(a) Critical conditions

(b) Special cases

FIGURE 3.11   Critical conditions of tilt-buckling.

An alternate solution is possible if we use the fourth-order differential equation, Eq. (18c). The solution to this equation is given by Eq. (19), or

$$w(x) = A_1 \sin kx + A_2 \cos kx + A_3 x + A_4 \tag{19}$$

Referring to Fig. 3.10 and placing the origin at the fixed end with x increasing toward the free end, we obtain the boundary conditions

$$w(0) = 0$$

$$w_{,x}(0) = 0$$

$$w_{,xx}(L) = 0$$

$$-\left[EIw_{,xxx}(L) + Pw_{,x}(L)\right] = -P\left[-\frac{w(L)}{a}\right]$$

Use of the boundary conditions leads to the following characteristic equation

$$\begin{vmatrix} 0 & 1 & 0 & 1 \\ k & 0 & 1 & 0 \\ \sin kL & \cos kL & 0 & 0 \\ \sin kL & \cos kL & \left(\frac{L}{a}+1\right) & 1 \end{vmatrix} = 0$$

The expansion of this determinant is:

$$\tan kL = kL\left(1 + \frac{a}{L}\right) \tag{53}$$

Tilt-buckling is also possible for other systems. Biezeno and Grammel (1956) discuss a number of cases of tilt-buckling in torsion and in bending (see Fig. 3.12).

Consider a shaft of torsional rigidity GJ, length L, fixed at one end, and carrying a rigid disc of diameter a, loaded as shown in Fig. 3.12a. When the disc is in the tilted position, the applied torque is $aP \sin \varphi$. The torque transmitted to the shaft is $(GJ)(\varphi/L)$. From equilibrium considerations

$$GJ\frac{\varphi}{L} = aP \sin \varphi \tag{54}$$

It is clear that a bifurcation point exists at $\varphi = 0$ and $P = GJ/aL$ (see Fig. 3.12b). Thus

$$P_{cr} = \frac{GJ}{aL} \tag{55}$$

Next, consider the beam shown in Fig. 3.12c. If $\ell_1 = L$, the governing differential equation becomes

$$EIw_{,xx} = -\frac{aP \sin \varphi}{L}x \tag{56}$$

Two integrations with respect to x and the use of the boundary conditions $w(0) = w(L) = 0$ yield

$$w = \frac{aP}{6EIL}\varphi x\left(L^2 - x^2\right) \tag{57}$$

Since $\varphi = -w_{,x}(L)$, then

$$\varphi = \frac{aPL}{3EI}\varphi \tag{58}$$

From this equation we see that a nontrivial solution exists if

$$P_{cr} = \frac{3EI}{aL} \tag{59a}$$

Similarly, the critical load for the more general case is given by

$$P_{cr} = \frac{3EI}{a} \cdot \frac{\ell_1 + \ell_2}{\ell_1^2 - \ell_1\ell_2 + \ell_2^2} \tag{59b}$$

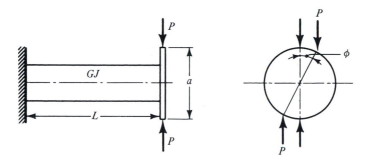

(a) Geometry of torsional shaft

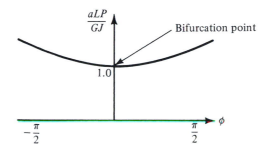

(b) Load-deflection curve for torsional tilt-buckling

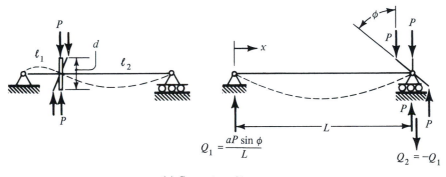

(c) Geometry of beams

FIGURE 3.12   Tilt-buckling of shafts and beams.

Note that the result given by Eq. (59a) can be derived from Eq. (59b) if we let $\ell_1 = L$ and $\ell_2 = 0$.

3.3.6   EFFECT OF TRANSVERSE SHEAR

One of the assumptions made in deriving the governing differential equations of beam bending (see Euler-Bernoulli assumptions) is that the effect of transverse shear on deformations is negligible. If this assumption is removed, the beam is called a Timoshenko beam, and the associated theory, Timoshenko theory; see Thomson (1993) and Timoshenko and Gere (1961).

Consider a beam element of length dx, as shown in Fig. 3.13a. The element deforms in a pure shear of magnitude γ and pure bending of magnitude $\psi_{,x}$, shown

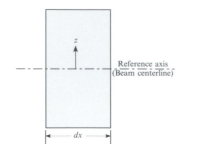

(a) Undeformed beam element

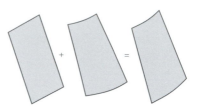

(b) Separate bending and shear deformations

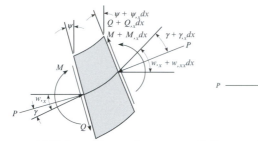

(c) Deformed beam element

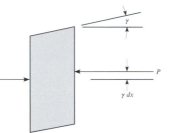

(d) Computation of bending moment caused by axial force

FIGURE 3.13   Geometry for shear corrections.

as decomposed in Fig. 3.13b the left and right faces of which rotate through angles ψ and $\psi + \psi_{,x}dx$, respectively as shown in Fig. 3.13c. The reference line rotates through (small) angles $w_{,x}$ and $w_{,x} + w_{,xx}dx$ at the left and right faces. Thus, the shear strain is

$$\gamma = w_{,x} - \psi \tag{60}$$

The resultant shear force is usually written as

$$Q = \frac{GA}{n}\left(w_{,x} - \psi\right) \tag{61}$$

where n is the so-called shear correction factor, included to account approximately for the nonuniformity of the shear strain. For a solid rectangular beam cross section $n = 1,2$; for a solid circular cross section $n = 1.11$; and for an I-beam, $n = 1.2A/A_f$ where A is the cross-sectional area and A_f is the area of the two flanges of the I-beam.

The bending stress at any position z relative to the midsurface is given by

$$\sigma_x = Ez\psi_{,x} \tag{62}$$

so that the bending moment can be written as

$$M = EI\psi_{,x} \tag{63}$$

From Fig. 3.13c, we can deduce two equilibrium equations

$$Q_{,x} - P\psi_{,x} = 0$$
$$M_{,x} + Q + P\gamma = 0 \tag{64}$$

The last term in the second equation arises from the relative position of the reference line on the left and right faces, as detailed in Fig. 3.13d. Note that the axial force is assumed to rotate with the section remaining normal to each face.

Considering now a simply supported column, the first of Eqs. (64) can be integrated to obtain

$$Q - P\psi = 0 \tag{65}$$

which, when substituted into the second, gives

$$M_{,x} + P(\gamma - \psi) = M_{,x} + Pw_{,x} = 0 \tag{66}$$

Again, for the simply-supported case, this equation can be integrated to obtain

$$M + Pw = EI\psi_{,x} + Pw = 0 \tag{67}$$

Combining Eq. (61) with this equation, one obtains

$$w_{,xx} + \frac{P}{EI}\left(1 + \frac{nP}{GA}\right) = 0 \tag{68}$$

Since the column is simply supported, the boundary conditions associated with the reduced-order equation, Eq. (68), are

$$w(0) = w(L) = 0 \tag{69}$$

It is easily concluded that the first eigenvalue is π and

$$P_{cr}\left(1 + \frac{nP_{cr}}{AG}\right) = \frac{\pi^2 EI}{L^2} = P_1 \tag{70}$$

Solving for P_{cr}, we obtain

$$P_{cr} = \frac{\sqrt{1 + 4nP_1/AG} - 1}{2n/AG} \tag{71a}$$

$$= \frac{2P_1}{1 + \sqrt{1 + 4nP_1/AG}} \tag{71b}$$

$$\approx P_1\left(1 - \frac{nP_1}{AG}\right) \tag{71c}$$

Note that equations (71a) and (71b) are equivalent.

Another expression is obtained for P_{cr} by a different approach; see Art. 2.17 of Timoshenko (1953). This expression is

$$P_{cr} = \frac{P_1}{1 + P_1/AG} \tag{72}$$

Note that Eqs. (71c) and (72) are approximately equal for $n = 1$.

3.4   THE KINETIC APPROACH

In using this approach, we are interested in the character of the motion (in the small) of the beam under constant P. In other words, if the column is compressed to any level of P, then the ends are made immovable (no additional axial deformation), and the column is given a small initial disturbance, what is the tendency of the system? If the system simply tends to oscillate about the undisturbed static equilibrium position, then the static equilibrium is stable. Before this can be accomplished, the differential equation governing the motion of a beam under the Euler-Bernoulli assumptions must be derived.

If m denotes the mass per unit length and if the effect of rotary inertia is neglected, then the equation of motion of the beam under constant axial load P is given by (see Churchill, 1958; Courant, 1953; and Thomson, 1965)

$$\left(EIw_{,xx}\right)_{,xx} + \bar{P}w_{,xx} + mw_{,tt} = 0 \tag{73}$$

If the ideal column is simply supported, then the solution of Eq. (73) is given by

$$w = f(t)\sin\frac{n\pi x}{L} \tag{74}$$

Note that the associated boundary conditions $w(0) = w(L) = w_{,xx}(0) = w_{,xx}(L) = 0$ are satisfied by this solution, Eq. (74).

Substitution into Eq. (80) yields

$$f_{,tt} + \left(\frac{n\pi}{L}\right)^2 \frac{1}{m}\left(\frac{n^2\pi^2 EI}{L^2} - P\right)f = 0 \tag{75}$$

and if

$$\omega_n^2 = \frac{1}{m}\left(\frac{n\pi}{L}\right)^2\left(\frac{n^2\pi^2 EI}{L^2} - P\right)$$

then

$$f_{,tt} + \omega_n^2 f = 0 \tag{76}$$

We see from Eq. (76) that, if $\omega_n^2 > 0$, then the motion is oscillatory, and if $\omega_n^2 < 0$, then the motion is diverging. Thus

$$P_{cr} = \frac{n^2\pi^2 EI}{L^2} \quad n = 1, 2, \ldots \tag{77}$$

and the smallest value corresponds to $n = 1$. Also note that the frequency of oscillations decreases as P approaches P_{cr}, while the frequency increases as the applied axial load increases in tension.

Another procedure that can be used here is as follows. Starting with Eq. (73), we may write the separated solution as

$$w(x, t) = g(x)e^{i\omega t} \tag{78}$$

where ω is the frequency of small oscillations. Use of this solution leads to the following ordinary differential equation

$$\left. \begin{aligned} EIg_{,xxxx} + Pg_{,xx} - m\omega^2 g = 0 \\[2ex] g_{,xxxx} + k^2 g_{,xx} - \frac{m\omega^2}{EI}g = 0 \end{aligned} \right\} \tag{79}$$

or

The general solution of this is

$$g(x) = \sum_{i=1}^{i=4} A_i e^{\lambda_i x} \tag{80}$$

where the A_i are constants and the λ_i are the roots of the biquadratic

$$\lambda^4 + k^2\lambda^2 - \frac{m\omega^2}{EI} = 0 \tag{81}$$

The λ_i roots are:

$$\left.\begin{array}{l} \lambda_{1,2} = \pm i\dfrac{k}{\sqrt{2}}\left(1 + \sqrt{1 + \dfrac{4m\omega^2}{k^4 EI}}\right)^{\frac{1}{2}} = \pm i\alpha \\[4mm] \lambda_{3,4} = \pm\dfrac{k}{\sqrt{2}}\left(-1 + \sqrt{1 + \dfrac{4m\omega^2}{k^4 EI}}\right)^{\frac{1}{2}} = \pm\beta \end{array}\right\} \tag{82}$$

The requirement of satisfying the prescribed boundary conditions leads to a set of four homogeneous algebraic equations in A_i. Since a nontrivial solution is sought, the determinant of the coefficient of the constants A_i must vanish. This is the characteristic equation.

To demonstrate this procedure, consider the case of a column with simply supported boundaries.

$$\begin{array}{r} g(0) = g(L) = 0 \\ g_{,xx}(0) = g_{,xx}(L) = 0 \end{array} \tag{83}$$

The characteristic equation for this case becomes

$$\begin{vmatrix} 1 & 1 & 1 & 1 \\ -\alpha^2 & -\alpha^2 & \beta^2 & \beta^2 \\ e^{-i\alpha L} & e^{i\alpha L} & e^{-\beta L} & e^{\beta L} \\ -\alpha^2 e^{-i\alpha L} & -\alpha^2 e^{i\alpha L} & \beta^2 e^{\beta L} & \beta^2 e^{\beta L} \end{vmatrix} = 0 \tag{84}$$

Subtraction of the first column from the second, the third from the fourth, and rearrangement of columns lead to the following

$$\begin{vmatrix} 1 & 1 & 0 & 0 \\ -\alpha^2 & -\beta^2 & 0 & 0 \\ e^{-i\alpha L} & e^{-\beta L} & e^{i\alpha L} - e^{-i\alpha L} & e^{\beta L} - e^{-\beta L} \\ -\alpha^2 e^{-i\alpha L} & \beta^2 e^{-\beta L} & -\alpha^2(e^{i\alpha L} - e^{-i\alpha L}) & \beta^2(e^{\beta L} - e^{-\beta L}) \end{vmatrix} = 0 \tag{85}$$

Expanding the above, we obtain

$$\left(\beta^2 + \alpha^2\right)^2 \left(e^{-i\alpha L} - e^{-i\alpha L}\right)\left(e^{\beta L} - e^{-\beta L}\right) = 0 \tag{86}$$

and since

$$\left.\begin{array}{l} e^{i\alpha L} - e^{-i\alpha L} = 2i\sin\alpha L \\ e^{\beta L} - e^{-\beta L} = 2\sinh\beta L \end{array}\right\} \tag{87}$$

then Eq. (86) becomes

$$4\left(\alpha^2 + \beta^2\right)^2 \sin\alpha L \sinh\beta L = 0 \tag{88}$$

It is easily seen that

$$4\left(\alpha^2 + \beta^2\right)^2 \sinh\beta L \neq 0 \tag{89}$$

Thus

$$\sin \alpha L = 0 \tag{90}$$

The solution of Eq. (90) leads to

$$\alpha L = n\pi \tag{91}$$

Replacing the expression for α and squaring both sides, we obtain

$$1 + \sqrt{1 + \frac{4m\omega^2}{k^4 EI}} = \frac{2n^2\pi^2}{L^2 k^2}$$

or

$$1 + \frac{4m\omega^2}{k^4 EI} = \frac{4n^4\pi^4}{L^4 k^4} - \frac{4n^2\pi^2}{L^2 k^2} + 1 \tag{92}$$

from which

$$\omega^2 = \frac{EI}{m}\left(\frac{n\pi}{L}\right)^2\left[\left(\frac{n\pi}{L}\right)^2 - k^2\right] \tag{93}$$

We see from Eq. (93) that the motion ceases to be oscillatory as $\omega \to 0$; consequently, the static equilibrium point corresponding to $k(P)$ ceases to be stable. Thus

$$P = \frac{n^2\pi^2 EI}{L^2}$$

and the smallest load corresponds to $n = 1$ or

$$P_{cr} = \frac{\pi^2 EI}{L^2}$$

3.5   ELASTICALLY SUPPORTED COLUMNS

In most structural configurations, columns are supported by other structural elements which provide elastic types of restraints at the ends of the columns. These restraints are similar to spring supports of the rotational (lb-inches per radian) as well as the extensional type (lbs per inch). In many cases, by knowing the structural configuration supporting the column, we can accurately estimate the intensity (spring constant) of the corresponding spring. In this section, the characteristic equation for a spring-supported column is derived and solutions are presented for a number of special cases.

In Fig. 3.14, the column of length L has a constant stiffness EI. It is spring-supported at both ends and is loaded axially by a compressive load P. The buckling equation is, as before,

$$w_{,xxxx} + k^2 w_{,xx} = 0 \quad k^2 = \frac{P}{EI} \tag{94}$$

and the boundary conditions are given by

$$\text{at } x = 0: \quad -\left[EIw_{,xxx} + Pw_{,x}\right] = \overline{\alpha}_0 w$$

$$EIw_{,xx} = \overline{\beta}_0 w_{,x} \tag{95a}$$

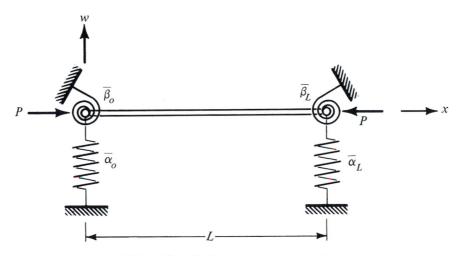

FIGURE 3.14 Elastically supported column.

$$\text{at } x = L: \quad -\left[EIw_{,xxx} + Pw_{,x}\right] = -\overline{\alpha}_L w \tag{95b}$$

$$EIw_{,xx} = -\overline{\beta}_L w_{,x}$$

If we introduce the new parameters α_0, α_L, β_0, and β_L

$$\left.\begin{array}{ll} \alpha_0 = \dfrac{\overline{\alpha}_0}{EI} & \beta_0 = \dfrac{\overline{\beta}_0}{EI} \\[2mm] \alpha_L = \dfrac{\overline{\alpha}_L}{EI} & \beta_L = \dfrac{\overline{\beta}_L}{EI} \end{array}\right\} \tag{96}$$

the buckling equation and the associated boundary conditions become

$$w_{,xxxx} + k^2 w_{,xx} = 0 \tag{97a}$$

$$w_{,xxx}(0) + k^2 w_{,x}(0) + \alpha_0 w(0) = 0 \tag{97b}$$

$$w_{,xx}(0) - \beta_0 w_{,x}(0) = 0 \tag{97c}$$

$$w_{,xxx}(L) + k^2 w_{,x}(L) - \alpha_L w(L) = 0 \tag{97d}$$

$$w_{,xx}(L) + \beta_L w_{,x}(L) = 0 \tag{97e}$$

Thus the problem has been reduced to an eigen-boundary-value problem.

The solution to the differential equation is given by Eq. (19), and the use of Eqs. (97a) leads to the following four linear homogeneous algebraic equations in the A_i's ($i = 1, 2, 3, 4$):

$$\left.\begin{array}{l} \alpha_0 A_2 + k^2 A_3 + \alpha_0 A_4 = 0 \\[1mm] \beta_0 k A_1 + k^2 A_2 + \beta_0 A_3 = 0 \\[1mm] (\alpha_L \sin kL) A_1 + (\alpha_L \cos kL) A_2 - (k^2 - \alpha_L L) A_3 + \alpha_L A_4 = 0 \\[1mm] (k\beta_L \cos kL - k^2 \sin kL) A_1 - (\beta_L k \sin kL + k^2 \cos kL) A_2 + \beta_L A_3 = 0 \end{array}\right\} \tag{98}$$

The requirement of the existence of a nontrivial solution leads to the vanishing of the determinant

$$
\begin{vmatrix}
0 & \alpha_1 & k^2 & \alpha_1 \\
\beta_0 k & k^2 & \beta_L & 0 \\
\alpha_L \sin kL & \alpha_L \cos kL & \alpha_L L - k^2 & \alpha_L \\
(k\beta_L \cos kL - k^2 \sin kL) & -(k\beta_L \sin kL + k^2 \cos kL) & \beta_L & 0
\end{vmatrix} = 0 \quad (99)
$$

If we denote the quantity kL by u, the characteristic equation is given by the following transcendental equation:

$$
\begin{aligned}
&\left\{ -(\alpha_0 + \alpha_L)\frac{u^6}{L^6} + \left[\beta_0\beta_L(\alpha_0 + \alpha_L) + \alpha_0\alpha_L L \right]\frac{u^4}{L^4} \right. \\
&\left. + \alpha_0\alpha_L(\beta_0 + \beta_L - \beta_0\beta_L L)\frac{u^2}{L^2} \right\} \sin u \\
&+ \left[(\alpha_0 + \alpha_L)(\beta_0 + \beta_L)\frac{u^5}{L^5} - \alpha_0\alpha_L L(\beta_0 + \beta_L)\frac{u^3}{L^3} \right. \\
&\left. - 2\alpha_0\alpha_L\beta_0\beta_L\frac{u}{L} \right] \cos u + 2\alpha_0\alpha_L\beta_0\beta_L\frac{u}{L} = 0
\end{aligned}
\quad (100)
$$

Solution of this transcendental equation (the smallest positive root) yields u_{cr}, from which we can compute P_{cr}:

$$
P_{cr} = \frac{u_{cr}^2 EI}{L^2} \quad (101)
$$

A number of special cases are reported below. These are (1) both ends pinned; (2) both ends clamped; (3) one end clamped and the other free; (4) one end clamped and the other pinned; and (5) both ends free against translation, constrained against rotation.

1. *Both Ends Pinned.* This case is characterized by $\alpha_0 = \alpha_L = \infty$ and $\beta_0 = \beta_L = 0$. If every term of Eq. (100) is divided by $\alpha_0\alpha_L$, and if we let $1/\alpha_0$, $1/\alpha_L$, β_0, and β_L approach zero, then

$$
L\frac{u^4}{L^4}\sin u = 0 \quad (102)
$$

From Eq. (102) $u_{cr} = \pi$ and

$$
P_{cr} = \frac{\pi^2 EI}{L^2}
$$

2. *Both Ends Clamped.* This case is characterized by $\alpha_0 = \alpha_L = \beta_0 = \beta_L = \infty$. If we divide Eq. (100) by $\alpha_0\alpha_L\beta_0\beta_L$ and take the limit as $1/\alpha_0$, $1/\alpha_L$, $1/\beta_0$, and $1/\beta_L$ approach zero, then

$$
-L\frac{u^2}{L^2}\sin u - 2\frac{u}{L}\cos u + 2\frac{u}{L} = 0 \quad (103)
$$

Since $u/L \neq 0$, then

$$\frac{u}{2}\sin u - (1 - \cos u) = 0$$
$$\frac{u}{2}2\sin\frac{u}{2}\cos\frac{u}{2} - 2\sin^2\frac{u}{2} = 0 \tag{104}$$
$$\left(\frac{u}{2}\cos\frac{u}{2} - \sin\frac{u}{2}\right)\sin\frac{u}{2} = 0$$

The smallest root is obtained from $\sin u/2 = 0$, or $u_{cr} = 2\pi$, from which

$$P_{cr} = \frac{4\pi^2 EI}{L^2}$$

3. One End Clamped, the Other Free. This special case is characterized by either $\alpha_0 = \beta_0 = 0$ and $\alpha_L = \beta_L = \infty$, or $\alpha_0 = \beta_0 = \infty$ and $\alpha_L = \beta_L = 0$. For either case, the characteristic equation reduces to

$$\frac{u^5}{L^5}\cos u = 0$$

from which $u_{cr} = \pi/2$ and $P_{cr} = \pi^2 EI/4L^2$.

4. One End Clamped, the Other Pinned. This case is characterized by either

$$\alpha_0 = \beta_0 = \alpha_L = \infty \quad \text{and} \quad \beta_L = 0 \quad \text{or} \quad \alpha_0 = \alpha_L = \beta_L = \infty \quad \text{and} \quad \beta_0 = 0$$

The characteristic equation reduces to

$$\left(\frac{u}{L}\right)^2\sin u - L\left(\frac{u}{L}\right)^3\cos u = 0 \tag{105}$$

Since $(u/L) \neq 0$, then

$$\tan u = u \tag{106}$$

The smallest root of Eq. (106) leads to the critical condition or

$$u_{cr} = 4.493 \quad \text{and} \quad P_{cr} = 20.19\frac{EI}{L^2}$$

5. Both Ends Free Against Translation, Constrained Against Rotation. This last special case is characterized by $\alpha_0 = \alpha_L = 0$ and $\beta_0 = \beta_L = \infty$. First divide Eq. (82) by $\beta_0\beta_L(\alpha_0 + \alpha_L)$. Then, since $1/\beta_0 = 1/\beta_L = 0$ and

$$\lim_{\substack{\alpha_0 \to 0 \\ \alpha_L \to 0}} \frac{\alpha_0\,\alpha_L}{\alpha_0 + \alpha_L} = 0$$

the characteristic equation reduces to

$$\frac{u^4}{L^4}\sin u = 0 \tag{107}$$

from which $u_{cr} = \pi$ and $P_{cr} = \pi^2 EI/L^2$.

3.6   CRITICAL SPRING STIFFNESS

To clearly demonstrate the meaning of critical spring stiffness, consider the following example (see Fig. 3.15). The column is pinned at the left end and

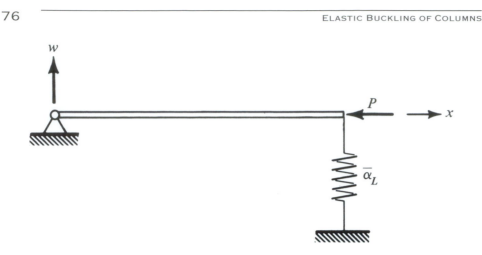

FIGURE 3.15   Critical spring stiffness model.

supported with an extensional spring at the loaded right end. From the discussion of the previous section, this case is characterized by $\alpha_0 = \infty$, $\beta_0 = \beta_L = 0$, and α_L. The characteristic equation becomes

$$\left(\frac{u}{L}\right)^4 \left[-\left(\frac{u}{L}\right)^2 + \alpha_L L\right] \sin u = 0 \tag{108}$$

Since $u/L \neq 0$, then

$$\left(-k^2 + \alpha_L L\right) \sin kL = 0$$

Thus either

$$P_{cr} = \overline{\alpha}_L L$$

or

$$P_{cr} = \frac{\pi^2 EI}{L^2} = P_E$$

Notice that when for $\overline{\alpha}_L$ is very small, $P_{cr} = \overline{\alpha}_L L < P_E$, but as $\overline{\alpha}_L$ increases, P_{cr} increases until $P_{cr} = \overline{\alpha}_L L = P_E$. This can happen when $\alpha_L = \pi^2 EI/L^3$. Any further increase in $\overline{\alpha}_L$ will yield $\overline{\alpha}_L L > P_E$, which implies that $P_{cr} = P_E$. This means that for $\overline{\alpha}_L > \pi^2 EI/L^3$, the column will always buckle in an Euler mode, and therefore there is no need to make the spring any stiffer than $\overline{\alpha}_L = \pi^2 EI/L^3$ because no increase in the critical load can result from it. Then the value $\pi^2 EI/L^3$ is called a critical spring stiffness.

Another case where there exists a critical spring stiffness is shown in Fig. 3.16. Consider the spring to act at the middle of the bar and the bar to deflect in a symmetric mode. From symmetry, the vertical reactions are $Q/2 = \alpha\delta/2$ (Fig. 3.16) and the reduced-order equation for the range $0 < x < L/2$ is

$$EIw_{,xx} + Pw = \frac{Q}{2}x \tag{109}$$

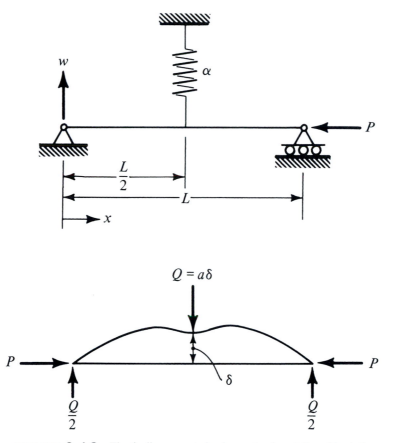

FIGURE 3.16   Elastically supported columns (spring at the midpoint).

The associated boundary and auxiliary conditions are

$$
\left.
\begin{array}{r}
w(0) = 0 \\[2mm]
w_{,x}\left(\dfrac{L}{2}\right) = 0 \\[2mm]
w\left(\dfrac{L}{2}\right) = \delta
\end{array}
\right\}
\tag{110}
$$

The solution is

$$
w = \frac{Q}{2P}\left(x - \frac{1}{k}\frac{\sin kx}{\cos kL/2}\right)
\tag{111}
$$

where

$$
k^2 = \frac{P}{EI}
$$

or

$$
w = \frac{\alpha\delta}{2P}\left(x - \frac{1}{k}\frac{\sin kx}{\cos kL/2}\right)
\tag{112}
$$

Using the auxiliary condition, we obtain

$$\delta = \frac{\alpha\delta}{2Pk}\left(\frac{kL}{2} - \tan\frac{kL}{2}\right) \tag{113}$$

The requirement of the existence of a nontrivial solution leads to the following characteristic equation:

$$1 = \frac{\alpha L}{4P}\left[1 - \frac{\tan(kL/2)}{kL/2}\right]$$

or

$$-\frac{16EI}{\alpha L^3}\left(\frac{kL}{2}\right)^2 = -1 + \frac{\tan(kL/2)}{kL/2} \tag{114}$$

We see from Eq. (114) that $(kL/2)_{cr}$ and consequently P_{cr} may be calculated for any given value of α. This may be done either numerically or graphically (see Fig. 3.17). Note from Fig. 3.15 that as $\alpha \to 0$, $(kL/2) \to \pi/2$ and $P_{cr} \to \pi^2 EI/L^2$ as expected. Furthermore, as $\alpha \to \infty$, $(kL/2)_{cr} \to 4.493$ and $P_{cr} \to (20.19)4EI/L^2$. However, if the bar were to buckle in an antisymmetric mode (with respect to $L/2$). then

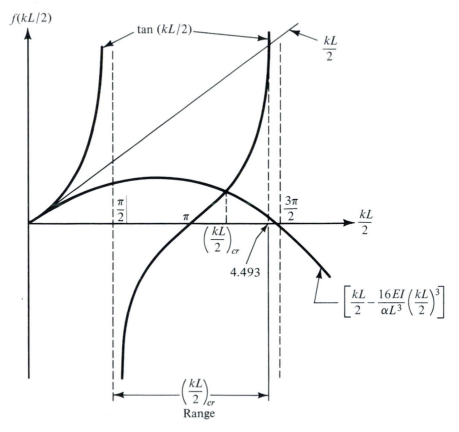

FIGURE 3.17   Critical conditions for spring-supported columns (at the midpoint).

$P_{cr} = 4\pi^2 EI/L^2$; therefore there is no need for the spring to be any stiffer than $16\pi^2 EI/L^3$ (corresponding to $kL/2 = \pi$). This maximum value of spring stiffness, α, required for the bar to carry the maximum possible axial load ($= 4\pi^2 EI/L^2$) is called critical spring stiffness, or

$$\alpha_{cr} = \frac{16\pi^2 EI}{L^3} \tag{115}$$

If the elastic support is applied at a distance ℓ_1 from the left end of the bar, then the characteristic equation is

$$-\frac{\sin k\ell_1 \sin k\ell_2}{Pk \sin kL} + \frac{\ell_1 \ell_2}{L} - \frac{1}{\alpha} = 0 \tag{116}$$

where

$$\ell_2 = L - \ell_1$$

For details, see Art. 2.6 of Timoshenko (1961).

3.7   ELASTICA THEORY FOR BEAMS

The nonlinear theory of beams goes back to Kirchhoff and Clebsch, as discussed by Love (1944). This theory is geometrically exact, which means that there are no small-angle assumptions in the treatment of finite rotation of the cross-sectional frame. Although elastica theory has been extended in various ways during the last 50 years, it was originally intended for beams undergoing extension, twist, and bending in two directions. The approximation in the formulation of the theory is in the constitutive law that relates the cross-sectional stress resultants to the generalized strains. Asymptotic methods have been used to show that Kirchhoff-Clebsch theory is asymptotically exact to three-dimensional elasticity for the case of $h/\ell \to 0$ where h is a cross-sectional characteristic dimension and ℓ is the characteristic wavelength of the deformation along the beam axis of centroids. This observation is only true if the beam cross section is assumed to be closed and to have only one characteristic length, thus excluding for now the important case of beams with thin-walled, open sections. For this reason we will devote Section 3.8 to a specialized derivation for that case, not a geometrically exact treatment however.

Kirchhoff-Clebsch theory can be derived either from vector mechanics or from the principle of minimum total potential energy. In the latter case, however, it is necessary to define the variations of the generalized strains in terms of appropriate virtual displacement and rotation measures in such a way that those variations are independent of the way the displacement of the axis of cross-sectional centroids and the finite rotation of the cross-sectional frame are expressed. The resulting equations are the same either way and are said to be "intrinsic"—meaning that they too are independent of the way the beam displacement and rotation measures are chosen.

3.7.1   EQUILIBRIUM EQUATIONS

Consider a beam slice (a segment of infinitesimal length) loaded by distributed forces and moments along its length and by concentrated forces and moments at the ends. Love (1944) shows that the resulting vector equations can be written out in scalar form as

$$F_1' + F_3 K_2 - F_2 K_3 + f_1 = 0$$
$$F_2' - F_3 K_1 + F_1 K_3 + f_2 = 0$$
$$F_3' + F_2 K_1 - F_1 K_2 + f_3 = 0$$
$$M_1' - K_3 M_2 + K_2 M_3 + m_1 = 0 \qquad (117)$$
$$M_2' - (1 + \varepsilon) F_3 + K_3 M_1 - K_1 M_3 + m_2 = 0$$
$$M_3' + (1 + \varepsilon) F_2 - K_2 M_1 + K_1 M_2 + m_3 = 0$$

These equations can also be written in a convenient matrix form as

$$F' + \widetilde{K} F + f = 0$$
$$M' + \widetilde{K} M + (1 + \varepsilon) \widetilde{e}_1 F + m = 0 \qquad (118)$$

where $(\ )'$ denotes the derivative with respect to the axial coordinate x_1 along the beam reference line, often chosen as the locus of the centroids or of the shear centers, F is the column matrix of stress resultant components measured in the deformed beam cross-sectional frame with $F_i = \mathbf{F} \cdot \mathbf{B}_i$, M is the column matrix of stress moment components measured in the deformed beam cross-sectional frame with $M_i = \mathbf{M} \cdot \mathbf{B}_i$, K is the column matrix of deformed beam curvature components measured in the deformed beam cross-sectional frame with $K_i = \mathbf{K} \cdot \mathbf{B}_i$, f is the column matrix of applied distributed force components measured in the deformed beam cross-sectional frame with $f_i = \mathbf{f} \cdot \mathbf{B}_i$, m is the column matrix of applied distributed force components measured in the deformed beam cross-sectional frame with $m_i = \mathbf{m} \cdot \mathbf{B}_i$, $e_1 = \lfloor 1 \quad 0 \quad 0 \rfloor^T$, ε is the stretching strain of the beam reference line, and $(\ )_{ij} = -e_{ijk}(\ )_k$. For example, for the column matrix K, the associated matrix $\widetilde{K}$ is the antisymmetric matrix

$$\widetilde{K} = \begin{bmatrix} 0 & -K_3 & K_2 \\ K_3 & 0 & -K_1 \\ -K_2 & K_1 & 0 \end{bmatrix} \qquad (119)$$

where $K = k + \kappa$, k is the column matrix of initial twist and curvature components measured in the cross-sectional frame of the undeformed beam, with $k_i = \mathbf{k} \cdot \mathbf{b}_i$, and κ is a measure of the elastic twist and bending. Note that the cross-sectional frame $\mathbf{b}_i$ has $\mathbf{b}_1$ tangent to the reference line of the undeformed beam and $\mathbf{b}_2$ and $\mathbf{b}_3$ perpendicular to it, in the plane of the reference cross section of the undeformed beam along the cross-sectional coordinate lines x_2 and x_3. If the undeformed beam is free of initial twist and curvature, then $k_i = 0$.

The cross-sectional frame base vectors $\mathbf{B}_i$ are defined such that $\mathbf{B}_1$ is tangent to the reference line of the deformed beam and $\mathbf{B}_2$ and $\mathbf{B}_3$ are perpendicular to it and lie parallel to the deformed beam reference cross-sectional plane. The material points of the reference cross section of the undeformed beam have small displacement in the deformed beam cross-sectional frame. This displacement is referred to as warping, and it has components both in and out of the cross-sectional plane. The warping is determined when the constitutive law is found and is of the order of $h\varepsilon^*$ where $\varepsilon^* = \max(\varepsilon, h\kappa_1, h\kappa_2, h\kappa_3)$ is the magnitude of the maximum strain in the beam.

3.7.2 CONSTITUTIVE LAW AND STRAIN ENERGY

A linear version of the constitutive law associated with the theory of Eqs. (118) is appropriate for beams with closed cross sections and can be written as

$$\left\{ \begin{array}{c} F_1 \\ M_1 \\ M_2 \\ M_3 \end{array} \right\} = \left[\begin{array}{cccc} S_{11} & S_{12} & S_{13} & S_{14} \\ S_{12} & S_{22} & S_{23} & S_{24} \\ S_{13} & S_{23} & S_{33} & S_{34} \\ S_{14} & S_{24} & S_{34} & S_{44} \end{array} \right] \left\{ \begin{array}{c} \varepsilon \\ \kappa_1 \\ \kappa_2 \\ \kappa_3 \end{array} \right\} \tag{120}$$

where S_{ij}, for $i, j = 1, 2, 3$, and 4 are the cross-sectional stiffness constants. They may vary along the beam (i.e. as a function of x_1) and depend on the initial twist k_1 and the initial curvature components k_2 and k_3. The corresponding strain energy per unit length can be written as

$$\Psi = \frac{1}{2} \left\{ \begin{array}{c} \varepsilon \\ \kappa_1 \\ \kappa_2 \\ \kappa_3 \end{array} \right\}^T \left[\begin{array}{cccc} S_{11} & S_{12} & S_{13} & S_{14} \\ S_{12} & S_{22} & S_{23} & S_{24} \\ S_{13} & S_{23} & S_{33} & S_{34} \\ S_{14} & S_{24} & S_{34} & S_{44} \end{array} \right] \left\{ \begin{array}{c} \varepsilon \\ \kappa_1 \\ \kappa_2 \\ \kappa_3 \end{array} \right\} \tag{121}$$

For example, consider a homogeneous, isotropic, prismatic beam with reference line along the coincident shear center and centroid, and with centroidal, cross-sectional principal axes along x_2 and x_3. For this case the elastic constants S_{ij} are all zero except for $S_{11} = EA$, $S_{22} = GJ$, $S_{33} = EI_2$, and $S_{44} = EI_3$, where E is the Young's modulus, G the shear modulus, A the cross-sectional area, J the Saint-Venant torsional constant, I_2 the area moment of inertia about x_2, and I_3 the area moment of inertia about x_3. Nonzero values of initial twist, initial curvature, and offsets between the reference line and shear center and/or centroid will bring coupling terms into the model.

3.7.3 KINEMATICAL EQUATIONS

When needed, the kinematical equations can be written in a variety of ways. One of the beautiful aspects of the equilibrium and constitutive equations, Eqs. (118) and (120), respectively, is that they are independent of the coordinate systems in which the displacement variables are expressed and of the definition of the finite rotation variables used to express the change of orientation between $\mathbf{b}_i$ and $\mathbf{B}_i$. This change of orientation may be defined in terms of the 3×3 matrix of direction cosines C, elements of which are

$$C_{ij} = \mathbf{B}_i \cdot \mathbf{b}_j \tag{122}$$

The three generalized strains κ_i may now be defined in terms of the antisymmetric matrix

$$\widetilde{\kappa} = -C'C^T + C\widetilde{k}C^T - \widetilde{k} \tag{123}$$

The quantities κ_i are sometimes referred to as moment strains because the partial derivative of the strain energy per unit length, as written in Eq. (121), with respect to κ_i is M_i. However, for convenience one may choose to express C in terms of three finite rotation variables, represented here generically in terms of a 3×1 column matrix of rotation measures θ. Thus, C can be expressed as $C(\theta)$, and the moment strains can be written as

$$\kappa = R\theta' + Ck - k \tag{124}$$

where R and C depend on the angular displacement parameters of the formulation. For example, for a formulation in terms of Rodrigues parameters, applied to non-linear beam kinematics by Hodges (1987), C and R become

$$C = \frac{(1 - \frac{1}{4}\theta^T\theta)I - \tilde{\theta} + \frac{1}{2}\theta\theta^T}{1 + \frac{1}{4}\theta^T\theta}$$

$$R = \frac{I - \frac{1}{2}\tilde{\theta}}{1 + \frac{1}{4}\theta^T\theta} \tag{125}$$

where I is the 3×3 identity matrix. This formulation is noted to be especially helpful because of the simplicity of these expressions. Depending on the context, the symbol θ instead may be used to refer to a column matrix of orientation angles, in which case C and R are more complicated functions of the angles. In any case, for the elastica theory as presented here, the three rotational variables θ_i for $i = 1$, 2, and 3, are not independent. This interdependence is addressed below after introducing displacement variables.

The position vector from a fixed point to a point on the deformed beam reference line can be written as

$$\mathbf{R} = \mathbf{r} + \mathbf{u} \tag{126}$$

where $\mathbf{r}$ is the position vector from a fixed point to the corresponding point on the undeformed beam reference line. Thus, $\mathbf{u}$ becomes the displacement vector of the reference line. One convenient way to define the displacement variables is to let

$$u_i = \mathbf{u} \cdot \mathbf{b}_i \tag{127}$$

where $\mathbf{b}_i$ is the orthonormal triad associated with the undeformed beam cross-sectional frame.

The unit vector tangent to the deformed beam reference line becomes

$$\frac{\partial \mathbf{R}}{\partial s} = \mathbf{B}_1 \tag{128}$$

where s is the running arc-length coordinate along the deformed beam reference line and, by definition of the stretching strain measure, $s' = 1 + \varepsilon$. Substituting Eq. (126) into Eq. (128), one obtains

$$\mathbf{R}' = \mathbf{r}' + \mathbf{u}' = (1 + \varepsilon)\mathbf{B}_1 \tag{129}$$

Since

$$\mathbf{r}' = \mathbf{b}_1$$
$$\mathbf{b}'_i = \mathbf{k} \times \mathbf{b}_i \tag{130}$$

and making use of the fixed length of $\mathbf{B}_1$ as unity, the stretching strain can be written as

$$\varepsilon = \sqrt{\left(e_1 + u' + \tilde{k}u\right)^T \left(e_1 + u' + \tilde{k}u\right)} - 1 \tag{131}$$

Now we address the interdependence of the three rotational variables. One way to approach this is to take an alternative expression for $\mathbf{R}'$, such that

$$\mathbf{R}' = \mathbf{r}' + \mathbf{u}' = (1 + \gamma_{11})\mathbf{B}_1 + 2\gamma_{12}\mathbf{B}_2 + 2\gamma_{13}\mathbf{B}_3 \tag{132}$$

Here $2\gamma_{12}$ and $2\gamma_{13}$ are shear angles in a more general theory. Letting $\gamma = \lfloor \gamma_{11} \ 2\gamma_{12} \ 2\gamma_{13} \rfloor^T$ we can again make use of Eqs. (126) to obtain

$$\gamma = C\left(e_1 + \tilde{k}u + u'\right) - e_1 \tag{133}$$

and constrain γ so that the shear angles are zero, viz.,

$$\gamma = \varepsilon e_1 \text{ or } \widetilde{e}_1 \gamma = 0 \tag{134}$$

In the event that Rodrigues parameters are used for the finite rotation, one can use Eqs. (133) and (134) to obtain expressions for θ_2 and θ_3, given by

$$\theta_2 = \frac{\theta_1 C_{12} - 2C_{13}}{1 + C_{11}}$$
$$\theta_3 = \frac{\theta_1 C_{13} + 2C_{12}}{1 + C_{11}} \tag{135}$$

with θ_1 remaining as an independent torsional variable, and where elements of the first row of C can be expressed as

$$C^T e_1 = \frac{e_1 + \widetilde{k}u + u'}{1 + \varepsilon} \tag{136}$$

and for small strain $\gamma_{11} = \varepsilon$. Three scalar equations are implied by Eq. (136), given by

$$C_{11} = \frac{1 + u_1' + k_2 u_3 - k_3 u_2}{1 + \varepsilon}$$
$$C_{12} = \frac{u_2' + k_3 u_1 - k_1 u_3}{1 + \varepsilon} \tag{137}$$
$$C_{13} = \frac{u_3' + k_1 u_2 - k_2 u_1}{1 + \varepsilon}$$

One may find the remaining elements of C from substituting the expressions for θ_2 and θ_3 in Eqs. (135) into the first of Eqs. (135), recalling that θ_1 is the independent torsional variable.

In later Chapters we will need to make certain approximations by perturbing about particular deformed states. Denoting the state about which the perturbations are made by the $(^-)$ and the perturbations by $(\hat{\ })$, we will base these approximations on

$$C \approx \left(I - \widetilde{\hat{\theta}} + \frac{1}{2}\widetilde{\hat{\theta}}\widetilde{\hat{\theta}} \right) \overline{C}$$
$$\gamma = (\overline{\varepsilon} + \hat{\varepsilon})e_1 \tag{138}$$
$$K = \overline{K} + \hat{\kappa}$$

where $\hat{\theta}$ is a column matrix containing the measure numbers of an infinitesimal rotation vector. In the sequel we shall need to keep only first-order perturbations in most cases. However, in a few places where we need to develop the energy, we will make use of perturbations up through second order in $\hat{\theta}$. One can use Eq. (123) to eliminate C' so that

$$C' = C\widetilde{\kappa} - \widetilde{K}C \tag{139}$$

which can be used along with Eqs. (134) and (138) to obtain approximate expressions valid through second order for $\hat{\varepsilon}$, $\hat{\theta}_2$, $\hat{\theta}_2$, and $\hat{\kappa}$, given by

$$\hat{\varepsilon} = \hat{\phi}_1 + \frac{1}{2(1 + \overline{\varepsilon})}\left(\hat{\phi}_2^2 + \hat{\phi}_3^2 \right)$$
$$\hat{\theta}_2 = -\hat{\phi}_3\left(1 - \hat{\phi}_1 \right) + \frac{1}{2}\hat{\theta}_1\hat{\phi}_2$$
$$\hat{\theta}_3 = \hat{\phi}_2\left(1 - \hat{\phi}_1 \right) + \frac{1}{2}\hat{\theta}_1\hat{\phi}_3 \tag{140}$$
$$\hat{\kappa} = \left(I - \widetilde{\hat{\theta}} \right)\hat{\theta}' + \widetilde{K}\hat{\theta} + \frac{1}{2}\widetilde{\hat{\theta}}\widetilde{\hat{\theta}}\overline{K}$$

where

$$\hat{\phi} = \overline{C}\left(\hat{u}' + \widetilde{k}\hat{u}\right) \tag{141}$$

Simplified versions of these second-order approximations are used in Section 8.2 for the case of a beam subjected to an axial force and a twisting moment and in Section 11.6 for a beam subjected to a bending moment that is constant in time.

For the particular case of small displacement and rotation variables, the kinematical equations can be linearized about the state of zero deformation and rewritten as

$$
\begin{aligned}
C &= I - \widetilde{\theta} \\
\kappa &= \theta' + \widetilde{k}\theta \\
\gamma &= u' + \widetilde{k}u + \widetilde{e}_1\theta \\
\varepsilon &= u_1' + k_2 u_3 - k_3 u_2 \\
\theta_2 &= -u_3' - k_1 u_2 + k_2 u_1 \\
\theta_3 &= u_2' + k_3 u_1 - k_1 u_3
\end{aligned}
\tag{142}
$$

3.7.4 EXAMPLE OF USING INTRINSIC EQUATIONS—EULER COLUMN

Consider a cantilevered beam loaded by an axial compressive force $-P\mathbf{b}_1$ as shown in Fig. 3.18. The unit vectors $\mathbf{B}_i$ are shown for a deformation in the plane of least flexural rigidity, here denoted as the $\mathbf{b}_1$-$\mathbf{b}_2$ plane. The pre-buckling deformation is only a compressive strain along the $\mathbf{b}_1$ direction, and the deformed state during buckling is governed by the reduced constitutive law

$$
\left\{ \begin{array}{c} F_1 \\ M_3 \end{array} \right\} = \left[\begin{array}{cc} EA & 0 \\ 0 & EI_3 \end{array} \right] \left\{ \begin{array}{c} \varepsilon \\ \kappa_3 \end{array} \right\} \tag{143}
$$

The pre-buckling state is, by inspection,

$$
\begin{aligned}
\overline{F} &= -Pe_1 = EA\overline{\varepsilon}e_1 \\
\overline{K} &= \overline{\kappa} = \overline{M} = \overline{\theta} = 0
\end{aligned}
\tag{144}
$$

This state clearly satisfies the equilibrium equations, Eqs. (118). Denoting small perturbations of the pre-buckling quantities by ($\hat{\ }$), we then write them as

$$
\begin{aligned}
F &= \overline{F} + \hat{F} = -Pe_1 + \hat{F} \\
M &= \overline{M} + \hat{M} = \hat{M} \\
K &= \overline{K} + \hat{\kappa} = \hat{\kappa}
\end{aligned}
\tag{145}
$$

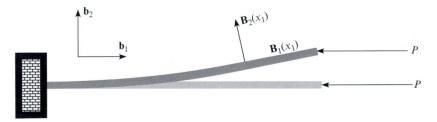

FIGURE 3.18 Schematic of an axially-loaded beam

and then substitute them into Eqs. (118). To get the buckling equations, we drop all terms of second degree and higher in the ˆ terms. The buckling equations are then

$$\hat{F}' + P\tilde{e}_1\hat{\kappa} = 0$$
$$\hat{M}' + \left(1 - \frac{P}{EA}\right)\tilde{e}_1\hat{F} = 0 \tag{146}$$

Noting that $\hat{M} = \hat{M}_3 e_3$ and $\hat{\kappa} = \hat{\kappa}_3 e_3$, one can write the only non-trivial scalar buckling equations as

$$\hat{F}_2' - \frac{P\hat{M}_3}{EI_3} = 0$$
$$\hat{M}_3' + \left(1 - \frac{P}{EA}\right)\hat{F}_2 = 0 \tag{147}$$

with boundary conditions $\hat{F}_2(0) = \hat{M}_3(\ell) = 0$. The first boundary condition implies that $\hat{M}_3'(0) = 0$. The two Eqs. (147) can be combined into one, which is

$$\hat{M}_3'' + \frac{P\left(1 - \frac{P}{EA}\right)}{EI_3}\hat{M}_3 = 0 \tag{148}$$

Since the constitutive law assumes that the strain is small compared to unity, one should neglect P/EA compared to unity, so that

$$\hat{M}_3'' + \frac{P}{EI_3}\hat{M}_3 = 0 \tag{149}$$

with boundary conditions $\hat{M}_3'(0) = \hat{M}_3(\ell) = 0$. The solution is

$$\hat{M}_3 = \cos\left(\sqrt{\frac{P}{EI_3}}x_1\right) \tag{150}$$

with

$$P_{\text{cr}} = \frac{\pi^2 EI_3}{4\ell^2} \tag{151}$$

This illustration has served to show how one can apply elastica theory to the buckling of an Euler column. The treatment is simpler than a traditional displacement-based analysis. In Chapters to come, we will apply it to more complex problems.

3.8   BUCKLING OF THIN-WALLED BEAM-COLUMNS

When thin-walled, prismatic beams with open sections are loaded with a compressive axial force, they may buckle in either bending *or in torsion*. The torsional buckling phenomenon requires a set of equations that takes into account the effect of axial force on the effective torsional stiffness, and the open cross section requires that the Vlasov effect (see Section 3.8.1) be taken into account. Neither of these effects are accounted for in the preceding Sections, so a derivation of governing equations for this problem will now be undertaken.

3.8.1 VLASOV THEORY FOR THIN-WALLED, PRISMATIC BEAMS WITH OPEN CROSS SECTION

Vlasov beam theory is a refined theory of beams that addresses effects associated with thin-walled, prismatic beams of open cross sections. The present adaptation of Vlasov theory[1] is based on the following assumptions:

1. The beam is spanwise uniform and prismatic;
2. The beam is slender, such that $a \ll \ell$, where a is a characteristic cross-sectional dimension and ℓ is a characteristic wavelength of elastic deformation along the beam;
3. The beam is thin-walled, such that $h \ll a$ and $h \ll R$, where h is a characteristic wall thickness and R is a characteristic radius of curvature of the midsurface.

Assumption 2 applies to beam theory in general, including elastica theory. Assumption 3 introduces multiple length scales in the cross-sectional domain. Additional details are discussed by Hodges (2006).

Four beam (1-D) variables are introduced, which correspond to displacement of the cross section as a rigid body: $u_i(x_1)$ is the translation of the cross section at x_1 in the x_i direction; and $\theta(x_1)$ is a rotation of the cross section at x_1 about x_1. The coordinates x_i are Cartesian (see Fig. 3.19). For the sake of convenience a curvilinear system of coordinates is introduced with s and ξ being the contour and through-the-thickness coordinates respectively; $\mathbf{r} = x_i \mathbf{b}_i$ is a position vector of the shell midsurface, vectors are denoted with bold letters.

In the general application of dimensional reduction to problems of elasticity one starts with a 3-D representation and reduces the theory to 2-D or 1-D. Here, because we are starting with a thin-walled beam structure, we start with a 2-D (shell) representation and move to a 1-D (beam) representation. Thus, the strain energy of a beam with cross section S is approximated in the following fashion:

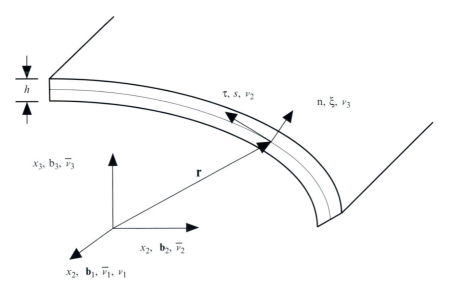

FIGURE 3.19   Configuration and coordinate system

---

[1] The author gratefully acknowledges the unpublished class notes of Dr. V. Volovoi in 1997 for the development in this section.

$$2U = \int_0^\ell \int_S \int_{-h/2}^{h/2} \sigma_{ij}\varepsilon_{ij}d\xi \, ds \, dx_1$$

$$\approx \int_0^\ell \int_S \left(N_{\alpha\beta}\gamma_{\alpha\beta} + M_{\alpha\beta}\rho_{\alpha\beta}\right) ds \, dx_1 \tag{152}$$

Greek indices for shell variables run from 1 to 2. Summation is implied for repeated indices, and the meaning of corresponding strain measures and stress resultants will be explained below.

Without going into too much detail, we will use the variational-asymptotic method in our derivation. This is a rigorous procedure that uses small parameters that are inherent to the problem. An iterative procedure is invoked which allows one to calculate approximations to the 3-D strain energy. In our case the main small parameter is a/ℓ. The so-called "zeroth-order" approximation is obtained by setting the small parameter to zero. Thus, the truncated strain energy is degenerate: it has a null space comprised of four rigid-body motions of the cross section. These four null space modes serve to introduce the four 1-D variables. Next, perturbations of such "rigid" displacements, w_i, are examined. Only the leading terms with respect to the small parameters are retained. The resulting expressions for strains are substituted into the strain energy. Finally, the functional thus obtained is minimized with respect to the unknowns w_i. It is important to emphasize that *the cross section is not considered rigid*, and general inplane displacements are present. Non-rigid inplane displacements (in curvilinear coordinates) are denoted as w_2 and w_3. These "warping" displacements are small, but they do contribute to the strain energy of the beam and should not be neglected. In many textbooks this warping is taken into consideration by correcting the constitutive law. That approach can lead to a certain amount of confusion since, as shown below, the warping displacements themselves do not have to be calculated explicitly to obtain the correct 1-D strain energy.

Diplacement Field

The following notation is used

$$(\dot{\ }) \equiv \frac{d(\ )}{ds}; \quad (\ )' \equiv \frac{d(\ )}{dx_1}$$

$$\boldsymbol{\tau} = \dot{\mathbf{r}} = \dot{x}_2\mathbf{b}_2 + \dot{x}_3\mathbf{b}_3$$

$$\mathbf{n} = \boldsymbol{\tau} \times \mathbf{b}_1 = \dot{x}_3\mathbf{b}_2 - \dot{x}_2\mathbf{b}_3 \tag{153}$$

$$r_\tau = \boldsymbol{\tau} \cdot \mathbf{r} = x_2\dot{x}_2 + x_3\dot{x}_3$$

$$r_n = \mathbf{n} \cdot \mathbf{r} = x_2\dot{x}_3 - x_3\dot{x}_2$$

$$R = \dot{x}_2/\ddot{x}_3 = -\dot{x}_3/\ddot{x}_2$$

The displacements in the curvilinear system, v_i, are expressed in terms of the displacements in the Cartesian system, $\bar{v}_i$, as

$$v_1 = \bar{v}_1$$

$$v_2 = \bar{v}_2\dot{x}_2 + \bar{v}_3\dot{x}_3 \tag{154}$$

$$v_3 = \bar{v}_2\dot{x}_3 - \bar{v}_3\dot{x}_2$$

This leads to an expression for the displacement field in the cross-sectional plane given by

$$v_2 = u_2 \dot{x}_2 + u_3 \dot{x}_3 + \theta r_n + w_2$$
$$v_3 = u_2 \dot{x}_3 - u_3 \dot{x}_2 - \theta r_\tau + w_3 \tag{155}$$

The rigid portion of the inplane displacements is illustrated in Fig. 3.20 in which one sees a displacement from the point M_o to M being comprised of rigid displacements from M_o to M' and from M_o to M''. The former is associated with u_2 and u_3 while the latter is caused by a rigid-body rotation θ about the origin O. The displacement variables are defined at the point O.

The axial displacement has the form

$$v_1 = u_1 + \hat{w}_1 + w_1 \tag{156}$$

The quantity $\hat{w}_1$ is larger (i.e., of a lower order) than the displacements w_1, w_2, and w_3, and must be determined before we can go any further. At this point $\hat{w}_2$ and $\hat{w}_3$ are zero.

Shell Theory

The strain energy of shells is determined by two strain measures: membrane, $\gamma_{\alpha\beta}$, and bending, $\rho_{\alpha\beta}$. For cylindrical shells expressions for those two types of measures have the form

$$\gamma_{11} = v_{1,1} \qquad\qquad \rho_{11} = v_{3,11}$$
$$2\gamma_{12} = v_{1,2} + v_{2,1} \qquad \rho_{12} = v_{3,12} + \frac{1}{4R}\left(v_{1,2} - 3\, v_{2,1}\right) \tag{157}$$
$$\gamma_{22} = v_{2,2} + \frac{v_3}{R} \qquad \rho_{22} = v_{3,22} - \left(\frac{v_2}{R}\right)_{,2}$$

It is noted that letting R tend to infinity, one obtains the more familiar expressions for plates. In order to find an equation for v_1, we substitute Eqs. (155) and (156) into the expression for $2\gamma_{12}$, yielding

$$2\gamma_{12} = \hat{w}_{1,2} + \underline{u_2' \dot{x}_2 + u_3' \dot{x}_3} + \underline{\underline{\theta' r_n}} + w_{1,2} \tag{158}$$

The underlined terms must be cancelled out by a proper choice of $\hat{w}_{1,2}$. This in effect eliminates them from the strain energy by a process sometimes referred to as killing these terms. Were the single-underlined terms not killed, the 1-D strain energy would

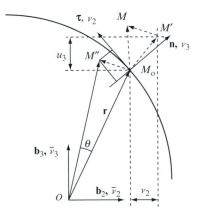

FIGURE 3.20   Rigid inplane displacements

depend on u'_α, the rigid-body rotations of the section, which cannot be tolerated because a rigid-body rotation of the entire beam should not result in any strain. Were the double-underlined term not killed it would produce too large a torsional rigidity, but only for open cross sections. So, we can write

$$\hat{w}_{1,2} = -u'_2 \dot{x}_2 - u'_3 \dot{x}_3 - \theta' r_n \tag{159}$$

After these terms are killed, only $w_{1,2}$ remains in γ_{12}, and ρ_{12} provides the torsional rigidity.[1] The axial displacement can now be obtained via integration with respect to circumferential coordinate, so that

$$v_1 = u_1(x_1) - u'_2 x_2 - u'_3 x_3 - \theta' \int_{s_0}^{s} r_n ds + w_1 \tag{160}$$

where the coefficient of θ' in Eq. (160), namely

$$\eta(s) \equiv \int_{s_0}^{s} r_n ds \tag{161}$$

is called the *sectorial coordinate* and is a solution of a classical Saint-Venant torsional problem within the shell approximation. The sectorial coordinate $\eta(s)$ must be continuous around the contour, and the origin of the variable s should be chosen so that $\int_S \eta(s) ds = 0$.

Semi-inversion

Let us now find the contribution of w_i to the 1-D strain energy. In order to achieve this, the constitutive relationships must be introduced, viz.,

$$\begin{Bmatrix} N_{11} \\ N_{12} \\ N_{22} \\ M_{11} \\ M_{12} \\ M_{22} \end{Bmatrix} = \begin{bmatrix} A & B \\ B & D \end{bmatrix} \begin{Bmatrix} \gamma_{11} \\ 2\gamma_{12} \\ \gamma_{22} \\ \rho_{11} \\ 2\rho_{12} \\ \rho_{22} \end{Bmatrix} \tag{162}$$

where, for isotropic materials, we have

$$A = h\mu \begin{bmatrix} 1+\sigma & 0 & \sigma \\ 0 & 1 & 0 \\ \sigma & 0 & 1+\sigma \end{bmatrix} \tag{163}$$

$$B = 0 \qquad D = \frac{h^2}{12} A$$

with

$$\sigma \equiv \frac{\lambda}{(\lambda + 2\mu)} \tag{164}$$

and where μ and λ are the Lamé constants, which are expressed in terms of engineering material constants as

---

[1] For closed sections the situation is quite different: it is γ_{12} that provides the torsional rigidity, while the contribution from ρ_{12} can be neglected.

$$\mu \equiv G = \frac{E}{2(1+\nu)} \quad \lambda = \frac{E\nu}{(1-2\nu)(1+\nu)} \tag{165}$$

The traditional way of deriving equations for thin-walled, open-section beams[2] apparently originated with Reissner and Tsai (1972). The procedure includes invoking the hypotheses

$$N_{12} = 0 \quad N_{22} = 0 \quad M_{22} = 0 \tag{166}$$

Introducing the shell's strain energy per unit area, $\hat{\Psi}$, it is easily seen that Eq. (162) follows from

$$N_{12} = \frac{\partial \hat{\Psi}}{\partial(2\gamma_{12})}$$

$$N_{22} = \frac{\partial \hat{\Psi}}{\partial \gamma_{22}} \tag{167}$$

$$M_{22} = \frac{\partial \hat{\Psi}}{\partial \rho_{22}}$$

Thus, the assumption of Eqs. (166) is seen to be equivalent to minimizing the energy with respect to unknowns $2\gamma_{12}, \gamma_{22}$ and ρ_{22}. Once Eqs. (166) are obtained (either as an assumption, or as the result of a minimization procedure), it is convenient to "semi-invert" the constitutive relationship, yielding

$$\begin{Bmatrix} N_{11} \\ M_{11} \\ M_{12} \\ 2\gamma_{12} \\ \gamma_{22} \\ \rho_{22} \end{Bmatrix} = \begin{bmatrix} \overline{A} & \overline{B} \\ \overline{B} & \overline{D} \end{bmatrix} \begin{Bmatrix} \gamma_{11} \\ \rho_{11} \\ 2\rho_{12} \\ N_{12} \\ N_{22} \\ M_{22} \end{Bmatrix} \tag{168}$$

Using the results of Problems 15 and 16, the beam's strain energy per unit length is

$$2\Psi = \int_S \left(\overline{A}_{11}\gamma_{11}^2 + 4\overline{A}_{13}\gamma_{11}\rho_{12} + 4\overline{A}_{33}\rho_{12}^2 \right) ds$$

$$= \int_S \left(Eh\gamma_{11}^2 + \frac{Gh^3}{3}\theta'^2 \right) ds \tag{169}$$

Now, substituting the displacement field variables we've obtained so far into γ_{11}, one finds that

$$\gamma_{11} = u_1'(x_1) - u_2''x_2 - u_3''x_3 - \theta'' \eta \tag{170}$$

For a doubly-symmetric cross section, with the origin of the coordinate system at the intersection of the two planes of symmetry, one finds

$$2\Psi = EAu_1'^2 + EI_{33}u_2''^2 + EI_{22}u_3''^2 + GJ\theta'^2 + E\Gamma\theta''^2 \tag{171}$$

where

$$A = hp \quad J = \frac{ph^3}{3} \quad \Gamma = \int_S \eta^2 ds$$

$$I_{33} = h \int_S x_2^2 ds \quad I_{22} = h \int_S x_3^2 ds \tag{172}$$

---

[2] For closed sections the situation is different because constraints of single-valuedness have to be imposed on the displacement field, leading to a more complicated procedure.

and p is the arc-length of the contour. Here A is the cross-sectional area, J is the Saint-Venant torsional constant, I_{22} and I_{33} are the area moments of inertia, and Γ is the warping rigidity.

For a general cross section the strain energy can still be written in the form of Eq. (171), but the displacement must be defined at the "main pole" of the cross section, called the shear center. The shear center has coordinates relative to O given by

$$
\begin{aligned}
a_2 &= \frac{h}{I_{22}} \int_S \eta_O x_3 ds \\
a_3 &= -\frac{h}{I_{33}} \int_S \eta_O x_2 ds
\end{aligned}
\tag{173}
$$

where η_O is a sectorial coordinate with respect to the centroid of the cross section. Like $\eta(s)$, the sectorial coordinate $\eta_O(s)$ must be continuous around the contour; but, unlike $\eta(s)$, its line integral over S need not be zero. In Cartesian coordinates the inplane displacements caused by rigid-body motion can then be expressed as

$$
\bar{v}_2 = u_2 - (x_3 - a_3)\theta \quad \bar{v}_3 = u_3 + (x_2 - a_2)\theta
\tag{174}
$$

Differential Equations

The strain energy from Eq. (171) leads to the following system of 1-D differential equations in the presence of distributed lateral forces:

$$
\begin{aligned}
EAu_1'' &= 0 \\
EI_{33}u_2'''' &= q_2 \\
EI_{22}u_3'''' &= q_3 \\
E\Gamma\theta'''' - GJ\theta'' &= m_1
\end{aligned}
\tag{175}
$$

This is the standard form of Vlasov's theory for a thin-walled, isotropic beam with open cross section. In order to apply this model to buckling under compressive loads, however, certain nonlinear effects must be brought into the analysis.

3.8.2 TORSIONAL-FLEXURAL BUCKLING UNDER A COMPRESSIVE AXIAL LOAD

The equations of equilibrium, Eqs. (175), were derived about the undeformed state. Another way to put it is that in the derivation the difference between the undeformed and deformed states is taken as negligible. This is appropriate for linear theory, but is not applicable for buckling. In order to apply our theory to buckling we will modify these equations to take into account the initial stresses in the pre-buckling state of the structure, while continuing to ignore the distinctions between the undeformed and deformed configurations.

To do so, let us consider a system of axial forces that is applied in compression to the ends of a beam $\bar{\sigma}_{11}(x_2, x_3)$. It should be noted that the ends of the beam are free to move toward each other, else the problem is not well-posed. If this system of forces on the end cross sections is linear with respect to each of the Cartesian coordinates x_2 and x_3, then it can be verified that $\sigma_{11}(x_1, x_2, x_3) = \bar{\sigma}_{11}(x_2, x_3)$ is an exact solution of the 3-D elasticity problem assuming no other applied tractions. Thus,

$$\bar{\sigma}_{11}(x_2, x_3) = -\frac{P}{A} + \frac{M_2}{I_{22}}x_3 - \frac{M_3}{I_{33}}x_2 \tag{176}$$

All other stress components are zero. For example, such a system of forces can model a concentrated, noncentroidally applied force P; see Fig. 3.21. Here moment components M_2 and M_3 are induced by P and are taken about the centroidal axis.

Let us now consider a projection of the deformed state onto the x_1-x_2 plane (see Fig. 3.22). Clearly, this leads to an equation of equilibrium of the form

$$Q_2 dx_1 ds = \bar{\sigma}_{11} h ds \frac{dx_1}{\rho_2} \qquad \frac{1}{\rho_2} \approx \bar{v}_2'' \tag{177}$$

Here ρ_2 is the curvature of $\mathbf{\Pi}(\mathbf{r})(s, x_1)$, the projection of the radius vector $\mathbf{r}$ onto the x_1-x_2 plane. A similar expression can be obtained for the x_1-x_3 plane. Canceling out $ds\, dx_1$ renders

$$Q_2 = \bar{\sigma}_{11} h \bar{v}_2'' \qquad Q_3 = \bar{\sigma}_{11} h \bar{v}_3'' \tag{178}$$

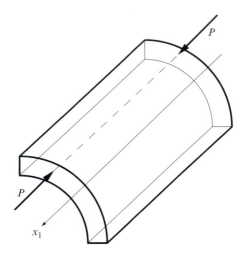

FIGURE 3.21   Application of a concentrated load

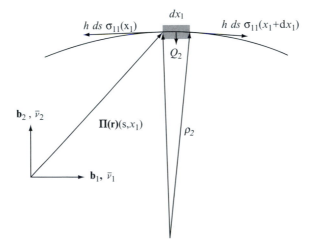

FIGURE 3.22   Projection of the deformed state onto the x_1-x_2 plane

Now we can integrate these expressions along the contour and obtain expressions for the distributed lateral forces and torsional moment, given by

$$q_2 = \int_S Q_2 ds \qquad q_3 = \int_S Q_3 ds$$

$$m_1 = \int_S [Q_3(x_2 - a_2) - Q_2(x_3 - a_3)] ds$$

(179)

Substituting expression for $\bar{\sigma}_{11}$ from Eq. (176) into Eq. (179) yields

$$q_2 = -Pu_2'' - (a_3 P + M_2)\theta''$$
$$q_3 = -Pu_3'' + (a_2 P - M_3)\theta''$$
$$m_1 = -(a_3 P + M_2)u_2'' + (a_2 P - M_3)u_3''$$
$$+ \left(-\rho^2 P + 2\beta_3 M_2 - 2\beta_2 M_3\right)\theta''$$

(180)

where

$$\rho^2 = \frac{I_{22} + I_{33}}{A} + a_2^2 + a_3^2$$

$$\beta_2 = \frac{\Psi_3}{2I_{33}} - a_2 \qquad \beta_3 = \frac{\Psi_2}{2I_{22}} - a_3$$

(181)

$$\Psi_2 = h\int_S \left(x_3^3 + x_3 x_2^2\right) ds \qquad \Psi_3 = h\int_S \left(x_2^3 + x_2 x_3^2\right) ds$$

We can now rewrite Eqs. (175) as

$$EAu_1'' = 0$$
$$EI_{33}u_2''' + Pu_2'' + (Pa_3 + M_2)\theta'' = 0$$
$$EI_{22}u_3''' + Pu_3'' + (Pa_2 - M_3)\theta'' = 0$$
$$E\Gamma\theta''' + (Pa_3 + M_2)u_2'' - (Pa_2 - M_3)u_3''$$
$$+ \left(\rho^2 P - 2\beta_3 M_2 + 2\beta_2 M_3 - GJ\right)\theta'' = 0$$

(182)

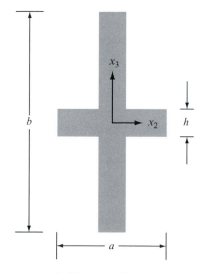

FIGURE 3.23   Cruciform cross section

As an example, we consider a beam with a cruciform cross section. The equations decouple for this case, so that buckling involving only torsion is possible, the governing equation for which is

$$E\Gamma\theta''' + (\rho^2 P - GJ)\theta'' = 0 \tag{183}$$

Thus, the torsional buckling of a beam with cruciform cross section depends on $E\Gamma$, the warping rigidity; GJ, the Saint-Venant torsional rigidity; and ρ, the polar radius of gyration about the shear center. The boundary conditions are more complicated than in the usual case of Saint-Venant torsion because of the fourth-order term in the equation. The choices at $x = 0$ and $x = \ell$ are

essential boundary conditions   natural boundary conditions

$$\theta = 0 \qquad \text{or} \qquad E\Gamma\theta''' + (\rho^2 P - GJ)\theta' = 0 \tag{184}$$

$$\theta' = 0 \qquad \text{or} \qquad \theta'' = 0$$

where the essential boundary conditions, i.e. those on θ and θ', are appropriate for zero rotation and zero warping, respectively, and the natural boundary conditions are for the cases of zero twisting moment and zero longitudinal stress, respectively.

For example, for a restraint and the boundary that does not allow rotation θ but does allow the ends of the beam to freely warp, the boundary conditions are $\theta(0) = \theta(\ell) = \theta''(0) = \theta''(\ell) = 0$. For a load applied at the centroid, the critical load is

$$P_{cr} = \frac{E\Gamma\lambda^2 + GJ}{\rho^2} \tag{185}$$

where $\lambda = \pi/\ell$. Whether a beam with a specific cross-sectional geometry buckles in torsion or in one of the two bending directions depends on which critical load is lowest.

PROBLEMS

1. Calculate the critical load for an ideal column of length l and the following boundary conditions:
   (a) $w(0) = w(\ell) = 0$
       $w_{,xx}(0) = w_{,x}(\ell) = 0$
   (b) $w(0) = 0$, $w_{,x}(\ell) = 0$
       $w_{,xx}(0) = 0$, $w_{,xxx}(\ell) = 0$
2. An ideal column is pinned at one end and fixed to a rigid bar of length a at the other end. The second end of the rigid bar is pinned on rollers (see figure). Find the critical load and discuss the extreme cases ($a \to 0$ and $a \to \infty$).
3. Find the critical condition for a system similar to the one in Problem 2, with the exception that the left end of the ideal column is clamped (see figure). Discuss the extreme cases.

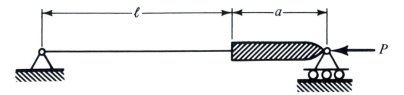

FIGURE P3.2

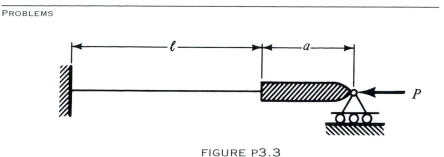

FIGURE P3.3

4. A simply supported imperfect elastic bar carries horizontal thrust P at each end, with eccentricities e_1 and e_2. The initial line of centroids is curved and given by

$$w_0 = \sum_{n=1}^{\infty} a_n \sin \frac{n\pi x}{L}$$

Find the expression for the deflection w and the critical P value deduced from this.

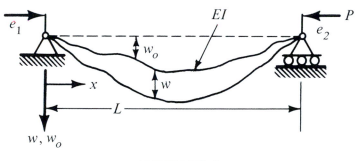

FIGURE P3.4

5. An ideal column of length L is pinned at A and built in at B on a rigid disc of radius R, which is supported by an immovable frictionless pin at its center (see figure). Derive the characteristic equation and the expression for P_{cr}. Discuss the extreme cases $(R \to 0$ and $R \to \infty)$.

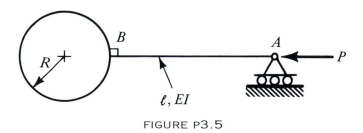

FIGURE P3.5

6. Find an expression for the tilt-buckling load associated with the two systems depicted in Fig. P3.6 a and b.
7. A uniform disc of radius R rotates at constant angular velocity ω. A weightless elastic bar of length $L < R$ is fixed at one end and carries a mass m at the free end (see figure). Find the critical angular velocity at which buckling will occur.

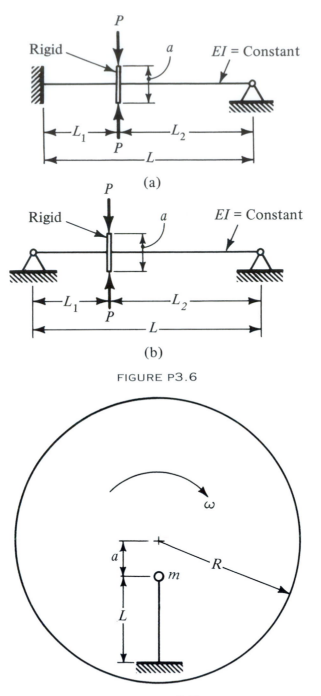

FIGURE P3.6

FIGURE P3.7

8. A column is loaded by a tensile force T at a small angle φ to the vertical (see part a of the figure). Show by deriving and solving the differential equation that

$$\delta = \ell \tan \varphi \left(1 - \frac{\tan h\gamma\ell}{\gamma\ell} \right)$$

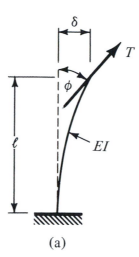

(a)

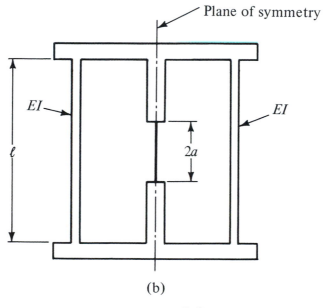

(b)

FIGURE P3.8

where

$$\gamma^2 \approx \frac{T}{EI}$$

The testing machine sketched (b of the figure) is applying compressive loading $2P$ to the test specimen. Show that buckling of the machine is possible, and indicate how the vertical load varies when a/ℓ is made smaller. The machine columns are built in at both ends.

9. The top end of a flexible straight bar is attached by a stretched wire to a fixed point A. The initial tension T in the wire does not change appreciably when a small deflection δ occurs.

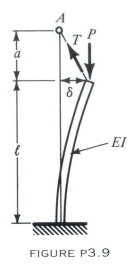

FIGURE P3.9

(a) Show that the critical values of P are determined by the equation

$$(k\ell)^2 = \frac{T\ell^2}{EI}\left(\frac{\ell}{a} - \frac{\tan k\ell}{ka}\right)$$

where

$$k^2 \approx \frac{P-T}{EI} \quad \text{and} \quad \sin\varphi \approx \frac{\delta}{a}$$

(b) Check this result by taking $T = 0$.

10. The vertical bar AB is supported by an extensional spring of stiffness α at A. Explain what is meant by the critical value of such a spring stiffness, and find an expression for it when the lower end B of the bar is pinned. Suppose that the elastic bar is built in at B. Show that there is no critical spring stiffness for this case.

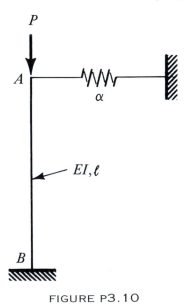

FIGURE P3.10

11. A slender elastic column of length L is pinned at the left end, while the right end is restrained by a vertical support and a linear rotational spring of stiffness α. Derive the characteristic equation, and indicate on a sketch the range of the first root for all spring stiffnesses (from 0 to ∞).

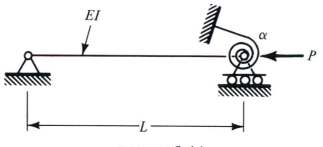

FIGURE P3.11

12. A flexible uniform light blade AB is pinned at A to a base. The spiral represents a linear rotational spring. The blade carries a mass particle m at B. The base is now made to spin (at ω), in the plane, about a center O between A and B, and the blade is consequently under tension $mc\omega^2$.

(a) Examine the possibility of a critical speed ω at which a slight deflection of the blade becomes possible. Show that this possibility exists if the equation

$$\frac{EI}{\alpha\ell} + \frac{\coth\gamma\ell}{\gamma\ell} = \frac{\ell}{\ell-c}\frac{1}{(\gamma\ell)^2}, \quad \gamma^2 \approx \frac{mc\omega^2}{EI}$$

has a real root for $\gamma\ell$.

(b) Consider the extreme case $\alpha \to \infty$ (clamped end). Prove that the "buckling" can occur when the center of rotation lies between A and B, but it can *not* occur when O is below A.

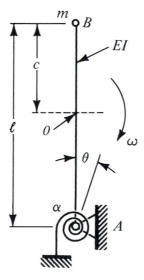

FIGURE P3.12

13.

(a) Find an expression for the critical spring stiffness for the system shown. The member of length *l* is flexible and is pinned at both ends. The member of length *a* is rigid and pinned at both ends.

(b) The left end of the flexible bar, instead of being pinned, is fixed into a block that can move horizontally (frictionless rollers). Show that the characteristic equation for a given value α is

$$1 - \frac{\tan k\ell}{k\ell} = \frac{(k\ell)^2}{\alpha \ell^3/EI - (\ell/a)(k\ell)^2}$$

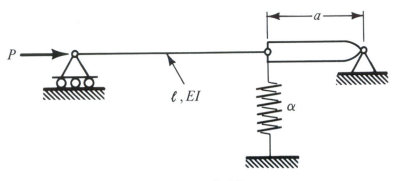

FIGURE P3.13

14. By use of the kinetic approach, find the critical compressive force for an ideal column with the following boundary conditions:
(α) Both ends fixed.
(β) One end fixed, the other pinned.
(γ) One end fixed, the other free.

15. Regarding the derivation in Section 3.8, show that the contribution of ρ_{11} to the strain energy can be neglected. Hint: $h \ll a$, and all elastic moduli are assumed to be of the same order of magnitude.

16. Regarding the derivation in Section 3.8, show that for isotropic materials $\overline{A}_{11} = hE$, $\overline{A}_{13} = 0$, $\overline{A}_{33} = \mu h^3/12$ and $\rho_{12} = \theta'$.

17. Analyze the axial-torsional buckling of a thin-walled cruciform section as shown in Fig. 3.23 in which the flange lengths (*a* and *b*) are of the same order and $h \ll a$, *b*. The axial compressive load is applied at the cross-sectional centroid, and the ends of the beam are free to move toward each other. Assume a clamped condition at one end with zero warping, and pinned conditions at the other end with freedom to warp. Determine values of the ratio *b/a* that dictate the mode of buckling to be torsional.

18. Analyze the axial-torsional buckling of a thin-walled channel section in which the flange lengths (*b*) are the same order as the width (*a*) and $h \ll a$, *b*. Assume simply supported boundary conditions and freedom to warp at both ends, that the axial compressive load is applied at the cross-sectional centroid, and the ends of the beam are free to move toward each other. Determine values of the ratio *b/a* that dictate the mode of buckling to be torsional.

REFERENCES

Biezeno, C. B. and Grammel, R. (1956). *Engineering Dynamics*. Vol. II. Part IV, Blackie and Son Ltd., London, p. 428.

Carslaw, H. S. and Jaeger, J. C. (1947). *Operational Methods in Applied Mathematics*. Dover Publications, Inc., New York, Chap. 11.

Churchill, R. V. (1958). *Operational Mathematics*. McGraw-Hill Book Co., New York, p. 252.

Courant, R. and Hilbert, D. (1953). *Methods of Mathematical Physics*. Vol. 1, Interscience Publishers, Inc., New York, p. 295.

Hodges, D. H. (1987). Finite rotation and nonlinear beam kinematics. *Vertica* 11, 297–307.

Hodges, D. H. (2006). *Nonlinear Beam Theory for Engineers*, American Institute of Aeronautics and Astronautics, Chapter 6.

Hoff, N. J. (1956). *The Analysis of Structures*. John Wiley & Sons, Inc., New York.

Love, A. E. H. (1944). *Mathematical Theory of Elasticity*. Dover Publications, New York, New York, 4th edition.

Reissner, E. and Tsai, W. T. (1972). Pure bending, stretching, and twisting of anisotropic cylindrical shells. *Journal of Applied Mechanics* 39, 148–154.

Shames, I. H. (1964). *Mechanics of Deformable Solids*. Prentice-Hall, Inc., Englewood Cliffs, N. J., Chap. 7.

Southwell, R. V. (1936). *An Introduction to the Theory of Elasticity*. Oxford at the Clarendon Press, p. 425.

Thomson, Wm. T. (1965). *Vibration Theory and Applications*. Prentice-Hall, Inc., Englewood Cliffs, N. J., p. 276.

Timoshenko, S. P. (1953). *History of Strength of Materials*. McGraw-Hill Book Co., New York, pp. 30–36.

Timoshenko, S. P. and Gere, J. M. (1961). *Theory of Elastic Stability*. McGraw-Hill Book Co., New York, p. 135.

Vlasov, V. Z. (1961). Thin-Walled Elastic Beams. National Science Foundation and Department of Commerce.

4

BUCKLING OF FRAMES

Frames of various types, especially the civil-engineering type, are widely used in structural configurations such as buildings and bridges. These frames are subjected to concentrated and distributed loads which, in many cases, may cause buckling of an element or group of elements of the frame. Because the members are rigidly connected to other members, flexural deformations in one element cause deformations in the neighboring elements. This results in a loss of flexural rigidity of the entire system. Knowledge of the critical condition is essential in the design of both simple and complex frames.

This chapter is intended to familiarize the student with buckling of some simple frames, and it presents a few of the methods that can successfully be used to arrive at the critical condition. A more complete presentation of the buckling analysis of frames may be found in the books of Bleich (1952) and Britvec (1973). Since one of the methods employed for the analysis of frames is based on the theory of beam-columns, a review section will first be presented (see also Timoshenko, 1961).

4.1 BEAM-COLUMN THEORY

A slender bar meeting the Euler-Bernoulli assumptions under transverse loads as well as an inplane compressive load (see Fig. 4.1) is called a *beam-column*. The equation governing the response of a beam-column was derived in Chapter 3:

$$\left(EIw_{,xx}\right)_{,xx}+\overline{P}w_{,xx} = q(x) + \sum_{i=1}^{n} Q_i\delta(x - x_i) + \sum_{j=1}^{m} C_j\eta\left(x - x_j\right) \tag{1}$$

The moment, M, and shear, V, at any station x are given by the following equations:

$$M = EIw_{,xx}$$
$$V = -\left[\left(EIw_{,xx}\right)_{,x}+\overline{P}w_{,x}\right] \tag{2}$$

The solutions to a number of problems are presented, and some of these solutions will be used in the buckling analysis of frames.

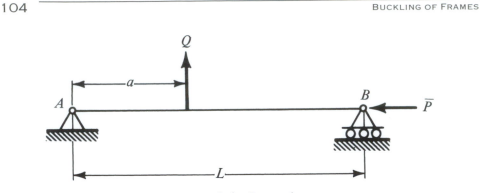

FIGURE 4.1 Beam-column.

4.1.1 BEAM-COLUMN WITH A CONCENTRATED LATERAL LOAD

A simply supported beam-column under the application of a concentrated lateral load, Q, at station $x = a$ is shown in Fig. 4.1. The bending stiffness, EI, of the beam-column is taken to be constant.

The governing differential equation and the proper boundary conditions are given by

$$EIw_{,xxxx} + \overline{P}w_{,xx} = Q\delta(x - a) \tag{3}$$

$$\begin{aligned} w(0) = w(L) = 0 \\ w_{,xx}(0) = w_{,xx}(L) = 0 \end{aligned} \tag{4}$$

If we now separate the interval $[0, L]$ into two regions $0 < x < a$ and $a < x < L$, and if we denote by $w^1(x)$ and $w^2(x)$ the displacements in the two intervals, respectively, the differential equations and proper boundary conditions are given by

$$EIw^1_{,xxxx} + \overline{P}w^1_{,xx} = 0 \tag{5}$$

$$EIw^2_{,xxxx} + \overline{P}w^2_{,xx} = 0 \tag{6}$$

$$\begin{aligned} w^1(0) = 0 \quad w^1_{,xx}(0) = 0 \\ w^2(L) = 0 \quad w^2_{,xx}(L) = 0 \end{aligned} \tag{7}$$

The solutions to Eqs. (5) and (6) are

$$w^1(x) = A_1 \sin kx + A_2 \cos ky + A_3 x + A_4 \tag{8}$$

$$w^2(x) = B_1 \sin kx + B_2 \cos kx + B_3 x + B_4 \tag{9}$$

where $k^2 = \overline{P}/EI$.

There are eight constants to be evaluated, A_i and B_i $(i = 1, 2, 3, 4)$. These constants may be evaluated by use of the boundary conditions, Eqs. (7), and the auxiliary conditions at $x = a$. The auxiliary conditions are based on the fact that, at $x = a$, the deflection, slope, and moment must be continuous and the shear is discontinuous by a known amount $(\Delta V = Q)$.

The auxiliary conditions are

$$w^1(a) = w^2(a)$$
$$w^1_{,x}(a) = w^2_{,x}(a)$$
$$w^1_{,xx}(a) = w^2_{,xx}(a) \tag{10}$$
$$-\left[EIw^1_{,xxx}(a) + \overline{P}w^1_{,x}(a)\right] = -\left[EIw^2_{,xxx}(a) + \overline{P}w^2_{,x}(a)\right] + Q$$

Use of the eight equations, Eqs. (7) and (10), leads to the following solution:

$$w^1(x) = \frac{Q\sin k(L-a)}{\overline{P}k\sin kL}\sin kx - \frac{Q}{\overline{P}}\left(1 - \frac{a}{L}\right)x \quad 0 \leq x \leq a$$
$$-w^2(x) = \frac{Q\sin ka}{\overline{P}k\sin kL}\sin k(L-x) - \frac{Qa}{\overline{P}}\left(1 - \frac{x}{L}\right) \quad a \leq x \leq L \tag{11}$$

By differentiation of Eq. (11), we obtain the following expressions for the slope and curvature (approximate):

$$w^1_{,x} = \frac{Q\sin k(L-a)}{\overline{P}\sin kL}\cos kx - \frac{Q(L-a)}{\overline{P}L} \quad 0 \leq x \leq a \tag{12}$$

$$w^2_{,x} = -\frac{Q\sin ka}{\overline{P}\sin kL}\cos k(L-x) + \frac{Qa}{\overline{P}L} \quad a \leq x \leq L \tag{13}$$

$$w^1_{,xx} = -\frac{Qk\sin k(L-a)}{\overline{P}\sin kL}\sin kx \quad 0 \leq x \leq a \tag{14}$$

$$w^2_{,xx} = -\frac{Qk\sin ka}{\overline{P}\sin kL}\sin k(L-x) \quad a \leq x \leq L \tag{15}$$

In the particular case for which $a = L/2$, the expressions for the deflection become

$$w^1 = \frac{Q\sin(kL/2)}{\overline{P}k\sin kL}\sin kx - \frac{Q}{2\overline{P}}x \tag{16a}$$

$$w^2 = \frac{Q\sin(kL/2)}{\overline{P}k\sin kL}\sin k(L-x) - \frac{QL}{2\overline{P}}\left(1 - \frac{x}{L}\right) \tag{16b}$$

The maximum deflection occurs at $x = L/2$, and the expression for it is

$$w\left(\frac{L}{2}\right) = \delta = \frac{Q}{2\overline{P}k}\left(\tan\frac{kL}{2} - \frac{kL}{2}\right) \tag{17}$$

In the absence of the inplane load $\overline{P}$, the expression for the maximum deflection (at $L/2$) is

$$\delta_0 = \frac{QL^3}{48EI} \tag{18}$$

If we rearrange the terms in Eq. (17) and make use of Eq. (18), the expression for the maximum deflection of the beam-column becomes

$$\delta = \delta_0 \frac{3(\tan u - u)}{u^3} \tag{19}$$

where $u = kL/2$. Introducing the following notation

$$\chi(u) = 3\frac{\tan u - u}{u^3} \tag{20}$$

the expression for δ becomes

$$\delta = \delta_0 \chi(u) \tag{21}$$

Numerical values of $\chi(u)$ are found in Appendix A of Ref. 1 for different values of u, and $\chi(u)$ is plotted versus u in Fig. 4.2 for $0 \le u < \pi/2$. The factor $\chi(u)$ in Eq. (21) gives the influence of the inplane load on the maximum deflection for the beam-column. From Fig. 4.2, we can decide up to what values of the axial thrust the neglect of its effect becomes unacceptable when we are interested in finding the deflection of this beam-column.

For this particular case ($a = L/2$), the rotation at $x = 0$ is given by

$$\frac{dw}{dx}\bigg|_{x=0} = \theta_A = \frac{QL^2}{16EI}\lambda(u) \tag{22}$$

where

$$\lambda(u) = \frac{2(1 - \cos u)}{u^2 \cos u} \tag{23}$$

Here again, the first factor in Eq. (22) denotes the slope at $x = 0$ in the absence of the inplane load $\overline{P}$, and $\lambda(u)$ gives the influence of inplane load $\overline{P}$. Numerical values of $\lambda(u)$ may be found in Appendix A of Bleich (1952). The plot of $\lambda(u)$ versus u resembles that of $\chi(u)$ versus u (see Fig. 4.2).

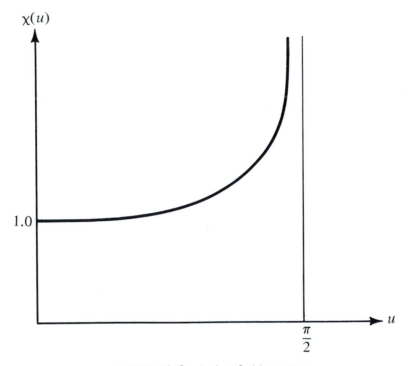

FIGURE 4.2   A plot of $\chi(u)$ versus u.

Finally, the expression for the bending moment at $x = L/2$ for this particular case is

$$M_{\max} = EI\left(\frac{d^2w}{dx^2}\right)_{x=L/2} = -\frac{QL}{4}\frac{\tan u}{u} \tag{24}$$

Note that as $\overline{P}$ approaches zero, u approaches zero, and since

$$\lim_{u\to 0}\frac{\tan u}{u} = 1 \tag{25}$$

the expression for the maximum bending moment for the beam is

$$M_{0_{\max}} = -\frac{QL}{4}$$

4.1.2 BEAM-COLUMN WITH TWO END-COUPLES

Consider the simply supported beam-column shown in Fig. 4.3 and loaded by two end-couples, M_A and M_B. The differential equation governing equilibrium is

$$EI\frac{d^4w}{dx^4} + \overline{P}\frac{d^2w}{dx^2} = 0 \tag{26}$$

The proper boundary conditions are

$$\begin{aligned} w(0) &= w(L) = 0 \\ EIw_{,xx}(0) &= M_A \\ EIw_{,xx}(L) &= M_B \end{aligned} \tag{27}$$

The solution to Eq. (26) is given by

$$w(x) = A_1\sin kx + A_2\cos kx + A_3x + A_4 \tag{28}$$

Use of the boundary conditions leads to the following expression for $w(x)$:

$$w(x) = \frac{M_A}{\overline{P}}\left[\frac{L-x}{L} - \frac{\sin k(L-x)}{\sin kL}\right] + \frac{M_B}{\overline{P}}\left(\frac{x}{L} - \frac{\sin kx}{\sin kL}\right) \tag{29}$$

Denoting by θ_A and θ_B the magnitudes of the rotation angles at A and B, respectively, we obtain

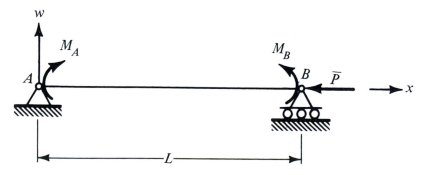

FIGURE 4.3   Beam-column loaded by two end-couples.

$$\theta_A = -\frac{dw}{dx}\Big|_{x=0} = \frac{M_A L}{3EI}\psi(u) + \frac{M_B L}{6EI}\varphi(u) \tag{30}$$

$$\theta_B = \frac{dw}{dx}\Big|_{x=L} = \frac{M_B L}{3EI}\psi(u) + \frac{M_A L}{6EI}\varphi(u) \tag{31}$$

where

$$\varphi(u) = \frac{3}{u}\left(\frac{1}{\sin 2u} - \frac{1}{2u}\right)$$

$$\psi(u) = \frac{3}{2u}\left(\frac{1}{2u} - \frac{1}{\tan 2u}\right) \tag{32}$$

$$u = \frac{kL}{2}$$

As before, the factors $\psi(u)$ and $\varphi(u)$ give the influence of the inplane load on the end rotations. This means that in the absence of the inplane load $\overline{P}$, the end rotations θ_{A_0} and θ_{B_0} are given by

$$\theta_{A_0} = \frac{M_A L}{3EI} + \frac{M_B L}{6EI}$$
$$\theta_{B_0} = \frac{M_B L}{3EI} + \frac{M_A L}{6EI} \tag{33}$$

4.1.3 SUPERPOSITION

Since beam-column problems are nonlinear problems, superposition of solutions does not hold in the usual manner. The results can be superimposed if and only if the axial load $\overline{P}$ is the same for two or more cases of different lateral loads. To demonstrate the point, consider a simply supported beam-column of length L loaded first by a lateral loading $q_1(x)$ and second by $q_2(x)$. Let the response of the system to the two loadings be denoted by $w^1(x)$ and $w^2(x)$, respectively. Then the equilibrium equations for the two problems are:

$$EIw^1_{,xxxx} + \overline{P}w^1_{,xx} = q_1(x) \tag{34}$$

$$EIw^2_{,xxxx} + \overline{P}w^2_{,xx} = q_2(x) \tag{35}$$

By addition, we obtain

$$EI\left(w^1 + w^2\right)_{,xxxx} + \overline{P}\left(w^1 + w^2\right)_{,xx} = q_1(x) + q_2(x) \tag{36}$$

Next, consider the case of simultaneous application of the loadings $q_1(x)$ and $q_2(x)$. For this case the equilibrium equation is:

$$EIw_{,xxxx} + \overline{P}w_{,xx} = q_1(x) + q_2(x) \tag{37}$$

By comparison of Eqs. (36) and (37), it is clear that superposition holds for this type of problem. Thus superposition holds for any number of transverse loadings (distributed and concentrated forces and applied moments) provided the inplane force is the same and the beam-column is supported in the same manner for all loading cases (see Fig. 4.4).

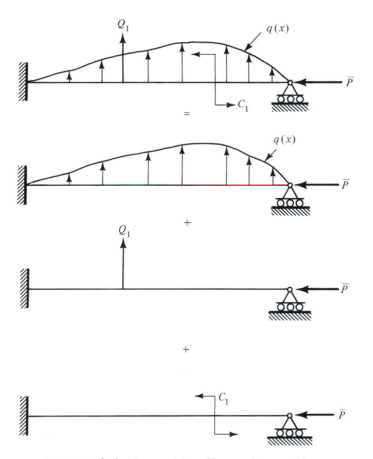

FIGURE 4.4   Superposition of beam-column problems.

4.2   APPLICATION OF BEAM-COLUMN THEORY TO THE BUCKLING OF ROTATIONALLY RESTRAINED COLUMNS

Consider a column which is supported against transverse translation at both ends but is restrained against rotation through rotational springs (see Fig. 4.5a). The problem here is to find $\overline{P}_{cr}$ as a function of the structural geometry (EI, L, β_0, and β_L).

Instead of this problem, we may consider the beam-column problem of Fig. 4.5b. According to the results of Section 4.1.2, the end rotations are given by

$$\theta_0 = \frac{M_0 L}{3EI} \psi(u) + \frac{M_L L}{6EI} \varphi(u) \tag{38}$$

$$\theta_L = \frac{M_0 L}{6EI} \varphi(u) + \frac{M_L L}{3EI} \psi(u) \tag{39}$$

By comparison of the beam-column problem to the original one, we may write

$$\begin{aligned} M_0 &= -\beta_0 \theta_0 \\ M_L &= -\beta_L \theta_L \end{aligned} \tag{40}$$

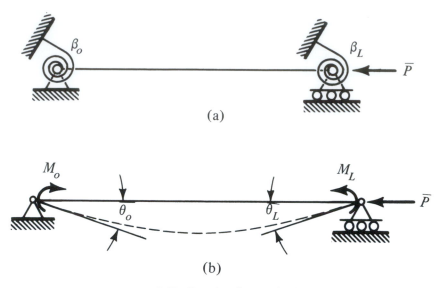

FIGURE 4.5   Rotationally restrained column.

Substitution into Eqs. (38) and (39) yields

$$M_0\left[\frac{1}{\beta_0}+\frac{L}{3EI}\psi(u)\right]+M_L\left[\frac{L}{6EI}\varphi(u)\right]=0$$

$$M_0\left[\frac{L}{6EI}\varphi(u)\right]+M_L\left[\frac{1}{\beta_L}+\frac{L}{3EI}\psi(u)\right]=0$$

(41)

These are two homogeneous linear algebraic equations in M_0 and M_L. For a nontrivial solution to exist (bifurcation), the determinant of the coefficients must vanish. Thus the characteristic equation is:

$$\left[\frac{1}{\beta_0}+\frac{L\psi(u)}{3EI}\right]\left[\frac{1}{\beta_L}+\frac{L\psi(u)}{3EI}\right]-\left[\frac{L\varphi(u)}{6EI}\right]^2=0$$

(42)

where $u=kL/2$ and $k^2=\overline{P}/EI$.

In the special case where $\beta_0=\beta_L=\beta$, Eq. (42) becomes

$$\frac{1}{\beta}+\frac{L\psi(u)}{3EI}\pm\frac{L\varphi(u)}{6EI}=0$$

(43)

From the first of Eqs. (41), it is seen that

$$M_L=-M_0\left[\frac{1}{\beta}+\frac{L\psi(u)}{3EI}\right]\left[\frac{L\varphi(u)}{6EI}\right]^{-1}$$

(44)

Therefore we see that the plus sign in Eq. (43) corresponds to the symmetric case ($M_0=M_L$) and the minus sign to the antisymmetric case ($M_0=-M_L$).

For the symmetric case, substitution for the expressions $\psi(u)$ and $\varphi(u)$ leads to the following characteristic equation:

$$\tan u=-\frac{2EI}{\beta L}u$$

(45)

Figure 4.6a shows that u_{cr}, depending on the value of β, lies between $\pi/2$ and π. When $\beta \to 0$, $u_{cr} \to \pi/2$ and $P_{cr} = \pi^2 EI/L^2$ (both ends simply supported). When $\beta \to \infty$, $u_{cr} \to \pi$ and $P_{cr} = 4\pi^2 EI/L^2$ (both ends clamped).

For the antisymmetric case, substitution for the expressions $\psi(u)$ and $\varphi(u)$ yields

$$\tan u = \frac{u}{1 + \dfrac{2EI}{\beta L} u^2} \tag{46}$$

From Fig. 4.6b we see that u lies between π and 4.493. When $\beta \to 0$, $u_{cr} \to \pi$ and $P_{cr} = 4\pi^2 EI/L^2$ (both ends simply supported). When $\beta \to \infty$, $u_{cr} \to 4.493$ and $P_{cr} = 4(4.493)^2 EI/L^2$ (both ends clamped).

Therefore, for this special case $(\beta_0 = \beta_L = \beta)$, the column will buckle in a symmetric mode $(\pi/2 < u_{cr} < \pi)$.

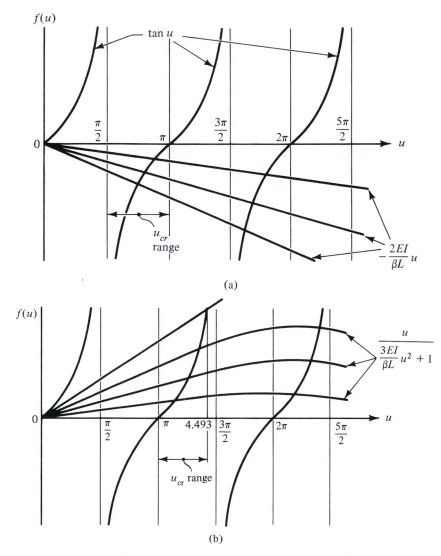

FIGURE 4.6 Critical conditions for rotationally restrained columns.

4.3   RECTANGULAR RIGID FRAMES

Consider the frame shown in Fig. 4.7 and note that, as load P increases quasistatically from zero, it is possible to reach some PQ combination for which the frame will buckle (bifurcation). It is also clear that buckling may be caused by the existence of only P or Q and that the mode of failure in any case can be either symmetric or antisymmetric (see Figs. 4.8a and 4.8b).

In this particular problem, each member is elastically restrained against rotation at the ends because of the rigid connection to the adjacent member. Therefore the method described in the previous section may be applied, provided the rotational spring constant can be expressed in terms of the structural geometry of the adjacent members.

Symmetric and antisymmetric buckling are treated separately in the following sections.

4.3.1   SYMMETRIC BUCKLING

If we decompose the frame shown in Fig. 4.7, and if the frame is assumed to buckle in a symmetric mode, then the bending moments at the four corners are all equal (see Fig. 4.9).

Noting that $\theta_A = \theta_B$, from Eq. (38)

$$\theta_A = -\frac{Mb}{3(EI)_1}\psi(u_1) - \frac{Mb}{6(EI)_1}\varphi(u_1) \tag{47}$$

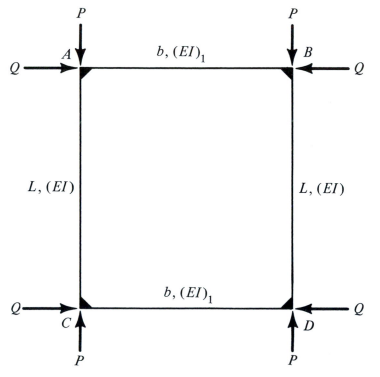

FIGURE 4.7   Geometry of a rigid frame.

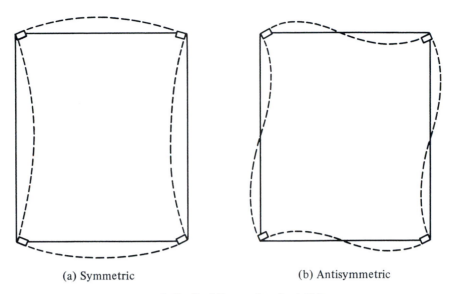

(a) Symmetric   (b) Antisymmetric

FIGURE 4.8   Buckling modes of a rigid frame.

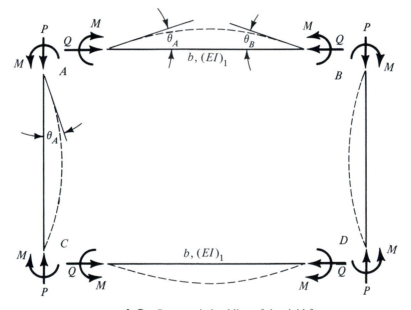

FIGURE 4.9   Symmetric buckling of the rigid frame.

where

$$u_1 = k_1 \frac{b}{2} \quad \text{and} \quad k_1^2 = \frac{Q}{(EI)_1}$$

From Eq. (47) we can obtain

$$\theta_A = -\frac{Mb}{2(EI)_1} \cdot \frac{\tan u_1}{u_1} \tag{48}$$

From Eq. (40), we obtain the expression for β. This equation is applicable because of the directions of moments and rotations as applied to AC (see Figs. 4.9 and 4.5b).

$$\beta^{-1} = -\frac{\theta_A}{M} = \frac{b}{2(EI)_1}\frac{\tan u_1}{u_1} \tag{49}$$

Use of this expression for β in Eq. (43) for the vertical member (AC) yields (symmetric buckling)

$$\frac{b}{2(EI)_1}\frac{\tan u_1}{u_1} + \frac{L}{3EI}\psi(u) + \frac{L}{6EI}\varphi(u) = 0 \tag{50}$$

where

$$u = \frac{kL}{2} \quad \text{and} \quad k^2 = \frac{P}{EI}$$

Thus the characteristic equation becomes

$$\frac{L}{2EI}\frac{\tan u}{u} + \frac{b}{2(EI)_1}\frac{\tan u_1}{u_1} = 0 \tag{51}$$

or

$$\frac{\tan u}{u} = -\frac{EI}{(EI)_1}\frac{b}{L}\frac{\tan u_1}{u_1}$$

In the special case for which $(EI)_1/b = EI/L$, then

$$\frac{\tan u}{u} = -\frac{\tan u_1}{u_1} \tag{52}$$

The solution to this equation is plotted in Fig. 4.10, and it represents the boundary between the stable and unstable regions.

Consider also the special case for which $Q = 0$. For this case, since

$$\lim_{u_1 \to 0}\frac{\tan u_1}{u_1} = 1$$

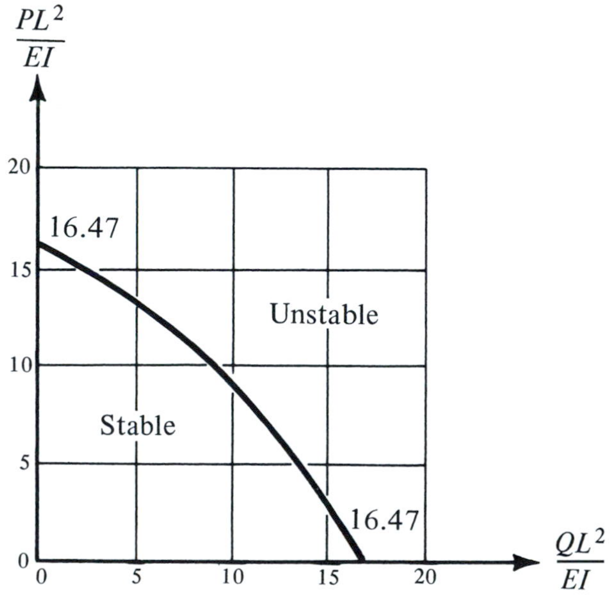

FIGURE 4.10 Critical conditions for a rigid square frame of constant stiffness.

the characteristic equation becomes

$$\frac{\tan u}{u} = -\frac{EIb}{(EI)_1 L} \tag{53}$$

Furthermore, if $EI = (EI)_1$ and $b = L$, then

$$\tan u = -u \tag{54}$$

The smallest root of this equation is 2.029; therefore

$$u_{cr} = \left(\frac{kL}{2}\right)_{cr} = 2.029$$

and

$$P_{cr} = 16.47 \frac{EI}{L^2} \tag{55}$$

4.3.2 ANTISYMMETRIC BUCKLING

If the frame buckles in an antisymmetric mode, then (see Fig. 4.11)

$$\theta_A = -\frac{Mb}{3(EI)_1}\psi(u_1) + \frac{Mb}{6(EI)_1}\varphi(u_1) \tag{56}$$

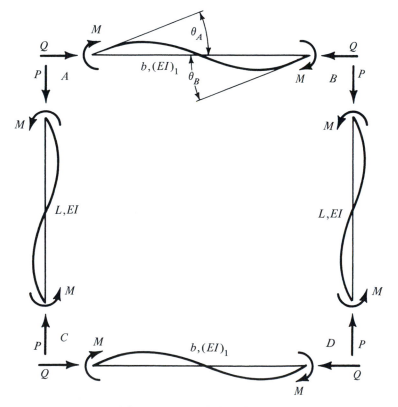

FIGURE 4.11 Antisymmetric buckling of the rigid frame.

From Eq. (56) and Eq. (40), we obtain

$$\frac{1}{\beta} = \frac{b}{6(EI)_1} \cdot \frac{3}{u_1}\left(\frac{1}{u_1} - \cot u_1\right) \tag{57}$$

Substitution of Eq. (57) into Eq. (43) yields the characteristic equation for antisymmetric buckling:

$$\frac{b}{(EI)_1}\frac{1}{u_1}\left(\frac{1}{u_1} - \cot u_1\right) = -\frac{L}{EI}\frac{1}{u}\left(\frac{1}{u} - \cot u\right) \tag{58}$$

Let us next consider a few special cases. First consider the case of $Q = 0$. Recognizing that $\psi(0) = \varphi(0) = 1$, then

$$\frac{1}{\beta} = \frac{b}{6(EI)_1} \tag{59}$$

The characteristic equation for this case is obtained if we substitute the expression for β, Eq. (59), into Eq. (43):

$$\frac{1}{u}\left(\frac{1}{u} - \cot u\right) = -\frac{EIb}{3(EI)_1 L} \tag{60}$$

Furthermore, if we assume that the frame is square and of constant stiffness $[L = b,\ EI = (EI)_1]$, then

$$\frac{1}{u} - \cot u = -\frac{u}{3} \tag{61}$$

The solution of this transcendental equation yields $u_{cr} > \pi$, and the critical load is higher than the corresponding symmetric mode critical load, Eq. (55).

Finally, if the frame is square with constant stiffness but $Q \neq 0$, then the characteristic equation becomes

$$\frac{1}{u_1}\left(\frac{1}{u_1} - \cot u_1\right) = -\frac{1}{u}\left(\frac{1}{u} - \cot u\right) \tag{62}$$

For this case, a plot similar to that in Fig. 4.10 may be generated. Since the intercepts are higher than those corresponding to symmetric buckling, the constant-stiffness square frame will always buckle in a symmetric mode.

4.4   THE SIMPLY SUPPORTED PORTAL FRAME

Let us consider the portal frame shown in Fig. 4.12. We are interested in finding the smallest possible load (P_{cr}) which will cause the frame to buckle. To accomplish this, we must consider all possible modes of buckling, compute P_{cr} for each mode, and establish, through a comparison, P_{cr} and the corresponding mode. Note that the frame is symmetric. The different buckling modes are shown in Fig. 4.13. Also note that there is no possibility of a sway buckling mode when the horizontal bar buckles symmetrically.

First, the rotational elastic restraint provided to the vertical bars by the horizontal bar is the same as in the frame problem of Section 4.3 with $Q = 0$.

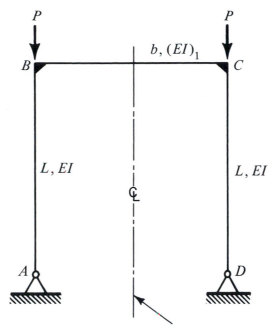

FIGURE 4.12   Simply supported portal frame.

$$\text{Symmetric}\quad \beta_B = \frac{2(EI)_1}{b}$$
$$\text{Antisymmetric}\quad \beta_B = \frac{6(EI)_1}{b} \tag{63}$$

The characteristic equation for the first two cases of Fig. 4.13, a and b, is given by Eq. (42) with $\beta_0 = \beta_A = 0$ and $\beta_L = \beta_B$, or

$$\left[\frac{1}{\beta_A} + \frac{L\psi(u)}{3EI}\right]\left[\frac{1}{\beta_B} + \frac{L\psi(u)}{3EI}\right] = \left[\frac{L\varphi(u)}{6EI}\right]^2 \tag{64}$$

Multiplying Eq. (64) by β_A and then setting $\beta_A = 0$, we have

$$\frac{1}{\beta_B} + \frac{L\psi(u)}{3EI} = 0 \tag{65}$$

Substitution of the expressions for $\psi(u)$, Eq. (32), and β_B, Eqs. (63), yields

$$\text{Symmetric (a)}\quad \frac{1}{2u} + (2u)\frac{EIb}{2(EI)_1 L} = \cot(2u)$$
$$\text{Antisymmetric (b)}\quad \frac{1}{2u} + (2u)\frac{EIb}{6(EI)_1 L} = \cot(2u) \tag{66}$$

It is shown qualitatively in Fig. 4.14 that the critical load for case (b) is higher than that for case (a). In addition, we see from this figure that $(2u)_{cr} > \pi$ for both cases.

As a special case of the characteristic equations, Eqs. (66), for this problem, we consider the horizontal bar to be extremely stiff. Then, $(EI)_1 \to \infty$ and the characteristic equation becomes

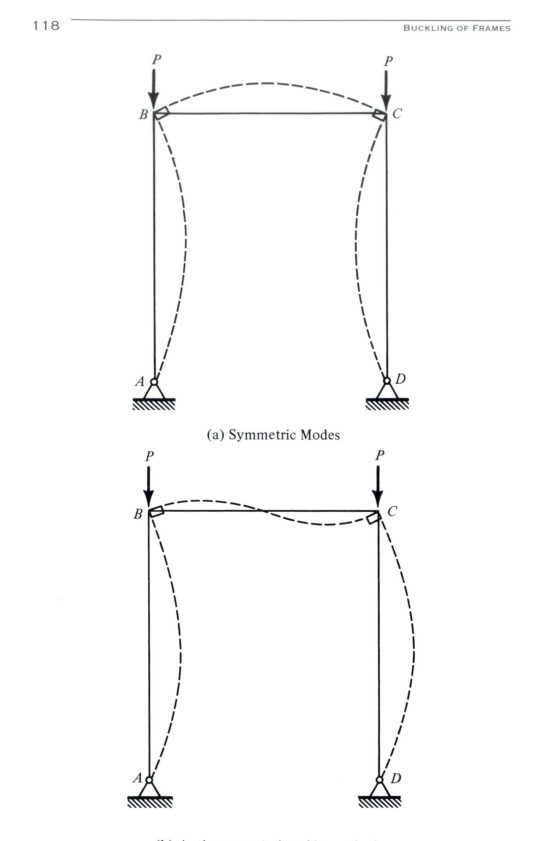

(a) Symmetric Modes

(b) Antisymmetric (no side motion)

FIGURE 4.13   Buckling modes for the simply supported portal frame.

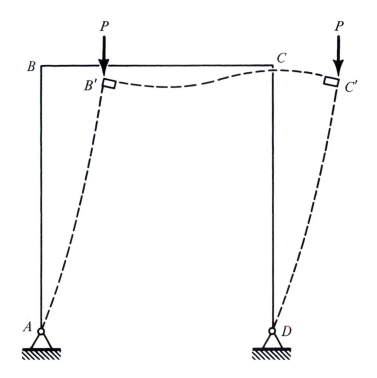

(c) Antisymmetric–sway buckling

FIGURE 4.13   **Cont'd.**

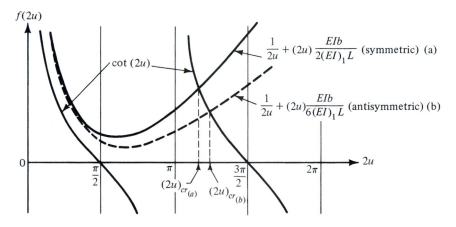

FIGURE 4.14   Critical conditions for cases (a) and (b).

$$\frac{1}{2u} = \cot(2u) \quad \text{or} \quad \tan(2u) = (2u) \tag{67}$$

Since, $2u = kL$, the above results in

$$P_{cr} = 20.19\frac{EI}{L^2}$$

This load represents the critical load for a column with one end fixed and the other simply supported.

The characteristic equation for the sway buckling case, Fig. 4.13c, cannot be obtained as a special case of Eq. (42) because point C is free to move in a direction normal to the column AB. Note that Eq. (42) represents the characteristic equation for a supported column with rotational end restraints (see Fig. 4.5a).

The characteristic equation for the case of sway buckling may be obtained if we consider the column shown in Fig. 4.15. Note that the rotational restraint provided by the horizontal bar in Fig. 4.14c is $6(EI)_1/b$. The column of Fig. 4.15 is a special case of the elastically supported column treated in Chapter 3. Therefore, the characteristic equation for this model is obtained from Eq. (97) of Chapter 3 with the following expressions for the spring constants

$$\alpha_0 = \infty \quad \alpha_L = 0 \quad \beta_0 = 0 \quad \beta_L = \frac{6(EI)_1/b}{EI}$$

In Eq. (97) of Chapter 3, the parameter u is defined by $u = kL$; therefore, wherever u appears, we must use $2u$.

Dividing Eq. (97) of Chapter 3 by α_0 and taking the limit as $1/\alpha_0 = \beta_0 = \alpha_L \to 0$, we have

$$-\frac{(2u)^6}{L^6} \sin(2u) + \frac{6EI_1}{bEI} \frac{(2u)^5}{L^5} \cos(2u) = 0$$

And finally

$$(2u) \tan 2u = \frac{6(EI)_1 L}{EIb} \tag{68a}$$

or

$$\tan(2u) = \frac{6(EI)_1 L/EIb}{(2u)} \tag{68b}$$

From Fig. 4.16 we see that $2u_{cr} < \pi/2$, and the critical load for the simply supported portal frame is characterized by Eq. (68b). Therefore, as the load P is increased quasistatically from zero, the frame will sway buckle when P reaches the value that satisfies Eq. (68b).

Assuming that $EI_1 L = EIb$, we obtain

$$\tan 2u = \frac{3}{u} \tag{69}$$

from which

FIGURE 4.15   Model for sway buckling of simply supported portal frames.

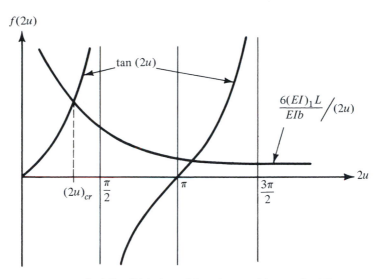

FIGURE 4.16   Critical conditions for case (c); sway buckling.

$$(2u)_{cr} = (kL)_{cr} = 1.350$$

and

$$P_{cr} = \frac{1.821 EI}{L^2}$$

4.5   ALTERNATE APPROACH

We have already demonstrated in Section 4.4 that we may use Eq. (97) of Chapter 3 to obtain the characteristic equation for a frame by reducing the problem to an elastically restrained column. This approach may be used for any frame once the amount of elastic restraint has been determined either by beam theory or beam-column theory. This idea will be demonstrated through the rigid frame and then applied to some additional cases.

4.5.1   RIGID FRAME

First consider the rigid frame shown in Fig. 4.7. First we reduce this problem to a column of length L, bending stiffness EI, and rotational restraints at the ends A and C of equal strength $(\bar{\beta}_A = \bar{\beta}_C$; see Fig. 4.17). Note that

$$\bar{\beta}_A = \bar{\beta}_C = \frac{2(EI)_1}{b} \cdot \frac{u_1}{\tan u_1}$$

from Eq. (49) for symmetric buckling (Fig. 4.9), and

$$\bar{\beta}_A = \bar{\beta}_C = \frac{[2(EI)_1/b]u_1}{(1/u_1) - \cot u_1}$$

from Eq. (57) for antisymmetric buckling (Fig. 4.11). To use Eq. (97) of Chapter 3, we must first recognize that wherever u appears in Eq. (97), we must use $2u$.

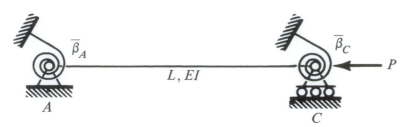

FIGURE 4.17   Column model for leg AC of the rigid frame.

Furthermore, the rotational restraint constants in Eq. (97) have been divided through by EI, or

$$\text{Symmetric} \quad \beta_0 = \beta_L = \frac{\overline{\beta}_A}{EI} = \frac{2(EI)_1}{bEI}\frac{u_1}{\tan u_1}$$

$$\text{Antisymmetric} \quad \beta_0 = \beta_L = \frac{\overline{\beta}_A}{EI} = \frac{[2(EI)_1/bEI]u_1}{(1/u_1) - \cot u_1} \tag{70}$$

For the model shown in Fig. 4.17, we have $\alpha_0 = \alpha_L = \infty$, and $\beta_0 = \beta_L$ given by Eq. (70).

We first divide Eq. (97) by $\alpha_0\alpha_L$ and take the limit as $1/\alpha_0 = 1/\alpha_L = 0$. This leads to the following characteristic equation for the column model of Fig. 4.17.

$$\left[L\left(\frac{2u}{L}\right)^3 + (2\beta_0 - \beta_0^2 L)\left(\frac{2u}{L}\right)\right]\sin(2u) - 2\beta_0\left[L\left(\frac{2u}{L}\right)^2 + \beta_0\right]\cos(2u) + 2\beta_0^2 = 0 \tag{71}$$

If we first express the trigonometric functions in terms of the single angle, we have

$$\left[L\left(\frac{2u}{L}\right)^3 + \left(2\beta_0 - \beta_0^2 L\right)\left(\frac{2u}{L}\right)\right]2\sin u \cos u - 2\beta_0 L\left(\frac{2u}{L}\right)^2(\cos^2 u - \sin^2 u)$$

$$+ 4\beta_0^2 \sin^2 u = 0 \tag{72}$$

Next, if we divide Eq. (72) through by $\cos^2 u$, we obtain the following quadratic equation in $\tan u$:

$$\left[2\beta_0^2 + \beta_0 L\left(\frac{2u}{L}\right)^2\right]\tan^2 u + \left[L\left(\frac{2u}{L}\right)^3 + (2\beta_0 - \beta_0^2 L)\frac{2u}{L}\right]\tan u - \beta_0 L\left(\frac{2u}{L}\right)^2 = 0 \tag{73}$$

The solution for $\tan u$ by the quadratic formula is

$$\tan u = -\frac{2u}{\beta_0 L} \tag{74a}$$

$$\cot u = \frac{1}{u} + \frac{2u}{\beta_0 L} \tag{74b}$$

It can be shown that Eq. (74a) corresponds to a symmetric mode, and therefore use of the corresponding expression for β_0 from Eqs. (70) yields

$$\frac{\tan u}{u} = -\frac{EIb}{(EI)_1 L}\cdot\frac{\tan u_1}{u_1} \tag{75}$$

This is the same as Eq. (51), as expected.

Similarly, Eq. (74b) corresponds to the antisymmetric mode, and substitution for β_0 yields the following expression [see Eq. (57)]:

$$\frac{1}{u}\left(\frac{1}{u} - \cot u\right) = -\frac{2}{\beta_0 L} = -\frac{bEI}{(EI)_1 L}\frac{1}{u_1}\left(\frac{1}{u_1} - \cot u_1\right) \tag{76}$$

4.5.2 THE CLAMPED PORTAL FRAME

Consider a portal frame similar to the one shown in Fig. 4.12 with the exception of having fixed supports at points A and D instead of simple supports. For this new problem, we must also consider all possible modes of failure. It is easily shown that the smallest load corresponds to sway buckling as demonstrated in Fig. 4.18. Therefore, we will only find P_{cr} for sway buckling. The column model that characterizes this mode of failure is shown in Fig. 4.19.

The characteristic equation for this model is obtained from Eq. (97) of Chapter 3 with

$$\alpha_0 = \infty, \quad \beta_0 = \infty, \quad \alpha_L = 0, \quad \text{and} \quad \beta_L = \frac{6(EI)_1}{bEI}$$

Note again that we must divide Eq. (97) by $\alpha_0\beta_0$, take the limit as $1/\alpha_0 = 1/\beta_0 = \alpha_L = 0$, and use $2u$ instead of u. The characteristic equation is

$$\tan 2u = -\frac{2u}{\beta_L L} = -\frac{2ubEI}{6L(EI)_1} \tag{77}$$

From Fig. 4.20 we see that, depending on the value of $bEI/L(EI)_1$, the critical value for $(2u)$ varies between $\pi/2$ and π as expected.

If $bEI = (EI)_1 L$, then from Eq. (77)

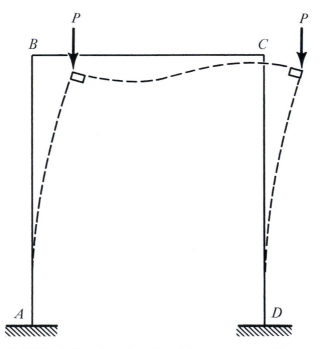

FIGURE 4.18 Sway buckling of the clamped portal frame.

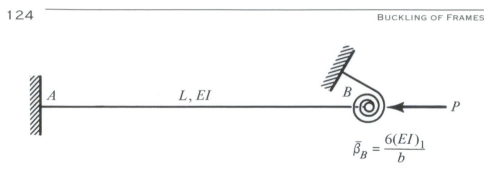

FIGURE 4.19   Model for sway buckling of clamped portal frames.

$$\tan 2u = -\frac{2u}{6}$$

and

$$(2u)_{cr} = (kL)_{cr} = 2.716$$

from which

$$P_{cr} = \frac{7.379EI}{L^2}$$

4.5.3   PARTIAL FRAMES

As a final application of the alternate approach, consider the partial frames of Fig. 4.21. The difference between these two partial frames is the support conditions at point A.

The elastic support provided by bar BC is a rotational spring with $\bar{\beta} = 4(EI)_{1/b}$, and an extensional spring, normal to the direction AB, with $\bar{\alpha} = (EA)_1/b$, where $(EA)_1$ is the extensional stiffness of bar BC. In most practical cases, $\bar{\alpha}$ is taken to be infinitely large. For such cases, the column models for the two partial frames are those shown in Fig. 4.22.

The characteristic equations for the two models are obtained from Eq. (97) of Chapter 3, with $\alpha_0 = \infty$, $\beta_0 = 0$, $\alpha_L = \infty$, $\beta_L = 4(EI)_1/bEI$ for case (a), and $\alpha_0 = \infty$, $\beta_0 = \infty$, $\alpha_L = \infty$, $\beta_L = 4(EI_1)/bEI$ for case (b). These equations are

$$\text{case (a)}\quad \left[L\left(\frac{2u}{L}\right)^4 + \beta_L\left(\frac{2u}{L}\right)^2\right]\sin 2u - L\beta_L\left(\frac{2u}{L}\right)^3\cos 2u = 0$$

or

$$\cot 2u = \frac{2u}{L\beta_L} + \frac{1}{2u}$$
$$\cot 2u = (2u)\frac{EIb}{4(EI)_1 L} + \frac{1}{2u} \tag{78}$$

$$\text{case (b)}\quad (1 - \beta_L L)\left(\frac{2u}{L}\right)^2\sin 2u - \left[L\left(\frac{2u}{L}\right)^3 + 2\beta_L\left(\frac{2u}{L}\right)\right]\cos 2u + 2\beta_L\left(\frac{2u}{L}\right) = 0$$

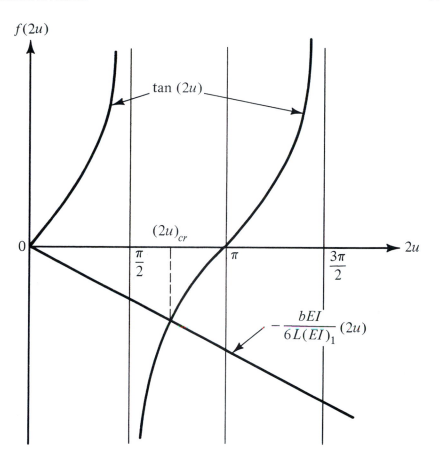

FIGURE 4.20   Critical conditions for sway buckling of clamped portal frames.

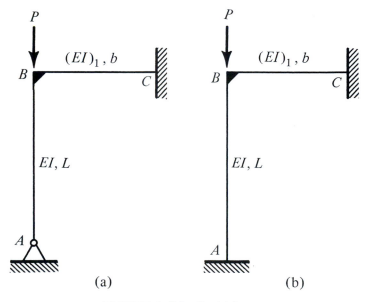

FIGURE 4.21   Partial frames.

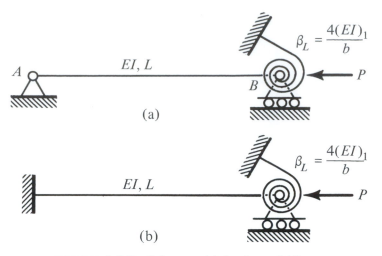

FIGURE 4.22   Column models for the partial frames.

or $\qquad (1 - \beta_L L)2u \sin 2u - (2u)^2 \cos 2u + 2\beta_L L(1 - \cos 2u) = 0$ $\qquad$ (79)

where $\beta_L = 4(EI)_1/bEI$.

If $(EI)_1 = EI$ and $L = b$, the characteristic equations become

$$\text{case (a)} \quad \cot 2u = \frac{2u}{4} + \frac{1}{2u} \qquad (80)$$

$$\text{case (b)} \quad -3(2u) \sin 2u - (2u)^2 \cos 2u + 8(1 - \cos 2u) = 0 \qquad (81)$$

For this particular condition and for case (a), $(2u)_{cr} = (kl)_{cr} = 3.829$ and

$$P_{cr} = \frac{14.66EI}{L^2}$$

4.6   NONLINEAR ANALYSIS*

4.6.1   INTRODUCTION

In recent years considerable attention has been given to stability of systems in the presence of initial imperfections. This attention is justified by the attempt to relate critical load conditions to load carrying capacity of the system. One of the major contributions to this problem is the initial post-buckling theory of Koiter (1967), who has shown that the effect of small initial imperfections is closely related to the initial post-buckling behavior of the corresponding perfect system. Koiter's initial post-buckling analysis is limited to systems that, in their ideally perfect state, exhibit a bifurcation point at the critical load. However, the bifurcation phenomenon is an exception rather than the rule. The majority of actual structural systems, if accurately modeled (mathematical model), exhibit limit point instability rather than

---

* From G.J. Simitses and A.N. Kounadis, "Buckling of Imperfect Rigid-Jointed Frames", ASCE Journal of the
  Engineering Mechanics Division, Vol. 104, No. 3, May/June 1978, pp.569-586. Reprinted with permission.

bifurcational buckling (with either stable or unstable branching). In many instances, the simplified modeling of a structural system, which leads to bifurcational buckling, is very attractive. This is so because a linearization of the governing equation for the prebuckling primary equilibrium state is possible, which in turn simplifies the estimation of the critical load. Usually, in such cases, the entire problem reduces to a linear eigen-boundary-value problem. On the contrary, for systems that exhibit limit point instability, a complete nonlinear analysis is required for the estimation of the critical load. The usual technique (Almroth et al., 1976) for determining the limit point consists of establishing the maximum on the load versus some characteristic displacement curve. This is accomplished by starting at some low level of the applied load, and obtaining the corresponding displacement by solving the nonlinear equilibrium equations in an exact (if possible) or an approximate manner. The procedure is repeated by step-increasing the applied load. The criterion for reaching the limit point (collapse load) is that convergence for the solution of the nonlinear equations cannot be obtained even for very small load increments. Such procedures for establishing critical loads (limit point) require computer solutions at many load levels and tend to be very expensive in terms of computer time. In addition, these procedures encounter numerical difficulties associated with convergence (Almroth et al., 1976).

This article presents a procedure for estimating critical loads for rigid-jointed frame structures (subject to instability including sway buckling). The procedure is applicable to both bifurcational and limit point instability problems.

Equilibrium and buckling equations are established for an n-bar frame, and the methodology is demonstrated through the analysis of two simple frames. The present analysis is based on linearly elastic behavior and nonlinear kinematic relations (moderate rotations), while the effect of transverse shear on deformation is neglected (Euler-Bernoulli beams). The buckling equations are derived by employing the perturbation technique (Bellman, 1969), which was demonstrated by Koiter (1966) and Sewell (1965) for buckling problems.

Since the procedure is demonstrated through frame-type configurations, the current state-of-the-art concerning stability analysis of such structural systems is briefly reviewed herein. Buckling analysis of rigid-jointed plane frameworks may be traced to Bleich (1919), Müller-Breslad (1908), and Zimmerman (1909, 1910, 1925). An excellent historical sketch is given in Bleich's textbook (1952), which also presents and discusses various methods for estimating critical conditions. In all these investigations, there are numerous simplifying or restrictive assumptions but, despite this, they lead to reasonable estimates of critical loads. In addition to the works mentioned so far, it is important to include the investigation by Chwalla (1938) of the buckling of a rigid one-story portal frame under a symmetric transverse load not applied at the joints of the horizontal bar. He has shown that sway buckling is possible, which appears as an unstable bifurcation from the nonlinear primary path. In obtaining, though, both the primary path and the bifurcation load, Chwalla employed linear equilibrium equations and linearly elastic behavior. In more recent years, similar problems have been studied by Barker, Horne, and Roderick (1947), Chilver (1956), Horne (1961), Livesley (1956), Merchant (1954, 1955) and others. Many of the aforementioned problems and procedures have been incorporated in books such as Horne (1961) and McMinn (1962). The stability of plane rectangular frames with sway under static and dynamic harmonic loads is investigated in Kounadis (1976) using the kinetic approach. In addition, there are several studies in the area of post-buckling analysis of specialized frameworks (Britvec, 1960; Britvec and Chilver, 1963;

Godley and Chilver, 1967) that enhance understanding of when and under what conditions critical loads can be related to load-carrying capacity. Finally, the book by Britvec (1973) presents many of the aforementioned analyses and procedures.

4.6.2 MATHEMATICAL FORMULATIONS

Consider a general frame composed of n straight, slender, constant, cross-sectional bars, which are rigidly connected to each other. Each bar is of length l_i, cross-sectional area A_i, cross-sectional second moment of area I_i, and it subscribes to a proper coordinate system (see Fig. 4.23) with displacement components ξ (along the length, x) and w (normal to the bar). The external loads applied to the frame may consist of concentrated loads and couples, as well as of distributed loads. The expression for the total potential of the frame is

$$U_r = \frac{1}{2} \sum_{i=1}^{n} \int_0^{l_i} \left[(EA)_i \left(\xi_{i,x} + \frac{1}{2} w_{i,x}^2 \right)^2 + (EI)_i w_{i,xx}^2 \right] dx + \Omega \tag{82}$$

in which $E =$ Young's modulus of elasticity and Ω is the potential of the external forces. Note that the behavior of the bars is assumed to be linearly elastic and the kinematic relations employed correspond to those of moderate rotations.

Equilibrium Equations

By employing the principle of the stationary value for the total potential, one may write the following equilibrium equations, when the loads are applied at the joints

$$\left(\xi_{i,x} + \frac{1}{2} w_{i,x}^2 \right)_{,x} = 0;$$

$$EI_i w_{i,xxxx} - EA_i \left(\xi_{i,x} + \frac{1}{2} w_{i,x}^2 \right) w_{i,xx} = 0;$$

$$i = 1, 2, \ldots, n \tag{83}$$

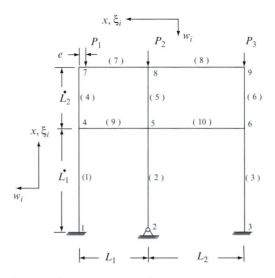

FIGURE 4.23   Typical multibay, multistory, rectangular rigid-jointed frame.

In the case for which the ith bar is loaded by a transverse distributed load, $q_i(x)$, concentrated forces, P_i^m, and couples $\overline{M}_i^n$ on $0 < x < l_i$, then the right-hand side of the second of Eqs. (83) is equal to

$$q_i(x) + \sum_m P_i^m \delta(x - x^m) - \sum_n \overline{M}_i^n \eta(x - x^n) \tag{84}$$

Note that δ and η are the Dirac-delta and doublet functions, respectively, and x^m and x^n denote the location of the concentrated forces and moments.

In addition, the associated boundary conditions for the ith bar are denoted by

$$\sum_{i=1}^n \left\{ \left[EA_i \left(\xi_{i,x} + \frac{1}{2} w_{i,x}^2 \right) \delta\xi_i \right] \Big|_0^{l_i} + \left(EI_i w_{i,xx} \delta w_{i,x} \right) \Big|_0^{l_i} + \right.$$
$$\left. \left[EA_i \left(\xi_{i,x} + \frac{1}{2} w_{i,x}^2 \right) w_{i,x} - EI_i w_{i,xxx} \right] \delta w_i \Big|_0^{l_i} + (\delta\Omega) \Big|_0^{l_i} \right\} = 0 \tag{85}$$

in which $(\delta\Omega)|_0^{l_i}$ exists only for bars whose end points are coincident with loaded frame joints. Finally, at each joint kinematic continuity must be imposed for the displacements ξ_i, w_i, $w_{i,x}$, and their admissible variations. Since in a general multi-story, multibay, rigid-jointed rectangular frame, there are three different (two-bar, three-bar, and four-bar joints) types of connections, the kinematic continuity equations and the natural boundary conditions are listed herein for each type separately (see Fig. 4.23). Consider the 10-bar frame, shown on Fig. 4.23. Each bar and each joint is identified by a number $[(i) = 1 - 10$ for the bars, and $j = 1 - 9$ for the joints]. The sign conventions are shown on the figure. Note that

$$l_1 = l_2 = l_3 = L_i^*; \quad l_4 = l_5 = l_6 = L_2^*; \quad l_7 = l_9 = L_1; \quad l_8 = l_{10} = L_2 \tag{86}$$

Also, note that e denotes a small load eccentricity (positive as shown).

A typical two-bar joint is characterized by joint 7, while a typical three-bar joint by joint 4, and a four-bar joint by joint 5. Kinematic continuity equations and natural boundary (joint equilibrium) equations are written herein for each one of the three typical joints. Note that in order to write the natural boundary conditions from Eqs. (85) one must use the kinematic continuity conditions at that joint.

Joint 7

The kinematic continuity conditions are

$$w_4(L_2^*) = \xi_7(L_1); \quad w_{4,x}(L_2^*) = w_{7,x}(L_1); \quad \xi_4(L_2^*) = -w_7(L_1) \tag{87}$$

The natural boundary conditions (joint equilibrium) are

$$(EI)_4 w_{4,xx}(L_2^*) + (EI)_7 w_{7,xx}(L_1) + P_1 e = 0; \quad (AE)_4 \left[\xi_{4,x}(L_2^*) + \frac{1}{2} w_{4,x}^2(L_2^*) \right]$$
$$+ (EI)_7 w_{7,xxx}(L_1) - (AE)_7 \left[\xi_{7,x}(L_1) + \frac{1}{2} w_{7,x}^2(L_1) \right] w_{7,x}(L_1) + P_1 = 0;$$
$$(AE)_4 \left[\xi_{4,x}(L_2^*) + \frac{1}{2} w_{4,x}^2(L_2^*) \right] w_{4,x}(L_2^*) - (EI)_4 w_{4,xxx}(L_2^*)$$
$$+ (AE)_7 \left[\xi_{7,x}(L_1) + \frac{1}{2} w_{7,x}^2(L_1) \right] = 0 \tag{88}$$

Joint 4

The kinematic continuity conditions are

$$w_1\left(L_1^*\right) = w_4(0) = \xi_9(L_1); \quad w_{1,x}\left(L_1^*\right) = w_{4,x}(0) = w_{9,x}(L_1);$$
$$\xi_1\left(L_1^*\right) = \xi_4(0) = -w_9(L_1) \tag{89}$$

The natural boundary conditions are

$$(EI)_1 w_{1,xx}\left(L_1^*\right) + (EI)_9 w_{9,xx}(L_1) - (EI)_4 w_{4,xx}(0) = 0; \; -(EI)_1\left[\xi_{1,x}\left(L_1^*\right).\right.$$

$$\left.+\frac{1}{2}w_{1,x}^2\left(L_1^*\right)\right] - (EI)_9 w_{9,xxx}(L_1) + (EA)_9\left[\xi_{9,x}(L_1) + \frac{1}{2}w_{9,x}^2(L_1)\right]w_{9,x}(L_1)$$

$$+ (EA)_4\left[\xi_{4,x}(0) + \frac{1}{2}w_{4,x}^2(0)\right] = 0; \; -(EI)_1 w_{1,xxx}\left(L_1^*\right) + (EA)_1\left[\xi_{1,x}\left(L_1^*\right).\right. \tag{90}$$

$$\left.+\frac{1}{2}w_{1,x}^2\left(L_1^*\right)\right]w_{1,x}\left(L_1^*\right) + (EA)_9\left[\xi_{9,x}(L_1) + \frac{1}{2}w_{9,x}^2(L_1)\right]$$

$$+ (EI)_4 w_{4,xxx}(0) - (EA)_4\left[\xi_{4,x}(0) + \frac{1}{2}w_{4,x}^2(0)\right]w_{4,x}(0) = 0$$

Joint 5

The kinematic continuity conditions are

$$w_2\left(L_1^*\right) = w_5(0) = \xi_9(0) = \xi_{10}(L_2); \; w_{2,x}\left(L_1^*\right) = w_{5,x}(0) = w_{9,x}(0)$$
$$= w_{10,x}(L_2); \; \xi_2\left(L_1^*\right) = \xi_5(0) = -w_9(0) = -w_{10}(L_2) \tag{91}$$

The natural boundary conditions are

$$-(EI)_9 w_{9,xx}(0) + (EI)_2 w_{2,xx}\left(L_1^*\right) + (EI)_{10} w_{10,xx}(L_2) - (EI)_5 w_{5,xx}(0) = 0;$$

$$-(EI)_{10} w_{10,xxx}(L_2) + (EA)_{10}\left[\xi_{10,x}(L_2) + \frac{1}{2}w_{10,x}^2(L_2)\right]w_{10,x}(L_2)$$

$$+ (EA)_5\left[\xi_{5,x}(0) + \frac{1}{2}w_{5,x}^2(0)\right] + (EI)_9 w_{9,xxx}(0) - (EA)_9\left[\xi_{9,x}(0).\right.$$

$$\left.+\frac{1}{2}w_{9,x}^2(0)\right]w_{9,x}(0) - (EA)_2\left[\xi_{2,x}\left(L_1^*\right) + \frac{1}{2}w_{2,x}^2\left(L_1^*\right)\right] = 0; \tag{92}$$

$$-(EI)_{5,xxx}(0) + (EA)_5\left[\xi_{5,x}(0) + \frac{1}{2}w_{5,x}^2(0)\right]w_{5,x}(0)$$

$$+ (EA)_9\left[\xi_{9,x}(0) + \frac{1}{2}w_{9,x}^2(0)\right] + (EI)_2 w_{2,xxx}\left(L_1^*\right) - (EA)_2\left[\xi_{2,x}\left(L_1^*\right)\right.$$

$$\left.+\frac{1}{2}w_{2,x}^2\left(L_1^*\right)\right]w_{2,x}\left(L_1^*\right) - (EA)_{10}\left[\xi_{10,x}(L_2) + \frac{1}{2}w_{10,x}^2(L_2)\right] = 0$$

Finally, for this particular illustrative example (Fig. 4.1) one must write the proper boundary conditions at the supports (joints 1, 2, and 3). These are, for joint 1

$$w_1(0) = 0; \quad w_{1,x}(0) = 0; \quad \xi_1(0) = 0 \tag{93}$$

for joint 2

$$w_2(0) = 0; \quad w_{2,xx}(0) = 0; \quad \xi_2(0) = 0 \tag{94}$$

and for joint 3

$$w_3(0) = 0; \quad w_{3,x}(0) = 0; \quad \xi_3(0) = 0 \tag{95}$$

Before outlining the solution methodology, it is convenient to obtain the general solution of the equilibrium equations for each bar, Eqs. (83). It is deduced from the first of Eqs. (83) that the axial force in the bar (denoted by S_i) is a constant, or

$$\xi_{i,x} + \frac{1}{2} w_{i,x}^2 = \pm \frac{S_i}{(AE)_i} \tag{96}$$

in which S_i = the magnitude of the axial force, and the positive and negative signs characterize tension and compression, respectively.

Depending on the sign of the axial force, the second of Eqs. (83) becomes a fourth-order, elliptic-type, ordinary differential equation (for compression) or a hyperbolic-type equation (for tension) for $S_i \neq 0$. Thus, one must differentiate between these two cases and write the appropriate general solution for each case. First, the following nondimensionalized parameters are introduced for convenience

$$k_i^2 = \frac{S_i l_i^2}{(EI)_i}; \; \lambda_i^2 = \frac{l_i^2 A_i}{I_i} \tag{97}$$

Then, the general solutions are obtained for $w_i(x)$ and $\xi_i(x)$. These are:

1. Compression in the ith bar

$$\left.\begin{aligned}
w_i(x) &= A_{i1} \sin \frac{k_i x}{l_i} + A_{i2} \cos \frac{k_i x}{l_i} + A_{i3}x + A_{i4} \\
\xi_i(x) &= -\frac{k_i^2}{\lambda_i^2}x + A_{i5} - \frac{1}{2}\int_0^x w_{i,x}^2 \, dx
\end{aligned}\right\} \tag{98}$$

2. Tension in the ith bar

$$\left.\begin{aligned}
w_i(x) &= A_{i1} \sinh \frac{k_i x}{l_i} + A_{i2} \cosh \frac{k_i x}{l_i} + A_{i3}x + A_{i4} \\
\xi_i(x) &= \frac{k_i^2}{\lambda_i^2}x + A_{i5} - \frac{1}{2}\int_0^x w_{i,x}^2 \, dx
\end{aligned}\right\} \tag{99}$$

Note that if a bar is also loaded transversely, the expression for $w_i(x)$, Eqs. (98), must contain the proper particular solution.

Regardless of tension or compression in the bar, there are six constants in Eqs. (98) or (99). These constants are: k_i, A_{ij}, $j = 1, 2, 3, 4, 5$. In addition, irrespective of the spatial distribution of the applied loads (concentrated forces at the joints only or in combination with distributed loads and other concentrated forces and couples), the response of an n-bar frame is known provided that the $6n$ constants can be evaluated. These $6n$ constants are

$$\left.\begin{aligned}
k_i: \quad & i = 1, 2, \ldots, n \\
A_{ij}: \quad & i = 1, 2, \ldots, n \text{ and } j = 1,2,3,4,5
\end{aligned}\right\} \tag{100}$$

The needed $6n$ equations are obtained from support conditions (three for each support) (see Eqs. 93, 94, and 95); from natural boundary conditions (equilibrium, three at each joint) (see Eqs. 88, 90, and 92); and from requiring kinematic continuity conditions at each joint [$3(p - 1)$], where p denotes the number of bars that are connected at the joint] (see Eqs. 87, 89, and 91). For an unbraced, multistory, multibay, rigid-jointed frame, the number of available conditions is $6n$. Note that if there is cross-bracing, one bar is added; therefore, six more constants are added, and six more kinematic continuity conditions (three at each end) must be satisfied.

In the particular case of Fig. 4.23, there are 10 bars ($n = 10$) and thus 60 constants must be determined for equilibrium response. The 60 needed equations come from the following: nine relations from the support conditions (three at each support), 18 natural boundary conditions from joints 4–9, and 33 kinematic continuity continuity conditions (nine from joint 5, six from each of joints 4, 6, and 8, and three from each of joints 7 and 9).

In closing this section there is a very important observation that is worth mentioning. Regardless of the position of a given bar, and regardless of the sign of the axial force in this given bar, application of the support, kinematic continuity, and natural boundary conditions yields only one nonlinear equation in the constants k_i, A_{ij}, $j = 1, 2, 3, 4, 5$. This is so because, when evaluating either Eqs. (98) or Eqs. (99), at $x = 0$, the resulting equations are linear in the constants; while when evaluating these equations at $x = l_i$, the resulting equations are linear except $\xi_i(l_i)$. Because of this linearity, it is possible to eliminate all constants, which appear in a linear fashion and end up with n number of nonlinear equations in the n nondimensionalized axial forces, k_i.

Buckling Equations

Regardless of whether the primary equilibrium path is linear or nonlinear, the buckling equations are needed in order to establish the level of the load at which there is loss of stability through the existence of either a bifurcation point or a limit point. In addition, through the buckling equations, one may establish the modes of deformation during the buckling process.

The buckling equations are derived by employing a perturbation method (Bellman, 1969; Roorda and Chilver, 1969; Sewell, 1965) based on the concept of the existence of an adjacent equilibrium position at either a bifurcation point or a limit point. The required steps are as follows: starting with the equilibrium equations and proper boundary conditions expressed in terms of displacements, perturb them by allowing small admissible changes in the displacement functions, make use of equilibrium at a point at which an adjacent equilibrium path is possible, and retain first-order terms in the admissible variations. The resulting inhomogeneous differential equations, for the case in consideration, are linear ordinary differential equations. These are the buckling equations.

Let $\xi_i^*(x)$ and $w_i^*(x)$ denote displacement components on the primary equilibrium path and $\overline{\xi}_i(x)$ and $\overline{w}_i(x)$ denote infinitesimally small but kinematically admissible displacement functions. Then by writing

$$\xi_i = \xi_i^* + \overline{\xi}_i; \quad w_i = w_i^* + \overline{w}_i \tag{101}$$

By substituting these expressions into the equilibrium equations and associated boundary terms, Eqs. (83) and (85), by recognizing that ξ_i^* and w_i^* satisfy primary path equilibrium, by performing the indicated operations, and by retaining first-order terms, one may derive the following set of buckling equations and associated boundary conditions

$$\left(\bar{\xi}_{i,_x} + w^*_{i,_x}\bar{w}_{i,_x}\right)_{,_x} = 0;$$

$$(EI)_i\bar{w}_{i,_{xxxx}} - (EA)_i\left(\xi^*_{i,_x} + \frac{1}{2}w^{*2}_{i,_x}\right)\bar{w}_{i,_{xx}} = (EA)_i\left(\bar{\xi}_{i,_x} + w^*_{i,_x}\bar{w}_{i,_x}\right)w^*_{i,_{xx}} \tag{102}$$

$$\sum_{i=1}^{n}\Bigg\langle \left[(EA)_i\left(\bar{\xi}_{i,_x} + w^*_{i,_x}\bar{w}_{i,_x}\right)\bar{\xi}_i\right]_0^{l_i} + \left[(EI)_i\bar{w}_{i,_{xx}}\bar{w}_{i,_x}\right]_0^{l_i} + \Bigg\{\Bigg[(EA)_i\Big(\xi^*_{i,_x}\cdot$$

$$+ \frac{1}{2}w^{*2}_{i,_x}\Big)\bar{w}_{i,_x} + (EA)_i\left(\bar{\xi}_{i,_x} + w^*_{i,_x}\bar{w}_{i,_x}\right)w^*_{i,_x} - (EI)_i\bar{w}_{i,_{xxx}}\Bigg]\bar{w}_i\Bigg\}\Bigg|_0^{l_i}\Bigg\rangle = 0 \tag{103}$$

Note that $\bar{\xi}_i$ and $\bar{w}_i$ must satisfy kinematic continuity conditions at every joint (see, e.g., Eqs. 88, 89 and 90).

Next Eqs. (102) may be rewritten by employing the primary state solution and by introducing the following

$$(EA)_i\left(\bar{\xi}_{i,_x} + w^*_{i,_x}\bar{w}_{i,_x}\right) = \bar{S}_i \tag{104}$$

Note that $\bar{S}_i$ may be positive or negative. Assuming that the ith bar is in compression at the instant of buckling, then Eqs. (102) becomes

$$\bar{\xi}_{i,_x} + w^*_{i,_x}\bar{w}_{i,_x} = \frac{\bar{S}_i}{(AE)_i};$$

$$\bar{w}_{i,_{xxxx}} + \frac{k_i^2}{l_i^2}\bar{w}_{i,_{xx}} = -\frac{\bar{S}_i}{(EI)_i}\left(\frac{k_i}{l_i}\right)^2\left(A_{i1}\sin\frac{k_ix}{l_i} + A_{i2}\cos\frac{k_ix}{l_i}\right) \tag{105}$$

The solution to the second of Eqs. (105) consists of complementary and particular parts and it is

$$\bar{w}_i(x) = \bar{A}_{i1}\sin\frac{k_ix}{l_i} + \bar{A}_{i2}\cos\frac{k_ix}{l_i} + \bar{A}_{i3}x + \bar{A}_{i4}$$

$$+ \frac{\bar{S}_il_ix}{2k_i(EI)_i}\left(A_{i2}\sin\frac{k_ix}{l_i} - A_{i1}\cos\frac{k_ix}{l_i}\right) \tag{106}$$

The expression for $\bar{\xi}_i$ is obtained from the first of Eqs. (105) and it is

$$\bar{\xi}_i(x) = \frac{\bar{S}_i}{(AE)_i}x + \bar{A}_{i5} - \int_0^x w^*_{i,_x}\bar{w}_{i,_x}dx \tag{107}$$

in which the integrand may be obtained from Eqs. (106) and the first of Eqs. (98). Note that if the member is in tension at the instant of buckling the solution for $\bar{w}_i$ will involve hyperbolic functions and the first of Eqs. (99) must be employed in Eq. (107). Regardless of this fact, there are six constants ($\bar{A}_{ij}$, $j = 1, 2, 3, 4, 5$, and $\bar{S}_i$) and the expressions for $\bar{w}_i$ and $\bar{\xi}_i$ (buckling modes) are linear functions of these constants.

In a similar manner as in the case of equilibrium, one may write the kinematic continuity, support, and natural boundary conditions in terms of the solutions for $\bar{w}_i$ and $\bar{\xi}_i$. It is important to observe here that these conditions are homogeneous and linear, and therefore their satisfaction leads to a system of homogeneous and linear algebraic equations in the constants $\bar{S}_i$, $\bar{A}_{ij}$ ($i = 1, 2, \ldots n$, and $j = 1, 2, \ldots 5$).

4.6.3　SOLUTION PROCEDURE

Given an n-bar, multistory, multibay plane frame, which is loaded by concentrated forces and moments (not necessarily at the joints) and distributed loads, the interest is in outlining a solution procedure for finding critical conditions. Critical condition, herein, refers to the level of the loads or combination of the loads for which the equilibrium of the system becomes unstable through either a bifurcation point or a limit point. Note that the governing equations, derived in the previous section, can easily be modified in order to accommodate small initial geometric imperfections.

Assuming that the interest lies in finding critical conditions, the solution procedure is as follows:

1. Through a simple and quick buckling analysis program, such as those in Hornef (1965), Kounadis (1977), and McMinn (1962), identify the bars (if any) that are in tension near the true buckling load.

2. On the basis of this first step, use Eqs. (98) for those bars in compression and Eqs. (99) for those in tension.

3. Employ the support, kinematic continuity, and natural boundary conditions to establish the $6n$ equations that signify equilibrium states for any level of the applied load.

4. Recognizing that $5n$ of the preceding equations are linear in the constants A_{ij} ($i = 1, 2, \ldots, n; j = 1, 2, \ldots, 5$), reduce the system to one of n nonlinear equations in k_i.

5. Since satisfaction of the kinematic continuity, support, and natural boundary conditions for the buckling solution leads to a system of $6n$ linear, homogeneous algebraic equations in $\overline{S}_i$ and $\overline{A}_{ij}$ (constants associated with the buckling modes), then the characteristic equation is established by requiring the determinant of the coefficients to vanish (for a nontrivial solution to exist). This step provides one more equation (nonlinear) in k_i, and the applied loads, and it holds true only at the critical equilibrium point (either bifurcation or limit point), which must correspond to the primary equilibrium path.

6. Steps 4 and 5 provide $(n + 1)$ nonlinear equations in k_i and the applied load. The simultaneous solution of these gives the critical load and the corresponding complete response (since knowledge of k_i implies knowledge of A_{ij} and together imply knowledge of $\xi_i(x)$, $w_i(x)$, and the associated stresses, strains, moments, etc.). In order to solve the system of $(n + 1)$ nonlinear equations, one must employ an efficient computing technique (Brush 1975; Merchant 1955; Timoshenko 1961) implemented on a high-speed digital computer. For a frame with small number of bars (two or three), the following scheme is devised by the writers. Assuming a two-bar frame, one must solve three nonlinear equations in k_i and λ_c (load parameter)

$$f_1(k_1, k_2, \lambda_C) = 0; \quad f_2(k_1, k_2, \lambda_C) = 0; \quad f_3(k_1, k_2, \lambda_C) = 0 \qquad (108)$$

If $\overline{k}_1$, $\overline{k}_2$, and $\overline{\lambda}_C$ characterize a point in the space (k_1, k_2, λ_C) that satisfies all three equations, Eqs. (108), then these values, also denote the minimum (which is zero) of the function F, in which

$$F = f_1^2 + f_2^2 + f_3^2 \qquad (109)$$

The mathematical search technique of Nelder and Mead (1964) is employed for finding this minimum. This approach is used in the illustrative examples contained herein.

Steps 1–6 provide the level of the load of the critical point, and the overall procedure is based on the assumption that the behavior of the material is linearly elastic. Since one can also find the complete response (equilibrium) of the system at that load level, it is possible to find out that the stresses are in the plastic range. In that case, if one is interested in finding out at which load level the stresses are in the

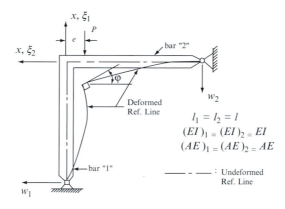

FIGURE 4.24   Geometry and sign convention, example 1.

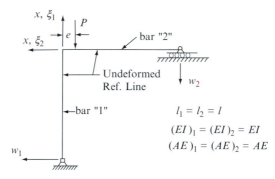

FIGURE 4.25   Geometry and sign convention, example 2.

plastic range, he may solve for the response by incrementing the applied load and solving for the n nonlinear equations, in a similar manner.

4.6.4   ILLUSTRATIVE EXAMPLES

Two illustrative examples have been chosen to demonstrate the solution methodology and procedure proposed in this article. The first example consist of a two-bar frame supported by immovable pins (see Fig. 4.24) and loaded as shown. The second one consist of a two-bar frame supported by an immovable pin at the end of the bar, which is parallel to the applied load (see Fig. 4.25) and by a movable (in the horizontal direction) pin at the end of the bar, which is normal (horizontal bar) to the applied load.

These examples were chosen because both have been analyzed before by different methods and provide excellent tests for the proposed methodology. The first example has been investigated theoretically and experimentally by Koiter (1966), Kounadis, Giri, and Simitses (1977), and Roorda (1965), while the second one was investigated by Huddleston (1967) for zero eccentricity.

Each example is analyzed and considered separately, but the procedure and the results are presented by making use of the following nondimensionalized parameters

$$\frac{x}{l_i} = \frac{x}{l} = \chi; \quad \frac{w_i}{l} = W_i; \quad \frac{\xi_i}{l} = \Xi_i; \quad \lambda^2 = \frac{A_i l_i^2}{I_i} = \frac{A l_i^2}{I}; \quad \bar{e} = \frac{e}{l};$$

$$\beta^2 = \frac{P l^2}{EI}; \quad \lambda_C = \frac{P}{P_{cl}}; \quad k_i^2 = \frac{S_i l_i^2}{(EI)_i}; \quad i = 1, 2$$

(110)

in which P_{cl} is the linear theory critical load.

Example 1.—For this particular example, the solution of the equilibrium equations is given by Eqs. (98), since both bars are in compression at the critical load. For this case, in view of Eqs. (110) the solution becomes

$$W_i(\chi) = A_{i1} \sin k_i\chi + A_{i2} \cos k_i\chi + A_{i3}\chi + A_{i4}$$

$$\Xi_i(\chi) = -\frac{k_i^2\chi}{\lambda^2} + A_{i5} - \frac{1}{2}\int_0^\chi W_{i,\chi}^2 \, d\chi; \quad i = 1, 2 \tag{111}$$

Note that $i = 1$ corresponds to the vertical bar and $i = 2$ to the horizontal bar (see Fig. 4.24).

Use of the kinematic continuity, Eqs. (87) natural, Eqs. (88) and support conditions, similar to Eqs. (91), in terms of the nondimensionalized parameters yields the following relations among the $2n$ (12) constants (k_i; A_{ij}, $i = 1, 2$, and $j = 1, 2, \ldots 5$). Thus

$$A_{i2} = A_{i4} = A_{i5} = 0; \quad i = 1, 2; \quad k_2^2 A_{23} - k_1^2 + \beta^2 = 0; \quad k_1^2 A_{13} + k_2^2 = 0;$$

$$A_{11} \, k_1 \cos k_1 - A_{21} \, k_2 \cos k_2 + A_{13} - A_{23} = 0;$$

$$A_{11} \, k_1^2 \sin k_1 + A_{21} \, k_2^2 \sin k_2 - \beta^2\bar{e} = 0; \tag{112}$$

$$\frac{k_2^2}{\lambda^2} = -A_{11} \sin k_1 - A_{13} - B_2(1); \quad \frac{k_1^2}{\lambda^2} = A_{21} \sin k_2 + A_{23} - B_1(1).$$

in which $B_1(\chi) = \dfrac{1}{2}\left[A_{13}^2\,\chi + 2A_{11}A_{13}\sin k_1\chi + \dfrac{k_1^2 A_{11}^2}{2}\left(\chi + \dfrac{\sin 2k_1\chi}{2k_1}\right)\right]$

$$\tag{113}$$

$$B_2(\chi) = \frac{1}{2}\left[A_{23}^2\,\chi + 2A_{21}A_{23}\sin k_2\chi + \frac{k_2^2 A_{21}^2}{2}\left(\chi + \frac{\sin 2k_2\chi}{2k_2}\right)\right]$$

Note that all of Eqs. (112), except the last two, are linear algebraic equations in terms of A_{ij} ($i = 1, 2$ and $j = 1, 2\ldots 5$) and therefore these constants can be expressed solely in terms of k_i, $i = 1, 2$. Thus, points on the equilibrium path for any value of the slenderness ratio, λ, eccentricity, $\bar{e}$, and level of the applied load, $\beta^2(\lambda_C)$ are characterized by the solution of the last two of Eqs. (112).

In summary, the expressions for the displacement components of equilibrium points are :

$$\Xi_i = -\frac{k_i^2}{\lambda^2}\chi - B_i(\chi); \quad W_i(\chi) = A_{i1} \sin k_i\chi + A_{i3}\chi; \quad i = 1, 2 \tag{114}$$

Once the solution is known, the complete response of the frame, at an equilibrium path position, is known.

Since it is possible, for a given problem, to have nonacceptable equilibrium paths that include a limit point, the development that follows is applicable only to the primary path. This implies that the sought solution (unstable point and corresponding critical condition) belongs to the primary equilibrium path, and thus it constitutes a realistic solution.

Next, in order to find P_{cr} (for all λ and $\bar{e}$), one must obtain the characteristic equation (from the solution to the buckling equations). By introducing the nondimensionalized parameters, by employing the general solution, Eqs. (106) and (107), to the buckling equations, and by making use of Eqs. (114) and all auxiliary conditions (kinematic continuity, etc.), one obtains a set of four linear homogeneous algebraic equations in $\overline{A}_{11}$, $\overline{A}_{21}$, and $\overline{S}_i$ $(i = 1, 2)$ (four equations in four unknowns), because enforcement of the auxiliary conditions yields $\overline{A}_{12} = \overline{A}_{14} = \overline{A}_{15} = 0$, and $\overline{A}_{13}$ have been expressed in terms of the remaining four. Thus, a nontrivial solution exists, if the determinant of the coefficients vanishes. This requirement yields the following characteristic equation:

$$\begin{vmatrix} a_{11} & a_{12} & a_{13} & a_{14} \\ a_{21} & a_{22} & a_{23} & a_{24} \\ a_{31} & a_{32} & a_{33} & a_{34} \\ a_{41} & a_{42} & a_{43} & a_{44} \end{vmatrix} = 0;$$

in which $a_{11} = \sin k_1$; $\quad a_{12} = A_{23} \sin k_2 + \dfrac{A_{21}}{2} k_2^2 \left(1 + \dfrac{\sin 2k_2}{2k_2}\right)$;

$$a_{13} = \lambda^2 \left(\frac{A_{11} \cos k_1}{2k_1} - \frac{A_{13}}{k_1^2} + \frac{A_{23} + A_{21} \sin k_2}{k_2^2}\right); \quad a_{14} = \lambda^2 \Bigg[\frac{1}{\lambda^2}$$

$$- \frac{1}{k_1^2} - \frac{A_{23}}{k_2^2}(A_{23} + A_{21} \sin k_2) + \frac{A_{21}^2}{4}\left(1 + \frac{\sin 2k_2}{4k_2} + \frac{\cos 2k_2}{2}\right) +$$

$$A_{21} A_{23} \frac{\cos k_2}{2k_2}\Bigg]; \quad a_{21} = k_1 \cos k_1; \quad a_{22} = -k_2 \cos k_2;$$

$$a_{23} = \lambda^2 \left[\frac{A_{11}}{2}\left(\frac{\cos k_1}{k_1} - \sin k_1\right) - \left(\frac{A_{13}}{k_1^2} + \frac{1}{k_2^2}\right)\right];$$

$$a_{24} = \lambda^2 \left[\frac{A_{21}}{2}\left(\sin k_2 - \frac{\cos k_2}{k_2}\right) + \frac{A_{23}}{k_2^2} - \frac{1}{k_1^2}\right]; \quad a_{31} = -A_{13} \sin k_1$$

$$- \frac{A_{11} k_1^2}{2}\left(2 + \frac{\sin 2k_1}{2k_1}\right); \quad a_{32} = 0; \quad a_{33} = -\lambda^2\Bigg[\frac{1}{\lambda^2} - \frac{1}{k_2^2} - \frac{A_{13}}{k_1^2}(A_{13} +$$

$$A_{11} \sin k_1) + \frac{A_{11}^2}{4}\left(1 + \frac{\sin 2k_1}{4k_1} + \frac{\cos 2k_1}{2}\right) + A_{11} A_{13} \frac{\cos k_1}{2k_1}\Bigg];$$

$a_{34} = 0$; $\quad a_{41} = 2k_1^2 \sin k_1$; $\quad a_{42} = 2k_2^2 \sin k_2$; $\quad a_{43} = \lambda^2 A_{11}(2 \sin k_1 + k_1 \cos k_1)$;

$a_{44} = \lambda^2 A_{21}(2 \sin k_2 + k_2 \cos k_2)$. $\hfill (115)$

Note that the last two of Eqs. (112) and Eqs. (115) constitute a system of three nonlinear equations in k_1, k_2, and β^2 (λ and $\bar{e}$ are fixed parameters). Also note that the A_{ij} values are expressed in terms of k_1, k_2, and β^2 by employing the remaining of Eqs. (112). The solution of this system yields the critical load, β_{cr}^2, and the response of the frame at the critical load.

The solution is accomplished by employing step 6 of the previous section, and numerical results have been generated. These results are presented graphically in Fig. 4.26 as critical load parameter ratio, $\lambda_c = P_{cr}/P_{cl} = \beta_{cr}^2/13.89$, versus eccentricity $\bar{e}$ for various values of the slenderness ratio ($\lambda = 40, 80, 120, \infty$). The $\lambda = 40$ value may be looked upon as a lower limit for linearly elastic behavior. It is observed that, for each λ value, there is a limiting value, of the eccentricity (positive) for which a limit point exists. For eccentricities below this critical value, the frame fails through the existence of a

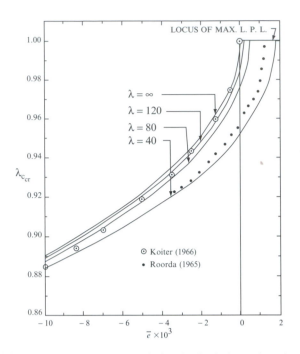

FIGURE 4.26   Effect of eccentricity on limit point loads for various slenderness ratios.

limit point. Contrary, for eccentricities higher than this critical value, no solution for the system is possible, which implies that there is neither a limit point nor a bifurcation point. For this range of eccentricities, the frame is always in stable equilibrium. Some of the results are presented in tabular form (Table 4.1). This table gives the limit point loads (λ_c or β_{cr}^2) and the corresponding values of k_1 and k_2 for $\lambda = 80$ and for various values of the eccentricity including the upper limit (critical eccentricity).

Example 2.—For this particular case, because of the movable support at the right extreme of the horizontal bar, the axial force, k_2, is zero and the solution to the equilibrium equations is

$$W_1 = A_{11}\sin k_1\chi + A_{12}\cos k_1\chi + A_{13}\chi + A_{14};$$

$$\Xi_1 = -\frac{k_1^2\chi}{\lambda^2} + A_{15} - \frac{1}{2}\int_0^\chi W_{1,\chi}^2\,d\chi; \tag{116}$$

$$W_2 = A_{21}\chi^3 + A_{22}\chi^2 + A_{23}\chi + A_{24}; \quad \Xi_2 = A_{25} - \frac{1}{2}\int_0^\chi W_{2,\chi}^2\,d\chi$$

Following the same procedure, as in the first example, the satisfaction of all auxiliary

TABLE 4.1   Critical Loads for $\lambda = 80$ and Various Eccentricities—Example 1

$\bar{e}$ (1)	$\lambda_c(=\beta_{cr}^2/13.89)$ (2)	β_{cr}^2 (3)	k_2^2 (4)	k_1^2 (5)
0.00047288	0.9997912	13.8871	0.001518461	13.8820
0.00000000	0.9753707	13.5479	0.09431323	13.6409
−0.00130000	0.9526637	13.2325	0.17481340	13.4195
−0.00250000	0.9390496	13.0434	0.22035590	13.2878
−0.00500000	0.9180633	12.7519	0.28645260	13.0866
−0.00700000	0.9048308	12.5681	0.32553400	12.9607
−0.01000000	0.8882145	12.3373	0.37176040	12.8035

Note: Eqs. (108) and (109) are satisfied with 10^{-10} accuracy.

conditions leads to a single nonlinear equation in k_1. This equation relates k_1 to all other parameters, λ, β^2, and $\bar{e}$ at points on the primary equilibrium path, and it is

$$\frac{k_1^2}{\lambda^2} = \frac{\beta^2 - k_1^2}{3} + \frac{k_1^2 + \beta^2(\bar{e} - 1)}{k_1} \cot k_1$$
$$- \frac{1}{4} \left[\frac{k_1^2 + \beta^2(\bar{e} - 1)}{k_1 \sin k_1} \right]^2 \left(1 + \frac{\sin 2k_1}{2k_1} \right)$$

(117)

Similarly the characteristic equation is

$$\frac{3}{2} \left[\frac{k_1^2 + \beta^2(\bar{e} - 1)}{k_1^2 \sin k_1} \right]^2 \left(\frac{k_1 \sin k_1}{2} + \frac{\sin k_1 \sin 2k_1}{4} + \frac{k_1^2 \cos k_1}{3} \right)$$
$$- \left[\frac{k_1 + \beta^2(\bar{e} - 1)}{k_1^2 \sin k_1} \right]^2 (2k_1 + \sin 2k_1) + 2 \cos k_1$$
$$- 2k_1 \sin k_1 \left(\frac{1}{3} + \frac{1}{\lambda^2} \right) = 0$$

(118)

The simultaneous solution of Eqs. (117) and (118) yields the critical load and the corresponding value k_1, and therefore the complete response of the system at the critical load. The constants A_{ij} ($i = 1, 2$, and $j = 1, 2, \ldots 5$) are related to k_1 through the following:

$$A_{12} = A_{13} = A_{14} = A_{15} = A_{22} = A_{24} = 0; \; A_{11} = \frac{k_1^2 + \beta^2(\bar{e} - 1)}{k_1^2 \sin k_1}$$

$$A_{21} = \frac{k_1^2 - \beta^2}{6}; \; A_{23} = \frac{k_1^2 + \beta^2(\bar{e} - 1)}{k_1^2} \cot k_1 + \frac{\beta^2 - k_1^2}{2}$$

(119)

$$A_{25} = \frac{k_1^2 + \beta^2(\bar{e} - 1)}{k_1^2} + \frac{1}{2} \int_0^1 W_{2,\chi}^2 \, d\chi \; k_2 = 0 \ldots,$$

in which the integral in the A_{25} expression can be expressed in terms of k_1, β^2, and $\bar{e}$.

Numerical results are generated and presented graphically in Fig. 4.27 as plots of $\lambda_C = \beta_{cr}^2/1.42155$ versus eccentricity for various values of the slenderness ratio ($\lambda = 40, 80, \infty$). As in the first example, here also there is a critical eccentricity, which is λ-dependent, up to which the frame fails through a limit point. For algebraically higher values of eccentricity, the frame is always in stable equilibrium with increasing load. From the generated results it is clear that, the linear theory buckling load and the result of Godley and Chilver (1967) are in excellent agreement for $\bar{e} = 0$, $\lambda \to \infty$. Note that linear theory and the approach employed in Huddleston

TABLE 4.2 Critical Loads for $\lambda = 40$ and Various Eccentricities—Example 2

$\bar{e}$ (1)	$\lambda_c (= \beta_{cr}^2/1.42155)$ (2)	β_{cr}^2 (3)	k_1^2 (4)
0.00187	0.99931	1.42058	1.41887
0.00000	0.93200	1.32489	1.35616
−0.00250	0.89683	1.27489	1.32407
−0.00500	0.87167	1.23912	1.30140
−0.00750	0.85123	1.21006	1.28315
−0.01000	0.83370	1.18515	1.26761

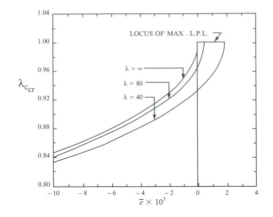

FIGURE 4.27   Effect of eccentricity and slenderness ratios on limit point loads, Example 2

(1967) do not account for the effect of the slenderness ratio. Some of the generated data are presented in tabular form (Table 4.2). This table gives the values of the critical load, β_{cr}^2, and the corresponding value of the nondimensionalized axial force in the vertical bar, k_1^2, for various eccentricities and $\lambda = 40$. In addition, λ_C is included in Table 4.2.

Additional applications can be found in Simitses (1981, 1982, 1984, 1986, 1990), Mohamed (1989, 1993), and Vlahinos (1986).

PROBLEMS

1. A horizontal column is rigidly attached to two vertical bars as shown. Find the expression for P_{cr} for buckling in the plane.
2. The horizontal column BE is rigidly attached to two vertical bars AC and FD.

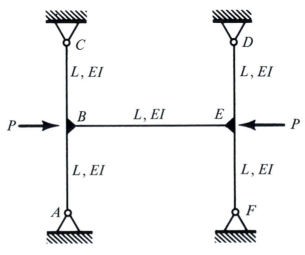

FIGURE P4.1

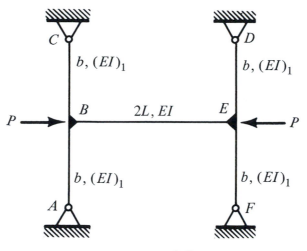

FIGURE P4.2

The flexural rigidities EI and $(EI)_1$ are for deflections normal to the plane of the figure. If the value of $(EI)_1$ is low, the critical load for such normal deflections can be raised by increasing it. Show that such an improvement is possible only until $(EI)_1$ reaches the value $\pi^2 b^3 \, EI/12L^3$. The torsional rigidity of AC and FD is to be treated as negligibly small.

3. A horizontal column AB and a vertical bar CD are rigidly attached at C. Derive the characteristic equation and compute the value of $(L^2/EI)P_{cr}$ for the special case of $EI = EI_1$ and $2b = L$.
4. Derive the characteristic equation for the cases shown in the figure.
5. A vertical column AB is rigidly attached to a flexible horizontal bar CB. B is a roller that can turn without friction on a smooth base. The load P remains vertical. Show that the characteristic equation is

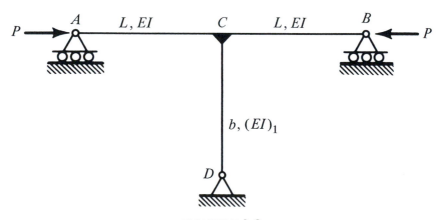

FIGURE P4.3

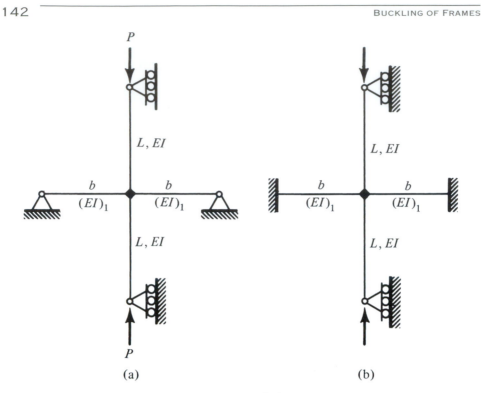

(a)   (b)

FIGURE P4.4

$$kL \tan kL = \frac{3(EI)_1 L}{EIb}, \quad k^2 = \frac{P}{EI}$$

Devise extreme cases to check this result.

6. Column AB and bar BC are identical and rigidly attached at B. The cross section is circular, with radius $R_0 \ll L$. Will the structure buckle in the plane of the figure or out of the plane?
7. Analyze the following partial frames for buckling in the plane of the figure.

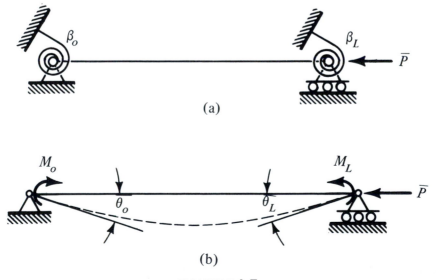

(a)

(b)

FIGURE P4.5

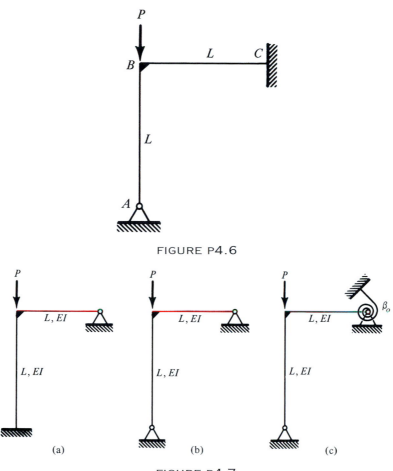

FIGURE P4.6

FIGURE P4.7

(a)   (b)   (c)

REFERENCES

Almroth, B. O., Meller, E., and Brogan, F. A. (1976). "Computer Solutions for Static and Dynamic Buckling of Shells." *Buckling of Structures*, International Union of Theoretical and Applied Mathematics Symposium, Cambridge, Mass., (1974). Springer-Verlag, Berlin, Germany.

Baker, J. F., Horne, M. R., and Roderick, J. W. (1949). The behaviour of continuous stanchions, *Proceedings of the Royal Society*, London, England, Vol. A-198, p. 493.

Ball, L. R. (1969). *Computational Solution of Nonlinear Operator Equations*. John Wiley and Sons, Inc., New York.

Bellman, R. (1969). *Perturbation Techniques in Mathematics. Physics and Engineering*. Holt. Rinehart, and Winston, New York.

Bleich, F. (1919). "Die Knickfestigkeit elasticher Stabrerbindungen." *Der Eisenbau*, Vol. 10, p. 27.

Bleich, F. (1952). *Buckling Strength of Metal Structures*. Edited by H. H. Bleich, McGraw-Hill Book Co., New York.

Britvec, S. J. (1960). "The Post-buckling Behavior of Frames," Thesis presented to Cambridge University, at Cambridge, England, in 1960, in partial fulfillment of the requirements for the degree of Doctor of Philosophy.

Britvec, S. J. (1973). *The Stability of Elastic Systems*. Pergamon Press, Inc., New York.

Britvec, S. J. and Chilver, H. A. (1963). Elastic buckling of rigidly-joined braced frames, *Journal of the Engineering Mechanics Division*, ASCE, Vol. 89, No. EM6, Proc. Paper 3736, pp. 217–255.

Brush, D. O. and Almroth, B. O. (1975). *Buckling of Bars, Plates, and Shells*. McGraw-Hill Book Co., New York.

Chilver, A. H. (1956). Buckling of a simple portal frame, *Journal of the Mechanics and Physics of Solids*, Vol. 5, p. 18.

Chwalla, E. (1938). "Die Stabilität lotrecht belasteter Rechteckrahmen," *Der Bauingenieur*, Vol. 19, p. 69.

Daniel, J. W. (1971). *The Approximate Minimization of Functionals*. Prentice-Hall, Inc., Englewood Cliffs, New Jersey.

Godley, H. R. M. and Chilver, H. A. (1967). The elastic post-buckling behaviour of unbraced frames, *International Journal of Mechanical Science*, Vol. 9, p. 323.

Horne, M. R. (1961). Stability of elastic-plastic structures, *Progress in Solid Mechanics*, North-Holland Publishing Co., Amsterdam, the Netherlands.

Horne, M. R. and Merchant, W. (1965). *The Stability of Frames*. Pergamon Press, London, England.

Huddleston, J. V. (1967). Nonlinear buckling and snap-over of a two-bar frame, *International Journal of Solids and Structures*, Vol. 3, pp. 1023–1030.

Koiter, W. T. (1966). Post buckling analysis of a simple two-bar frame, *Recent Progress in Applied Mechanics*, Almginst and Wiksell, Stockholm, Sweden.

Koiter, W. T. (1967). "On the Stability of Elastic Equilibrium," Thesis presented to the Polytechnic Institute, at Delft, the Netherlands, in 1945, in partial fulfillment of the requirements for the degree of Doctor of Philosophy (English translation *NASA TT-F-10833*).

Kounadis, A. (1976). "The Dynamic Procedure for the Study of Stability of Rectangular Frames Subjected to Static or Dynamic Loads." National Technical University Report, Athens, Greece.

Kounadis, A. N., Giri, J., and Simitses, G. J. (1977). Nonlinear stability analysis of an eccentrically loaded two-bar frame, *Journal of Applied Mechanics*, Vol. 44, Series E, No. 4, pp. 701–706.

Livesley, R. K. (1956). Application of electronic digital computers to some problems of structural analysis, *Structural Engineer*, Institution of Structural Engineers, London, England, Vol. 34, No. 6, pp. 161–167.

McMinn, S. J. (1962). *Matrices for Structural Analysis*. John Wiley and Sons, Inc., New York.

Merchant, W. (1954). The failure load of rigid-jointed frameworks as influenced by stability, *Structural Engineer*, Institution of Structural Engineers, London, England, Vol. 32, No. 7, pp. 185–190.

Merchant, W. (1955). Critical loads of tall building frames, *Structural Engineers*, Institution of Structural Engineers, London, England, Vol. 33, No. 3, pp. 84–89.

Mohamed, S. E. and Simitses, G. J. (1989). Buckling of flexibily-connected gabled frames, *International Journal of Non-Linear Mechanics*, Vol. 24, No. 5, pp. 353–364.

Mohamed, S. E. and Simitses, G. J. (1993). Stability and strength of rigid and semirigid plane frame-works, ASCE *J. of Aerospace Engineering*, Vol. 6, No. 2, pp. 186–198.

Müller-Breslau, H. (1908). *Die graphische Statik der Bau-Konstructionen*. Vol. II. 2, A. Kröner, Berlin, Germany.

Nelder, J. A. and Mead, R. (1964). A simplex method of function minimization, *Computer Journal*, Vol. 7, pp. 308–313.

Ortega, M. J. (1972). *Numerical Analysis*. Academic Press, New York.

Roorda, J. Feb. (1965). Stability of structures with small imperfections, *Journal of the Engineering Mechanics Division*, ASCE, Vol. 91, No. EMI, Proc. Paper 4230, pp. 87–106.

Roorda, J. and Chilver, H. A. (1969). "Frame Buckling: An Illustration of the Perturbation Technique." *Report No. 2*, Solid Mechanics Division, University of Waterloo, Canada.

Sewell, M. J. (1965). The static perturbation technique in buckling problems, *Journal of the Mechanics and Physics of Solids*, Vol. 13, p. 247.

Simitses, G. J. and Giri, J. (1984). Asymmetrically loaded portal frames, *Computers and Structures*, Vol. 19, No. 4, pp. 555–558.

Simitses, G. J. and Mohamed, S. E. (1990). Instability and collapse of flexibly-connected gabled frames, *International Journal of Solids and Structures*, Vol. 26, Nos. 9/10, pp. 1159–1171.

Simitses, G. J. and Vlahinos, A. S. (1982). Stability analysis of semi-rigidly connected simple frames, *Journal of Constructional Steel Research*, Vol. 2, No. 3, pp. 29–32.

Simitses, G. J. and Vlahinos, A. S. (1986). Sway buckling of unbraced multistory frames, *Computers and Structures*, Vol. 22, No. 6, pp. 1047–1054.

Simitses, G. J., Giri, J., and Kounadis, A. N. (1981). Nonlinear analysis of portal frames, *International Journal of Numerical Methods in Engineering*, Vol. 17, pp. 123–132.

Timoshenko, S. P. and Gere, J. M. (1961). *Theory of Elastic Stability*. McGraw-Hill Book Co., New York.

Vlahinos, A. S., Smith, C. V. and Simitses, G. J. (1986). A nonlinear solution scheme for multistory, multibay plane frames, *Computers and Structures*, Vol. 22, No. 6, pp. 1035–1045.

Zimmermann, H. (1909). "Die Knickfestigkeit des geraden Stabes mit mehreren Feldern." *Sitzungsberichte der preussischen Akademie der Wissenschaften*, p. 180.

Zimmermann, H. (1910). *Die Knickfestigkeit der Druckgurte offener Brücken*. W. Ernst and Sohn, Berlin, Germany.

Zimmermann, H. (1925). *Die knickfestigkeit der Stabrerbindungen*. W. Ernst und Sohn, Berlin, Germany.

5

THE ENERGY CRITERION
AND ENERGY-BASED
METHODS

5.1   REMARKS ON THE ENERGY CRITERION

As a basis for the energy criterion, we use the principle of the minimum total potential (see Appendix A). This principle states:

> Of all possible kinematically admissible deformation fields in an elastic conservative system, for a specified level of the external loads and the corresponding internal loads, only those that make the total potential assume a minimum value correspond to a stable equilibrium.

First of all, the system must be conservative for the principle to hold, which implies that the energy criterion holds only for such systems. The system is conservative if both the external and internal forces are conservative. Since we are dealing with an elastic system, the existence of a strain-energy density function (see Appendix A) implies that the internal forces are conservative. The external forces are conservative if the work done by these forces from state O to state I are independent of the path and depend only on the initial and final values of the kinematically admissible deformations (virtual displacements). The idea of a virtual displacement is discussed in detail in Appendix A.

This principle is an extension of the Lagrange-Dirichlet theorem to systems with infinitely many degrees of freedom. Although it is stated as a sufficiency condition for stable equilibrium, the energy criterion based upon this principle has been used as both a necessary and sufficient condition for stability.

In order to clearly state and apply the stability criterion, let $U_T[\bar{u}]$ be the total potential (functional) at an equilibrium position characterized by $\bar{u}$. Furthermore, let $U_T[\bar{u} + \varepsilon\bar{u}_1]$ be the total potential in the neighborhood of the equilibrium position, where $\bar{u}_1$ denotes kinematically admissible deformations and ε is a small nonzero constant. If we now expand the integrals and group them on the basis of powers of ε, then we may write

$$\Delta U_T = U_T[\bar{u} + \varepsilon \bar{u}_1] - U_T[\bar{u}]$$
$$= \varepsilon \delta U_T[\bar{u}, \bar{u}_1] + \varepsilon^2 \delta^2 U_T[\bar{u}, \bar{u}_1] \ldots \tag{1}$$

According to Eq. (1), δU_T, $\delta^2 U_T$, etc., denote the first, second, etc., variations in the total potential for kinematically admissible deformations.

For equilibrium, it is necessary that

$$\delta U_T = 0 \tag{2}$$

for all $\bar{u}_1$. For a relative minimum (stable equilibrium), it is necessary that

$$\delta^2 U_T \geq 0 \tag{3}$$

for all $\bar{u}_1$ (see Sagan, 1969). Note that, if the second variation is identically equal to zero for all $\bar{u}_i$, then no conclusion can be drawn and higher variations must be considered. For a relative minimum, since ε^3 can be positive or negative, the following two conditions must be satisfied

$$\delta^3 U_T \equiv 0$$
$$\delta^4 U_T \geq 0 \tag{4}$$

for all $\bar{u}_i$. These steps are continued if $\delta^4 U_T \equiv 0$ for all $\bar{u}_i$ (see Koiter, 1967).

If we assume that the second variation is not identically equal to zero, then the stability criterion requires that $\delta^2 U_T$ be positive definite for every nonvanishing virtual displacement. Similarly, $\delta^2 U_T$ is negative definite for instability. The implication here is that $\delta^2 U_T$ changes its character at the critical load. Consequently, the critical load is the least value for which $\delta^2 U_T$ ceases to be positive definite and becomes positive semidefinite. This means that there exists at least one nonvanishing virtual displacement (buckling mode) for which the second variation is zero.

The energy criterion has been used by Timoshenko (1956) and Trefftz (1933) in various modified forms. The formulation suggested by Timoshenko is based on the following arguments: In a position of stable equilibrium, the total potential is a minimum, and consequently the increment in the total potential, ΔU_T, for every small kinematically admissible deformation from the equilibrium position, is positive. In terms of the strain energy and potential of the external forces,

$$\Delta U_i + \Delta U_p > 0 \tag{5}$$

Since $\Delta U_p = -\Delta W_e$, where ΔW_e is the work done by the external forces during these small deviations, Eq. (5) becomes

$$\Delta U_i > \Delta W_e \tag{6a}$$

This inequality becomes an equality as the load is increased to its critical value, and

$$\Delta U_i = \Delta W_e \tag{6b}$$

We can demonstrate this concept by applying it to the column problem. The Trefftz criterion, generalized herein, is based on the argument that, when the critical load is reached, $\delta^2 U_T$ becomes positive semidefinite. This means that the minimum value of $\delta^2 U_T$ becomes zero for certain nonvanishing virtual displacements. Now, if we set $\delta^2 U_T = V$, we want V to possess a minimum (which is zero) for certain nonvanishing virtual displacements (buckling mode). Thus, the first variation of V must be zero, or the first variation of the second variation of U_T must vanish. This criterion is applied to the column problem in this chapter and to the shallow arch in Chapter 7.

5.2   TIMOSHENKO'S METHOD

Timoshenko's method is fully outlined and applied to a number of systems in Arts. 2.8–2.10 of Timoshenko (1956). This method, referred to by Timoshenko (1956) as the energy method, provides a shortcut to obtaining approximate but highly accurate values for the critical load. It avoids solving differential equations and becomes very useful when applied to systems with nonuniform stiffness, a case where the solution to the usual eigen-boundary-value problem is extremely difficult and in some cases impossible.

5.2.1   THE CANTILEVER COLUMN

Consider the cantilever shown in Fig. 5.1 under a constant directional thrust P applied quasistatically.

As the load is increased from zero, the work done by the force P is stored into the system as stretching strain energy. If we now allow a bending deformation, $w(x)$, which is very small so that it does not alter the stretching energy, the change in the total potential ΔU_T is given by

$$\Delta U_T = \Delta U_{i_B} + \Delta U_p \tag{7}$$

where ΔU_{i_B} is the bending strain energy and ΔU_p is the change in the potential of the external force.

$$\Delta U_{i_B} = \frac{1}{2} \int_0^L EI(w'')^2 dx \tag{8a}$$

$$\Delta U_p = -\Delta W_e = -P\Delta L = -\frac{1}{2}P \int_0^L (w')^2 dx \tag{8b}$$

According to Timoshenko's argument, the straight configuration is stable if $\Delta U_T > 0$ and unstable if $\Delta U_T < 0$. A critical condition is reached when $\Delta U_T = 0$.

The additional steps are to assume a form for the admissible bending deformation, $w(x)$ and perform the indicated operations in the method. Let

$$w(x) = A\left(1 - \cos\frac{\pi x}{2L}\right) \tag{9}$$

Note that A is an arbitrary constant and $1 - \cos(\pi x/2L)$ satisfies the kinematic boundary conditions at $x = 0$.

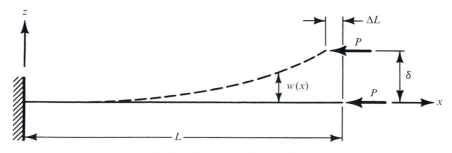

FIGURE 5.1   Geometry of the cantilever column.

$$\Delta U_{i_B} = \frac{\pi^4 EI A^2}{64 L^3}$$

and

$$\Delta U_p = -P \frac{\pi^2 A^2}{16 L}$$

From this expression we obtain that

$$P_{cr} = \frac{\pi^2 EI}{4 L^2}$$

which, of course, is the exact solution because the chosen deformation function happens to be the exact eigenfunction (see Chapter 3).

Next, let us use a different expression for $w(x)$:

$$w(x) = A x^2 \tag{10}$$

Then Timoshenko's method yields

$$P_{cr} = \frac{3 EI}{L^2}$$

which is higher, by approximately 21.3%, than the exact value.

Finally, we use for $w(x)$ the shape corresponding to the solution of a cantilever loaded transversely by a concentrated load at the free end:

$$w(x) = A x^2 (3L - x) \tag{11}$$

Timoshenko's method then yields

$$P_{cr} = 2.5 \frac{EI}{L^2}$$

This value is only 1.32% higher than the exact solution. The reason we get a better approximation in this case is because the expression for $w(x)$, given by Eq. (11), satisfies one of the natural boundary conditions at $x = L$, i.e., $M(L) = EI w''(L) = 0$. This condition is not satisfied by the expression for $w(x)$ given by Eq. (10).

5.2.2   THE SIMPLY SUPPORTED COLUMN

In a similar manner, according to Timoshenko's method, a critical condition is reached, when

$$\frac{1}{2} \int_0^L EI (w'')^2 dx = \frac{P}{2} \int_0^L (w')^2 dx \tag{12}$$

Assume that $w(x)$ is given by a function that is kinematically admissible, i.e.,

$$w(x) = A(L - x)x \tag{13}$$

Then by Eq. (12)

$$P_{cr} = 12 \frac{EI}{L^2}$$

This value is higher than the exact value by approximately 21.3%.

If we now choose a function that satisfies the natural boundary conditions as well, $M(0) = M(L) = 0$, we should expect the solution to improve.

Let

$$w(x) = A\left(L^3 x - 2Lx^3 + x^4\right)$$
$$w'(x) = A\left(L^3 - 6Lx^2 + 4x^3\right)$$
$$w''(x) = A\left(12x^2 - 12Lx\right)$$

Then by Eq. (12)

$$P_{cr} = 9.88 \frac{EI}{L^2}$$

which is approximately 0.13% higher than the exact value of the critical load.

5.2.3　THE RAYLEIGH AND TIMOSHENKO QUOTIENTS

When Timoshenko (Ref. 1) applied his method to the cantilever column, he did not use, for the strain energy, the expression given by Eq. (7). Instead he made use of the fact that (see Fig. 5.1)

$$M(x) = P[\delta - w(x)] \tag{14}$$

and

$$(w'') = \frac{M}{EI}$$

Then

$$
\begin{aligned}
\Delta U_{i_B} &= \frac{1}{2} \int_0^L \frac{M^2}{EI} dx \\
&= \frac{P^2}{2} \int_0^L \frac{(\delta - w)^2}{EI} dx
\end{aligned}
\tag{15}
$$

According to his method, when P approaches P_{cr}

$$\frac{1}{2} P_{cr}^2 \int_0^L \frac{(\delta - w)^2}{EI} dx = \frac{P_{cr}}{2} \int_0^L (w')^2 dx \tag{16}$$

If we now use Eq. (10) for w, and employ Eq. (16) to solve for P_{cr}, we have

$$P_{cr} = 2.5 \frac{EI}{L^2}$$

which is much closer to the exact value than what we got before ($3EI/L^2$).

Similarly, if the expression for w, given by Eq. (11), is used in Eq. (16)

$$P_{cr} = 2.4706 \frac{EI}{L^2}$$

This value, also, is closer to the exact value than what we got before. The improvement in the value, when using Eq. (16), occurs because the operation of differentiation indicated in Eq. (7) magnifies the error that exists when an approximate expression is used for $w(x)$.

We have seen so far that, as a consequence of Timoshenko's method, the expression for the critical load for any column may be written as

$$P = \frac{\int_0^L EI(w'')^2 dx}{\int_0^L (w')^2 dx} \tag{17}$$

This is called the integral or Rayleigh quotient, and it holds for all columns regardless of the boundary conditions. The name Rayleigh is used because a similar expression was derived and employed by Lord Rayleigh (Ref. 3) in the study of vibrations. Southwell (Ref. 4) outlines a procedure for finding buckling loads for a column which uses the Rayleigh quotient and refers to it as Rayleigh's method. Similar quotients may be derived for plates and certain shell configurations.

For the cantilever column, Eq. (16) is employed and the quotient becomes

$$P = \frac{\int_0^L (w')^2 dx}{\int_0^L \frac{(\delta - w)^2}{EI} dx} \tag{18}$$

Finally, if the column is simply supported, the reduced-order (second-degree) equation is applicable (see Chapter 3)

$$EIw'' + Pw = 0 \tag{19a}$$

and

$$w'' = -\frac{Pw}{EI} \tag{19b}$$

Substitution of this expression into the Rayleigh quotient yields

$$P = \frac{\int_0^L (w')^2 dx}{\int_0^L \frac{w^2}{EI} dx} \tag{20}$$

Quotients of the type given by Eqs. (18) and (20) are referred to as the Timoshenko quotient. Note that, when applicable, the Timoshenko quotient yields a closer approximation than the Rayleigh quotient.

Finally, when a column with elastic restraints at both ends is considered (see Fig. 3.14), the Rayleigh quotient becomes

$$P = \frac{\int_0^L EI(w'')^2 dx + \bar{\alpha}_0 w^2(0) + \bar{\alpha}_L w^2(L) + \bar{\beta}_0 [w'(0)]^2 + \bar{\beta}_L [w'(L)]^2}{\int_0^L (w')^2 dx} \tag{21}$$

Since the Timoshenko method leads to a quotient similar to that used by Lord Rayleigh, the method is often called the Rayleigh-Timoshenko method.

5.2.4   THE GENERAL RAYLEIGH-TIMOSHENKO METHOD

Starting with the concept of Timoshenko, we arrive at an integral quotient or the ratio of two functionals $I[u]$ and $J[u]$:

$$\lambda = \frac{I[u]}{J[u]} \tag{22}$$

We are interested in finding the function, u, which minimizes the quotient. If we use a series expression for u in the form of

$$u = \sum_{i=1}^{N} a_i g_i \tag{23}$$

where g_i are admissible functions and elements of a complete sequence (see Appendix A), the quotient becomes

$$\lambda = \frac{f_1(a_1, a_2, \ldots, a_n)}{f_2(a_1, a_2, \ldots, a_n)} \tag{24}$$

Now we must adjust the coefficients a_i in such a way that the ratio is a minimum. This requirement leads to

$$\frac{\partial \lambda}{\partial a_i} = \frac{1}{f_2^2}\left[\frac{\partial f_1}{\partial a_i} f_2 - \frac{\partial f_2}{\partial a_i} f_1\right] = 0 \tag{25}$$

Since f_2 is finite, use of Eq. (24) yields

$$\frac{\partial f_1}{\partial a_i} - \lambda \frac{\partial f_2}{\partial a_i} = 0, \quad i = 1, \ldots, N \tag{26}$$

This procedure is referred to by many authors, including Timoshenko (Ref. 3), as the Ritz procedure.

As an example of this procedure, consider the cantilever of Fig. 5-1. If we employ the Rayleigh quotient,

$$I[w] = \int_0^L EI(w'')^2 dx$$

and

$$J[w] = \int_0^L (w')^2 dx \tag{27}$$

Let

$$w_A(x) = \sum_{n=1}^{3} a_n x^{n+1} \tag{28}$$

Then

$$f_1(a_1, a_2, a_3) = I[w_A]$$

$$= 4EI\left[a_1^2 L + 3a_1 a_2 L^2 + \left(3a_2^2 + 4a_1 a_3\right)L^3 + 9a_2 a_3 L^4 + \frac{36}{5}a_3^2 L^5\right] \tag{29a}$$

and

$$f_2(a_1, a_2, a_3) = J[w_A] = \frac{4}{3}a_1^2 L^3 + 3a_1 a_2 L^4 + \left(\frac{9}{5}a_2^2 + \frac{16}{5}a_1 a_3\right)L^5$$

$$+ 4a_2 a_3 L^6 + \frac{16}{7}a_3^2 L^7 \tag{29b}$$

If we define $\lambda = PL^2/EI$, Eqs. (26) become

$$8L\left(1 - \frac{1}{3}\lambda\right)a_1 + 3(4 - \lambda)L^2 a_2 + 16\left(1 - \frac{1}{5}\lambda\right)L^3 a_3 = 0$$

$$3(4 - \lambda)L^2 a_1 + 6\left(4 - \frac{3}{5}\lambda\right)L^3 a_2 + 4(9 - \lambda)L^4 a_3 = 0 \qquad (30)$$

$$16\left(1 - \frac{1}{5}\lambda\right)L^3 a_1 + 4(9 - \lambda)L^4 a_2 + 32\left(\frac{9}{5} - \frac{1}{7}\lambda\right)L^5 a_3 = 0$$

Equations (30) represent a system of three linear homogeneous algebraic equations in a_1, a_2, and a_3. A nontrivial solution exists if the determinant of the coefficients vanishes.

$$\begin{vmatrix} 8\left(1 - \frac{1}{3}\lambda\right) & 3(4 - \lambda) & 4\left(1 - \frac{1}{5}\lambda\right) \\[2mm] 3(4 - \lambda) & 6\left(4 - \frac{3}{5}\lambda\right) & (9 - \lambda) \\[2mm] 4\left(1 - \frac{1}{5}\lambda\right) & (9 - \lambda) & 2\left(\frac{9}{5} - \frac{1}{7}\lambda\right) \end{vmatrix} = 0 \qquad (31)$$

If only one term is considered ($a_1 \neq 0$, $a_2 = a_3 = 0$), then $\lambda_{cr} = 3$ and $P = 3EI/L^2$ as before. If two terms are considered ($a_1 \neq 0$, $a_2 \neq 0$, and $a_3 \equiv 0$), then $\lambda_{cr} = 2.48596$, which is only 0.75% higher than the exact solution. When all three terms are considered (a computer program was employed), $\lambda_{cr} = 2.4677$, which is extremely close to the correct answer (2.4674).

Note that every approximation is higher than the exact solution, and as more terms are considered, we converge to the minimum of λ from above. This, as expected, is true for all problems for which the formulation is characterized by Eq. (22). This means that the value of λ obtained by the use of an approximate expression for u in Eq. (22) cannot be any smaller than the value of λ corresponding to the exact expression for u.

5.2.5 THE NONUNIFORM STIFFNESS COLUMN

Consider a simply supported column with a bending stiffness given by (see Fig. 5.2)

$$EI = EI_0\left(1 + \frac{I_1}{I_0}\sin\frac{\pi x}{L}\right) \qquad (32)$$

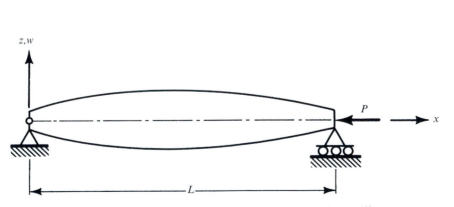

FIGURE 5.2   Simply supported column with nonuniform stiffness.

Furthermore, let the buckled deformation, $w(x)$, be approximated by

$$w_A(x) = A_1 \sin\frac{\pi x}{L} + A_3 \sin\frac{3\pi x}{L} \tag{33}$$

If we use the general Rayleigh-Timoshenko procedure as outlined in the previous section, then

$$f_1(A_1, A_3) = EI_0\left(\frac{\pi}{L}\right)^4 \frac{L}{2}\left[A_1^2\left(1+\frac{8}{3\pi}\frac{I_1}{I_0}\right) - 18A_1A_3\left(\frac{8}{15\pi}\right)\frac{I_1}{I_0}\right.$$
$$\left. + 81A_3^2\left(1+\frac{72}{35\pi}\frac{I_1}{I_0}\right)\right] \tag{34a}$$

$$f_2(A_1, A_3) = \left(\frac{\pi}{2}\right)^2 \frac{L}{2}\left(A_1^2 + 9A_3^2\right) \tag{34b}$$

Next, if we let $\lambda = P/P_{E_0}$, where $P_{E_0} = \pi^2 EI_0/L^2$, and employ Eqs. (26), we obtain the following system of equations in A_1 and A_3

$$\left[\left(1+\frac{8}{3\pi}\frac{I_1}{I_0}\right) - \lambda\right]A_1 - 9\left(\frac{8}{15\pi}\right)\frac{I_1}{I_0}A_3 = 0$$
$$-\frac{8}{15\pi}\frac{I_1}{I_0}A_1 + \left[9\left(1+\frac{72}{35\pi}\frac{I_1}{I_0}\right) - \lambda\right]A_3 = 0 \tag{35}$$

For a nontrivial solution to exist, the determinant of the coefficients must vanish. The expansion of the determinant yields the following quadratic equation in λ.

$$\lambda^2 - \lambda\left(10+\frac{2224}{105\pi}\frac{I_1}{I_0}\right) + 9\left(1+\frac{496}{105\pi}\frac{I_1}{I_0}+\frac{8192}{1575\pi^2}\frac{I_1^2}{I_0^2}\right) = 0 \tag{36}$$

From Eq. (36) we obtain λ_{cr}:

$$\lambda_{cr} = 5 + 3.371056\frac{I_1}{I_0} - 4\sqrt{1 + 1.261114\frac{I_1}{I_0} + 0.413813\frac{I_1^2}{I_0^2}} \tag{37}$$

Note that, if $w = A_1 \sin(\pi x/L)$ (one-term solution), then

$$\lambda_{cr} = 1 + \frac{8}{3\pi}\frac{I_1}{I_0} \tag{38}$$

The actual values for λ are shown in Table 5.1 for various values of I_1/I_0. At this point there are two basic questions that we must answer. First, "Since the exact value for λ_{cr} is not known, does the two-term solution yield a good approximation for λ_{cr}?" Second, "Since there is considerable difficulty in obtaining λ_{cr} for a column with nonuniform flexural stiffness, why bother to build and analyze such columns?" Another way of stating the second question is the following: "Is the critical load of a nonuniform stiffness column, $P_{NU_{cr}}$, higher than the critical load of a uniform stiffness column, $P_{U_{cr}}$, when the two columns are of equal weight?"

Before we answer these questions, we must consider the relation between the cross-sectional area, A, and the moment of inertia, I. If we let

$$I(x) = \alpha A^n(x) \tag{39}$$

where α is a constant, we note that, when $n = 1$, the width varies while the height remains constant; when $n = 2$, both the width and height vary in the same

TABLE 5.1 Critical loads for nonuniform and uniform columns of equal weight

$\frac{I_1}{I_0}$	$\lambda_{NU_{cr}}$		$\lambda_{U_{cr}}$ $n=1$	$\lambda_{U_{cr}}$ $n=2$	$\lambda_{NU_{cr}}/\lambda_{U_{cr_{n=1}}}$		$\lambda_{NU_{cr}}/\lambda_{U_{cr}n=2}$	
	One-Term Solution	Two-Term Solution			One-Term Solution	Two-Term Solution	One-Term Solution	Two-Term Solution
0	1.0000	1.0000	1.0000	1.0000	1.000	1.000	1.000	1.000
0.5	1.4244	1.4183	1.3183	1.3120	1.080	1.076	1.086	1.081
1.0	1.8488	1.8290	1.6366	1.6211	1.130	1.117	1.140$^+$	1.128
2.0	2.6977	2.6405	2.2732	2.2265	1.187	1.162	1.212	1.186
4.0	4.3953	4.2488	3.5465	3.4193	1.239	1.198	1.285	1.243
6.0	6.0930	5.8505	4.8197	4.6013	1.264	1.214	1.324	1.271
8.0	7.7906	7.4497	6.0930	5.7781	1.279	1.223	1.348	1.289
10.0	9.4883	9.0478	7.3662	6.9518	1.288	1.228	1.365	1.302
20.0	17.9765	17.0320	13.7324	12.7996	1.309	1.240	1.404	1.331
30.0	26.4648	25.0130	20.0986	18.6331	1.317	1.244	1.420	1.342
40.0	34.9530	32.9932	26.4648	24.4610	1.321	1.247	1.429	1.349
50.0	43.4413	40.9730	32.8310	30.2860	1.323	1.248	1.434	1.353
100.0	85.8826	80.8703	64.6620	59.3925	1.328	1.251	1.446	1.362
∞					1.333	1.253	1.460	1.373

proportion; and when $n = 3$, the height varies while the width remains constant. For a column with a rectangular cross section of height h and width b,

$$I = \frac{bh^3}{12} \quad \text{and} \quad A = bh$$

Now, for $n = 1$

$$\alpha = \frac{bh^3/12}{bh} = \frac{h^2}{12}$$

and for α to be a constant, h must be a constant.

Similarly, for $n = 2$,

$$\alpha = \frac{bh^3/12}{b^2h^2} = \frac{1}{12}\left(\frac{h}{b}\right)$$

and for α to be a constant, h/b must be a constant, which implies that the height and width vary proportionally.

Finally, for $n = 3$

$$\alpha = \frac{bh^3/12}{b^3h^3} = \frac{1}{12b^2}$$

and b must be a constant.

These conclusions are generally true for all symmetric cross-sections such as circular, elliptic, triangular, I-, and T-sections.

Returning to the two questions, we find the answer to the second question by comparing P_{cr} for the nonuniform geometry column with P_{cr} for the uniform column, provided the weights of the two are equal. However, since the two-term solution leads to a higher value for P_{cr} than the exact, this comparison is meaningful only if the two-term solution is a good approximation to the exact value of $P_{NU_{cr}}$. Let us first obtain the expressions for the volume (weight), $\overline{V}$, for the two columns.

For the uniform column

$$\overline{V} = \int_0^L A_U dx = A_U L \qquad (40)$$

For the nonuniform column

$$\overline{V} = \int_0^L A_{NU} dx \qquad (41)$$

Making use of Eqs. (32) and (39)

$$A_{NU} = \left[\frac{I_0}{\alpha} \left(1 + \frac{I_1}{I_0} \sin \frac{\pi x}{L} \right) \right]^{\frac{1}{n}} \qquad (42)$$

Substitution of Eq. (42) into Eq. (41) yields

$$\overline{V} = \left(\frac{\overline{I_0}}{\alpha} \right)^{\frac{1}{n}} \int_0^L \left(1 + \frac{I_1}{I_0} \sin \frac{\pi x}{L} \right)^{\frac{1}{n}} dx \qquad (43)$$

Let $\xi = \pi x / L$, then Eq. (43) becomes

$$\overline{V} = \left(\frac{I_0}{\alpha} \right)^{\frac{1}{n}} \frac{L}{\pi} \int_0^\pi \left(1 + \frac{I_1}{I_0} \sin \xi \right)^{\frac{1}{n}} d\xi \qquad (44)$$

For a simply supported column, $P_{U_{cr}}$ is given by the Euler load, or

$$P_{U_{cr}} = \frac{\pi^2 E I_U}{L^2}$$

Use of Eqs. (39) and (40) yields

$$P_{U_{cr}} = \frac{\pi^2 E}{L^2} \alpha \left(\frac{\overline{V}}{L} \right)^n \qquad (45)$$

Through the use of Eq. (44), Eq. (45) becomes

$$P_{U_{cr}} = \frac{\pi^2 E I_0}{L^2} \left[\int_0^\pi \left(1 + \frac{I_1}{I_0} \sin \xi \right)^{\frac{1}{n}} \frac{d\xi}{\pi} \right]^n \qquad (46)$$

From this equation

$$\lambda_{U_{cr}} = \left[\int_0^\pi \left(1 + \frac{I_1}{I_0} \sin \xi \right)^{\frac{1}{n}} \frac{d\xi}{\pi} \right]^n \qquad (47)$$

For $n = 1$

$$\lambda_{U_{cr}} = \frac{1}{\pi} \int_0^\pi \left(1 + \frac{I_1}{I_0} \sin \xi \right) d\xi = 1 + \frac{2}{\pi} \frac{I_1}{I_0}$$

There are no closed-form solutions for $n = 2$ and 3. Values of $\lambda_{U_{cr}}$ are presented in Table 5.1 for $n = 1$ and $n = 2$ for a large range of I_1/I_0. In addition, the ratios of $\lambda_{NU_{cr}}$ to $\lambda_{U_{cr}}$ are presented for the one- and two-term solutions.

A number of investigators have dealt with the shape of the optimum column (see Prager and Taylor, 1968; Simitses et al., 1972; and Tadjbakhsh and Keller, 1962). When there is no constraint on the stress level and the size of the area distribution, it is found that the optimum shape (for the simply supported column) starts with zero area at the ends and builds up to some maximum value at the center.

Furthermore, the ratio of the critical load, corresponding to the optimum column, to the critical load for a uniform geometry column of equal volume is given by 1.216, 1.333, and 1.410 for $n = 1$, 2, and 3, respectively. A study of the results of Table 5.1 in connection with the above conclusions suggests that:

1. Since Eq. (32) does not necessarily correspond to the optimum shape (even with $I_0 = 0$), then $\lambda_{NU_{cr}}/\lambda_{U_{cr}}$ should be smaller than 1.216 (for $n = 1$) and 1.333 (for $n = 2$). Because it is not, the two-term solution has not converged to the exact value and more terms are needed in Eq. (33).

2. The more uniform the column is (smaller I_1/I_0 values), the better the convergence is.

Finally, we may conclusively state that nonuniformity in stiffness, of the type expressed by Eq. (32), yields a stronger configuration than that of a uniform geometry of the same weight.

5.3   THE RAYLEIGH-RITZ METHOD

The Rayleigh-Ritz or simply the Ritz method is explained in Appendix A. As far as buckling problems are concerned, there are two possible applications of the method.

The first type of application concerns problems for which a Rayleigh quotient exists (columns, plates, cylindrical shells, etc.). In this case, if the total potential, or some characteristic functional such as $\delta^2 U_T$ according to the Trefftz criterion, is expressed in the form of $U_T = I[u] - \lambda J[u]$, where λ denotes the eigenvalues (the lowest of which corresponds to the critical load parameter), the method suggests that we express u in terms of a series of the type

$$u_A = \sum_{i=1}^{N} a_i g_i$$

where g_i are kinematically admissible functions. Then,

$$U_T = f_1(a_1, a_2, \ldots, a_N) - \lambda f_2(a_1, a_2, \ldots, a_N) \tag{48}$$

where

$$f_1 = I[u_A] \quad \text{and} \quad f_2 = J[u_A]$$

Requiring that U_T have a minimum leads to

$$\frac{\partial U_T}{\partial a_i} = \frac{\partial f_i}{\partial a_i} - \lambda \frac{\partial f_2}{\partial a_i} = 0 \quad \text{for} \quad i = 1, 2, \ldots, N \tag{49}$$

These equations are identical to Eqs. (26); thus, the Rayleigh-Ritz and the general Rayleigh-Timoshenko methods are identical. This is the reason that many authors call this particular application the Rayleigh-Ritz method as used by Timoshenko for buckling problems. Note that in this first type of application, the variation in $I - \lambda J$ with respect to u (keeping λ constant) leads to the same equations as the minimization of the Rayleigh quotient, given by Eq. (22).

Finally, for this type of application, convergence is guaranteed because we are dealing with a variational problem which satisfies the sufficiency conditions for a minimum. Some authors refer to this type of application as the Rayleigh-Ritz method, whereas when the method is applied to variational problems (stationary) that do not satisfy the sufficiency conditions for a minimum or a maximum, they call it simply the Ritz method. This distinction is not important. What is important is that there is no rigorous proof of convergence for this latter type of application, although the method has been used very successfully.

The second type of application does not depend on the existence of a Rayleigh quotient, and it is based on the stability criterion directly. If we express the deformation(s) by the finite series

$$u = \sum_{i=1}^{N} a_i g_i$$

where g_i are kinematically admissible functions, the total potential, $U_T[u]$ (functional) becomes a function of a_i, $U_T(a_i)$. For the equilibrium to be stable, the total potential must be a minimum, and the following conditions must be satisfied

$$\frac{\partial U_T}{\partial a_i} = 0 \quad i = 1, 2, \ldots, N \tag{50}$$

and

$$\begin{vmatrix} \dfrac{\partial^2 U_T}{\partial^2 a_1^2} & \dfrac{\partial^2 U_T}{\partial a_1 \partial a_2} & \cdots & \dfrac{\partial^2 U_T}{\partial a_1 \partial a_N} \\ \dfrac{\partial^2 U_T}{\partial a_2 \partial a_1} & \dfrac{\partial^2 U_T}{\partial a_2^2} & \cdots & \dfrac{\partial^2 U_T}{\partial a_2 \partial a_N} \\ \dfrac{\partial^2 U_T}{\partial a_N \partial a_1} & \dfrac{\partial^2 U_T}{\partial a_N \partial a_2} & \cdots & \dfrac{\partial^2 U_T}{\partial a_N^2} \end{vmatrix} > 0 \tag{51}$$

along with all its principal minors, such as

$$\frac{\partial^2 U_T}{\partial a_1^2} > 0, \quad \begin{vmatrix} \dfrac{\partial^2 U_T}{\partial a_1^2} & \dfrac{\partial^2 U_T}{\partial a_1 \partial a_2} \\ \dfrac{\partial^2 U_T}{\partial a_2 \partial a_1} & \dfrac{\partial^2 U_T}{\partial a_2^2} \end{vmatrix} > 0, \quad \text{etc.} \tag{52}$$

Equations (50) give us the equilibrium equations that relate the load to the displacement parameters a_i (generalized coordinates). They are N equations in $N + 1$ unknowns (a_i, $i = 1, 2, \ldots, N$, and the load parameter λ). From these equations, we may solve for the a_i's in terms of the load parameter. Knowing the equilibrium positions, we then proceed to study the stability or instability of these equilibrium positions by using the inequalities given by Eqs. (51) and (52). The value of the load parameter at which the equilibrium changes from stable to unstable is the critical value. Note at this point that, if the expressions in Eqs. (51) and (52) are identically equal to zero, no decision can be made about the stability or instability of this equilibrium position, and higher variations are needed.

This procedure will be demonstrated in the following application. Consider a simply supported column of uniform geometry as shown in Fig. 5.3. The kinematic and constitutive relations are

$$\varepsilon_{xx} = u_{,x} + \frac{1}{2} w_{,x}^2 - z w_{,xx}$$

$$\sigma_{xx} = E \varepsilon_{xx}$$

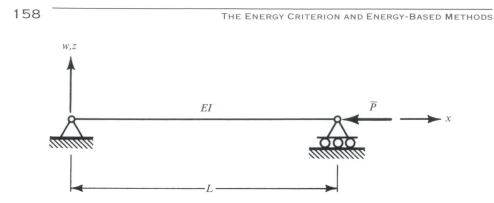

FIGURE 5.3   Geometry and sign convention for a simply supported column.

On the basis of these, the total potential is

$$U_T = \frac{1}{2}\int_0^L \left[EA\left(u_{,x} + \frac{1}{2}w_{,x}^2 \right)^2 + EIw_{,xx}^2 \right]dx + \bar{P}u(L) \tag{53}$$

Let us now use the following one-term approximations for $u(x)$ and $w(x)$:

$$u(x) = B_1 x$$
$$w(x) = C_1 \sin\frac{\pi x}{L} \tag{54}$$

Note that $u(0) = 0$ and $w(0) = w(L) = 0$, and the functions x and $\sin \pi x/L$ are kinematically admissible. Substitution of Eqs. (54) into Eqs. (53) yields

$$U_T = \frac{L}{2}\left\{ EA\left[B_1^2 + \frac{1}{2}\left(\frac{\pi}{L}\right)^2 B_1 C_1^2 + \frac{3}{32}\left(\frac{\pi}{L}\right)^4 C_1^4 \right] + \frac{1}{2}EI\left(\frac{\pi}{L}\right)^4 C_1^2 \right\} + \bar{P}LB_1 \tag{55}$$

For equilibrium

$$\frac{\partial U_T}{\partial B_1} = \frac{\partial U_T}{\partial C_1} = 0$$

$$L\left[EAB_1 + \frac{EA}{4}\left(\frac{\pi}{L}\right)^2 C_1^2 + \bar{P}L \right] = 0$$

$$\frac{L}{2}\left(\frac{\pi}{L}\right)^2\left[EAB_1 C_1 + \frac{3}{8}EA\left(\frac{\pi}{L}\right)^2 C_1^3 + EI\left(\frac{\pi}{L}\right)^2 C_1 \right] = 0 \tag{56}$$

The second derivatives are given by

$$\frac{\partial^2 U_T}{\partial B_1^2} = EAL$$

$$\frac{\partial^2 U_T}{\partial B_1 \partial C_1} = \frac{EAL}{2}\left(\frac{\pi}{L}\right)^2 C_1 \tag{57}$$

$$\frac{\partial^2 U_T}{\partial C_1^2} = \frac{EAL}{2}\left(\frac{\pi}{L}\right)^2 B_1 + \frac{9EAL}{16}\left(\frac{\pi}{L}\right)^4 C_1^2 + \frac{EIL}{2}\left(\frac{\pi}{L}\right)^4$$

If we let $P_E = \pi^2 EI/L^2$, the equilibrium equations, Eqs. (56), are

$$(EA)B_1 + \frac{EA}{4}\left(\frac{\pi}{L}\right)^2 C_1^2 = -\overline{P}$$

$$C_1\left[(EA)B_1 + \frac{3}{8}EA\left(\frac{\pi}{L}\right)^2 C_1^2 + P_E\right] = 0 \tag{58}$$

It is easily seen from Eqs. (58) that there are two possible solutions:

$$\text{(a)} \quad B_1 = -\frac{\overline{P}}{AE} \quad \text{and} \quad C_1 = 0$$

and

$$\text{(b)} \quad B_1 = \frac{-\overline{P} - 2(\overline{P} - P_E)}{AE} \quad \text{and} \quad C_1^2 = \frac{8}{AE}\left(\frac{L}{\pi}\right)^2(\overline{P} - P_E)$$

The corresponding deformation functions are

$$\text{(a)} \quad u(x) = -\frac{\overline{P}}{AE}x, \quad w(x) = 0$$

$$\text{(b)} \quad u(x) = -\frac{\overline{P} + 2(\overline{P} - P_E)}{AE}x$$

and

$$w(x) = \pm\left(\frac{8}{AE}\right)^{\frac{1}{2}}\left(\frac{L}{\pi}\right)(\overline{P} - P_E)^{\frac{1}{2}}\sin\frac{\pi x}{L}$$

The term $(8/AE)^{\frac{1}{2}}(L/\pi)(\overline{P} - P_E)^{\frac{1}{2}}$ represents the maximum deflection, δ (at $x = L/2$). From this, we may write the following two expressions for δ.

$$\frac{\delta}{L} = 2\sqrt{2}\left(\frac{\rho}{L}\right)\left(\frac{P}{P_E} - 1\right)^{\frac{1}{2}}$$

or

$$\frac{\delta}{\rho} = 2\sqrt{2}\left(\frac{P}{P_E} - 1\right)^{\frac{1}{2}}$$

where ρ is the radius of gyration of the cross-sectional area, $\rho^2 = I/A$. All of the equilibrium positions are shown, qualitatively, in Fig. 5.4.

The next problem is to determine the stability or instability of all the equilibrium positions. To this end, the two solutions are treated separately. First, let us consider the solution corresponding to the straight configuration $C_1 = 0$.

1. Making use of the expressions for the second partial derivatives, Eqs. (57), evaluated at $B_1 = -P/AE$ and $C_1 = 0$, we obtain the conditions for stability:

$$EAL > 0 \quad \text{and} \quad L\left(\frac{\pi}{L}\right)^2(-\overline{P} + P_E) > 0 \tag{59}$$

It is clear from these inequalities that the straight configuration is stable for $\overline{P} < P_E$ and unstable for $\overline{P} > P_E$, as expected.

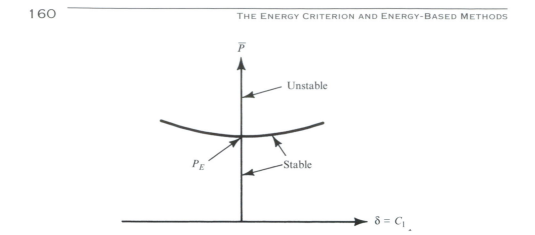

FIGURE 5.4 Equilibrium positions for the simply supported column.

2. Similarly, since $EAL > 0$, the condition for stability for the equilibrium positions characterized by $C \neq 0$ is

$$\begin{vmatrix} EAL & \dfrac{EAL}{2}\left(\dfrac{\pi}{L}\right)^2 C_1 \\ \dfrac{EAL}{2}\left(\dfrac{\pi}{L}\right)^2 C_1 & \dfrac{\pi^2}{2L}\left[EAB_1 + \dfrac{9}{8}EA\left(\dfrac{\pi}{L}\right)^2 C_1^2 + P_E\right] \end{vmatrix} > 0$$

Making use of the expressions for B_1 and C_1, and expanding the determinant, the above inequality becomes

$$\frac{\pi^2}{L}\left(\overline{P} - P_E\right) > 0$$

This inequality is definitely true, since these equilibrium positions (bent configuration) exist only if $P > P_E$.

The only question that remains to be answered is whether the position corresponding to $C_1 = 0$ and $\overline{P} = P_E$ (bifurcation point) is stable or unstable. If we expand the total potential about this position for small variations in B_1 and C_1, we obtain

$$\Delta U_T = \frac{1}{4!}\left(\frac{9}{8}\right)(EAL)\left(\frac{\pi}{L}\right)^4 (\delta C_1)^4$$

Clearly, $\Delta U_T > 0$ and this position is stable.

In this particular application of the Rayleigh-Ritz method, it must be pointed out that the analysis is exact for the discrete system and it is approximate for the continuous system. The results obtained for the column are exact as far as the critical load is concerned, mainly because the exact form for $u(x)$ was assumed for the straight configuration and because the expression for $w(x)$ is that of the linear eigenvalue problem. The results are also exact as far as the stability analysis is concerned. The only approximation involved is in the post-buckling curve for two reasons. First, because of the kinematic relations used, we do not expect the results to be applicable to bent configurations for which $w_{,x}^2 \ll 1$ does not hold. Second, even if $w_{,x}^2 \ll 1$, we do not necessarily have a good approximation for the bent configuration, and we must take more terms for the deformation functions to determine the convergence to the true deformation.

5.4   THE COLUMN BY THE TREFFTZ CRITERION

Consider the simply supported column, shown in Fig. 5.3. The kinematic and constitutive relations to be used are given by

$$\varepsilon_{xx} = u_{,x} + \frac{1}{2}w_{,x}^2 - zw_{,xx} \tag{60}$$

$$= \varepsilon_{xx}^0 - zw_{,xx}$$

$$\sigma_{xx} = E\varepsilon_{xx} \tag{61}$$

If we let

$$P = \int_A \sigma_{xx}dA \tag{62}$$

where A is the cross-sectional area, the total potential is given by

$$U_T = \frac{1}{2}\int_0^L \left(P\varepsilon_{xx}^0 + EIw_{,xx}^2\right)dx + \overline{P}u(L) \tag{63}$$

In terms of the displacements, the total potential becomes

$$U_T[u,\ w] = \frac{1}{2}\int_0^L \left[AE\left(u_{,x} + \frac{1}{2}w_{,x}^2\right)^2 + EIw_{,xx}^2\right]dx + \overline{P}u(L) \tag{64}$$

Let $\overline{u}(x)$ and $\overline{w}(x)$ denote positions of stable equilibrium, and let $\beta(x)$ and $\gamma(x)$ be kinematically admissible functions for $u(x)$ and $w(x)$, respectively. Then

$$U_T[\overline{u} + \varepsilon_1\beta,\ \overline{w} + \varepsilon_2\gamma] = \frac{1}{2}\int_0^L \left\{AE\left[\overline{u}_{,x} + \varepsilon_1\beta_{,x} + \frac{1}{2}\left(\overline{w}_{,x} + \varepsilon_2\gamma_{,x}\right)^2\right]^2 \right.$$

$$\left. + EI\left(\overline{w}_{,xx} + \varepsilon_2\gamma_{,xx}\right)^2\right\}dx + \overline{P}[u(L) + \varepsilon_1\beta(L)] \tag{65}$$

where ε_1 and ε_2 are small constants. After performing the indicated operations in the integrals and collecting like powers of the ε's, we have

$$U_T[\overline{u} + \varepsilon_1\beta,\ \overline{w} + \varepsilon_2\gamma] = \frac{1}{2}\int_0^L \left[EA\left(\overline{u}_{,x} + \frac{1}{2}\overline{w}_{,x}^2\right)^2 + EI\overline{w}_{,xx}^2\right]dx$$

$$+ \overline{P}u(L) + \varepsilon_1\left[\int_0^L EA\left(\overline{u}_{,x} + \frac{1}{2}\overline{w}_{,x}^2\right)\beta_{,x}dx + \overline{P}\beta(L)\right]$$

$$+ \varepsilon_2\int_0^L \left[EA\left(\overline{u}_{,x} + \frac{1}{2}\overline{w}_{,x}^2\right)\overline{w}_{,x}\gamma_{,x} + EI\overline{w}_{,xx}\gamma_{,xx}\right]dx \tag{66}$$

$$+ \frac{\varepsilon_1^2}{2}\int_0^L EA\beta_{,x}^2 dx + \varepsilon_1\varepsilon_2\int_0^L EA\overline{w}_{,x}\gamma_{,x}\ dx$$

$$+ \frac{\varepsilon_2^2}{2}\int_0^L \left[EA\overline{w}_{,x}^2\gamma_{,x}^2 + EA\left(\overline{u}_{,x} + \frac{1}{2}\overline{w}_{,x}^2\right)\gamma_{,x}^2 + EI\gamma_{,xx}^2\right]dx$$

$$+ \frac{\varepsilon_1\varepsilon_2^2}{2}\int_0^L EA\beta_{,x}\gamma_{,x}^2 dx + \frac{\varepsilon_2^3}{2}\int_0^L EA\overline{w}_{,x}\gamma_{,x}^3\ dx + \frac{\varepsilon_2^4}{8}\int_0^L EA\gamma_{,x}^4\ dx$$

Note that the terms on the right side of Eq. (66) that do not contain ε's denote $U_T[\bar{u}, \bar{w}]$. Furthermore, if we collect terms with like powers in ε, we may write Eq. (66) in the following form:

$$\Delta U_T = \delta U_T + \delta^2 U_T + \delta^3 U_T + \delta^4 U_T \tag{67}$$

Next, if we are interested in studying the stability of equilibrium positions corresponding to the straight configuration

$$\Delta U_T[\bar{u} + \varepsilon_1 \beta, \varepsilon_2 \gamma] = \delta U_T[\bar{u} + \varepsilon_1 \beta, \varepsilon_2 \gamma] + \delta^2 U_T[\bar{u} + \varepsilon_1 \beta, \varepsilon_2 \gamma] + \cdots \tag{68}$$

where

$$\delta U_T[\bar{u} + \varepsilon_1 \beta, \varepsilon_2 \gamma] = \varepsilon_1 \left[\int_0^L P\beta_{,x} dx + \bar{P}\beta(L) \right]$$

$$\delta^2 U_T[\bar{u} + \varepsilon_1 \beta, \varepsilon_2 \gamma] = \frac{\varepsilon_1^2}{2} \int_0^L EA\beta_{,x}^2 dx + \frac{\varepsilon_2^2}{2} \int_0^L \left[P\gamma_{,x}^2 + EI\gamma_{,xx}^2 \right] dx \tag{69}$$

Note that $P = EA\bar{u}_{,x}$ from Eqs. (60)–(62).

Equilibrium for the straight configuration is characterized by

$$\delta U_T[\bar{u} + \varepsilon_1 \beta, \varepsilon_2 \gamma] = 0 \tag{70}$$

This leads to $P_{,x} = 0$ or $P = $ constant. Use of the boundary condition at $x = L$ yields $P = -\bar{P}$. These equilibrium positions are stable if $\delta^2 U_T$ is positive definite for all $\beta(x)$ and $\gamma(x)$ functions.

According to the Trefftz criterion (see Section 5.1), when the critical load is reached, $\delta^2 U_T$ becomes positive semidefinite. From the second equation of Eqs. (69), we notice that the first term is positive for all $\beta(x)$ except zero. Therefore the second term must be positive for all $\gamma(x)$ for stability. Thus, $\delta^2 U_T$ becomes positive semidefinite when $\beta(x) = 0$ and

$$\delta \int_0^L \left(P\gamma_{,x}^2 + EI\gamma_{,xx}^2 \right) dx = 0$$

when $P = -\bar{P}$, or

$$\delta \int_0^L \left(EI\gamma_{,xx}^2 - \bar{P}\gamma_{,x}^2 \right) dx = 0 \tag{71}$$

This condition leads to the same eigen-boundary-value problem as the one in Section 3.3. Note that the variations in Eq. (71) are with respect to kinematically admissible functions.

Alternative Procedure. If we follow the approach used in Chapter 3, we notice that

$$\Delta U_T = \delta_\varepsilon U_T = \int_0^L \left[P\left(\delta u_{,x} + w_{,x}\delta w_{,x} + \frac{1}{2}\delta w_{,x}^2 \right) \right.$$

$$\left. + EI\left(w_{,xx}\delta w_{,xx} + \frac{1}{2}\delta w_{,xx}^2 \right) \right] dx + \bar{P}\delta u(L) = \int_0^L P\delta u_{,x}\, dx \tag{72}$$

$$+ \bar{P}\delta u(L) + \frac{1}{2}\int_0^L \left(P\delta w_{,x}^2 + EI\delta w_{,xx}^2 \right) dx = \delta U_T + \delta^2 U_T$$

According to this approach, δU_T is the same as the first of Eqs. (69). The difference in $\delta^2 U_T$ between the two approaches [see the second of Eqs. (69)] is the term

$$\int_0^L \frac{\varepsilon_1^2}{2} EA\beta(x) \, dx \tag{73}$$

This term, in the alternative approach, represents $\frac{1}{2}\int_0^L \delta P \, \delta u \, dx$ which is zero since the external and internal loads are kept constant during the virtual displacements δu and δw.

Therefore, again we have

$$\delta(\delta^2 U_T) = \delta\left[\int_0^L \left(EI\delta w_{,xx}^2 - \overline{P}\delta w_{,x}^2\right) dx\right] = 0$$

which is the same as Eq. (71).

It is important to note at this point that this particular form of the second variation is very attractive to the application of the Rayleigh-Ritz method, as demonstrated in the first type of application in Section 5.3.

$$\delta^2 U_T = 1 - \lambda J$$

where

$$I = \int_0^L EI\delta w_{,xx}^2 \, dx$$

$$\lambda = \overline{P} \tag{74}$$

$$J = \int_0^L \delta w_{,x}^2 \, dx$$

5.5   THE GALERKIN METHOD

The Galerkin method belongs in the class of approximate techniques for solving partial and ordinary differential equations. It was introduced by B. G. Galerkin (1941) in the study of rods and plates, and it has been extensively used ever since by many investigators not only of problems in solid mechanics but also in fluid mechanics, heat transfer, and other fields. Finlayson and Scriven (1966) give an extensive bibliography on the uses of the Galerkin method. In addition, they unify this method with other approximate techniques under the name of the Method of Weighted Residuals (MWR).

Before outlining and applying the method to a number of problems, we must state that the method is not necessarily restricted to problems for which the differential equations are Euler-Lagrange equations (derived from stationary principles), and thus, this method is more general than the Rayleigh-Ritz technique. When dealing with variational problems, the Galerkin and Ritz methods are closely related and under certain conditions completely equivalent (Singer, 1962).

The basic idea of the method is as follows: Suppose we are required to solve the differential equation

$$\mathcal{L}(u) = 0 \quad 0 \le x \le L \tag{75}$$

where $\mathcal{L}$ is a differential operator, operating on u, which is a function of a single independent variable x, subject to some boundary conditions. We seek an approximate solution, u_{appr}, in the form

$$U_{\text{appr}} = \sum_{i=1}^{N} a_i f_i(x) \tag{76}$$

where $f_i(x)$ are a certain sequence of functions, each of which satisfies all of the boundary conditions, but none of them, as a rule, satisfy the differential equation, and a_i are undetermined coefficients. We can consider the functions to be elements of a complete sequence. If the exact solution to the differential equation, Eq. (75), is denoted by $\bar{u}(x)$, then the operator, $\mathcal{L}$, operating on the difference $(u_{\text{appr}} - \bar{u})$ represents some kind of error or residual, $e(x)$,

$$e(x) = \mathcal{L}\left(u_{\text{appr}} - \bar{u}\right) = \mathcal{L}\left(u_{\text{appr}}\right) - \mathcal{L}(\bar{u}) = \mathcal{L}\left(u_{\text{appr}}\right) \tag{77}$$

If we substitute the series, Eq. (76), for u_{appr}, we have

$$e(x) = \mathcal{L}\left(\sum_{i=1}^{N} a_i f_i(x)\right) \tag{78}$$

Next we must choose the undetermined coefficients, a_i, such that the error is a minimum. To this end, we make the error orthogonal, in the interval $0 \le x \le L$, to some weighting functions. In the Galerkin method the weighting functions are the functions used in the series, $f_k(x)$, $k = 1, 2, \ldots, N$. This process leads to N integrals, called the Galerkin integrals

$$\int_0^L \left[\mathcal{L}\left(\sum_{i=1}^{N} a_i f_i(x)\right) f_k(x) \right] dx = 0 \quad k = 1, 2, \ldots, N \tag{79}$$

After performing the indicated operations, we have a system of N equations in N unknowns, a_i. The solution of this system is substituted into Eq. (76) to give the approximate solution to the problem. We obtain successive approximations by increasing N, and this gives us some idea about the convergence to the exact solution.

A number of questions and comments have been raised concerning choice of functions, convergence, and other particulars of the method. First, the choice of functions is not restricted by any means, but experience shows that, if the functions are elements of a complete sequence, convergence is improved. Furthermore, which complete sequence must be used depends on the particular problem. When there are certain symmetries to be satisfied, if the functions are so chosen beforehand, it eliminates a lot of unnecessary work. As far as the boundary conditions are concerned, the method, as originally developed and applied by Galerkin, requires that the chosen functions satisfy all of the boundary conditions. This requirement can be relaxed, as will be shown in Section 5.5.1. This can easily be done for variational problems (when the differential equation is an Euler-Lagrange equation), but it presents difficulties in all other problems.

In variational problems, we know precisely which boundary residuals or errors must be added and which must be subtracted from the Galerkin integral in order to relax the method. In nonvariational problems, the adding or subtracting of the boundary errors is based on mathematical convenience or the physics of the problem, and extreme care is required.

Second, convergence of the method has been and still is the subject of study for many mathematicians. Whenever the Galerkin and the Rayleigh-Ritz methods are equivalent, the convergence requirements and proofs for the Rayleigh-Ritz method imply convergence for the Galerkin method.

When the method is used in eigen-boundary-value problems, the Galerkin integrals lead to a system of N homogeneous algebraic equations in a_i. The requirement for a nontrivial solution leads to the vanishing of the determinant of the coefficients of the a_i, which is the characteristic equation.

5.5.1 THE METHOD DERIVED FROM STATIONARY PRINCIPLES

Although the Galerkin method may be used on all initial and boundary value problems, in the special case where it is applied to variational problems, it can be derived directly from the principle of the stationary value of the total potential. This is the case for all conservative problems of elastostatics.

To demonstrate this, consider the beam-column problem, Fig. 3.2, treated in Chapter 3. Let us start with Eq. (15) of Chapter 3. For convenience, let us eliminate the inplane component of deformation, through the use of the inplane equilibrium equation ($P_{,x} = 0$, which implies that $P =$ constant, and from the boundary conditions $P = \overline{P}$). With this, Eq. (15) becomes

$$\int_0^L \left[\left(EIw_{,xx}\right)_{,xx} - \overline{P}w_{,xx} - q(x) - \sum_{i=1}^n P_i\delta(x - x_i) + \sum_{j=1}^m C_j\eta\left(x - x_j\right) \right]\delta w dx$$

$$+ \left\{ \left[-\left(EIw_{,xx}\right)_{,x} + \overline{P}w_{,x} \right]_{x=L} - R_L \right\}\delta w(L)$$

$$- \left\{ \left[-\left(EIw_{,xx}\right)_{,x} + \overline{P}w_{,x} \right]_{x=0} - R_0 \right\}\delta w(0) + \left[\left(EIw_{,xx}\right)_{x=L} - \overline{M}_L \right]\delta w_{,x}(L)$$

$$- \left[\left(EIw_{,xx}\right)_{x=0}\overline{M}_0 \right]\delta w_{,x}(0) 0 \tag{80}$$

where δw denotes a virtual displacement.

From Eq. (80) we obtain the Euler-Lagrange equation and the associated boundary conditions.

D.E.

$$\left(EIw_{,xx}\right)_{,xx} - \overline{P}w_{,xx} - q(x) - \sum_{i=1}^n P_i\delta(x - x_i) + \sum_{j=1}^m C_j\eta\left(x - x_j\right) = 0 \tag{81}$$

Boundary Conditions

1. At $x = 0$

$$\begin{array}{cc} Either & or \\ -\left(EIw_{,xx}\right)_{,x}\overline{P}w_{,x} = R_0 & \delta w = 0 \\ EIw_{,xx} = \overline{M}_0 & \delta w_{,x} = 0 \end{array}$$

2. At $x = L$

$$\text{Either} \qquad\qquad or$$

$$-\left(EIw_{,xx}\right)_{,x} + \overline{P}w_{,x} = \overline{R}_L \qquad\qquad \delta w = 0$$

$$EIw_{,xx} = \overline{M}_L \qquad\qquad \delta w_{,x} = 0$$

Now let us suppose that for a given set of loads, $q(x)$, P_i, C_j, $\overline{P}$, we want to find the solution to the problem by employing Galerkin's method. We represent $w(x)$ by the series

$$w(x) = \sum_{m=1}^{N} a_m\, f_m(x) \tag{82}$$

where $f_m(x)$ satisfy all of the boundary conditions regardless of whether they are kinematic or natural. Then, the Galerkin integrals are

$$\int_0^L \left[\left(EI \sum_{m=1}^{N} a_m f_{m,xx} \right)_{,xx} - \overline{P} \sum_{m=1}^{N} a_m\, f_{m,zz} - q(x) - \sum_{i=1}^{n} P_i \delta(x - x_i) \right. $$
$$\left. + \sum_{j=1}^{m} C_j \eta\left(x - x_j\right) \right] f_k dx = 0 \quad k = 1, 2, 3, \ldots, N \tag{83}$$

These are N linear algebraic equations in a_m $(m = 1, 2, \ldots, N)$. We solve this system of equations for a_m, and we have the approximate solution by substituting these expressions for a_m into Eq. (82).

Another way of looking at the procedure is to directly associate it with Eq. (80). If the series representation for $w(x)$, Eq. (82), is substituted into Eq. (80), and if δw is taken to be $\delta a_k f_k(x)$, then we arrive at the same integrals as those given by Eqs. (83). Note that all the boundary terms vanish, and $\delta a_k \neq 0$ is taken outside the integral.

Next, suppose that the functions $f_m(x)$ in the series expressions for $w(x)$ satisfy only the kinematic boundary conditions. If we substitute the series into Eq. (80) we obtain

$$\delta a_k \int_0^L \left[\left(EI \sum_{m=1}^{N} a_m\, f_{m,xx} \right)_{,xx} - \overline{P} \sum_{m=1}^{N} a_m\, f_{m,xx} - q(x) - \sum_{i=1}^{n} P_i \delta(x - x_i) \right. $$
$$\left. + \sum_{j=1}^{m} C_j \eta\left(x - x_j\right) \right] f_k dx + \left\{ \left[-\left(EI \sum_{m=1}^{N} a_m\, f_{m,xx} \right)_{,x} \right. \right.$$
$$\left. \left. + \overline{P} \sum_{m=1}^{N} a_m\, f_{m,x} \right]_{x=L} - \overline{R}_L \right\} \delta a_k f_k(L) - \left\{ \left[-EI \sum_{m=1}^{N} a_m\, f_{m,xx} \right)_{,x} \right.\right.$$
$$\left. \left. + \overline{P} \sum_{m=1}^{N} a_m\, f_{m,x} \right]_{x=0} - \overline{R}_0 \right\} \delta a_k\, f_k(0) + \left[\left(EI \sum_{m=1}^{N} a_m\, f_{m,xx} \right)_{x=L} \right.$$
$$\left. - \overline{M}_L \right] \delta a_k\, f_{k,k}(L) - \left[\left(EI \sum_{m=1}^{N} a_m\, f_{m,xx} \right)_{x=0} - \overline{M}_0 \right] \delta a_k\, f_{k,x}(0) = 0 \tag{84}$$
$$k = 1, 2, \ldots, N$$

As before, since $\delta a_k \neq 0$, Eqs. (84) represent a system of N linear algebraic equations in $a_m, m = 1, 2, \ldots, N$. The solution yields a_m and, therefore, the approximate expression for $w(x)$.

Note that in this modification of the Galerkin method we have added the boundary errors or residuals to the original Galerkin integrals. In addition, since we have related the method to the principle of the stationary value of the total potential, the functions $f_m(x)$ must be kinematically admissible. Among other requirements (see Appendix A), they must satisfy the kinematic boundary conditions.

5.5.2 THE CLAMPED-FREE COLUMN

Consider a column of length L, which is clamped at $x = 0$ and free at $x = L$. This column is of uniform flexural stiffness, EI, and is loaded with axial load $\overline{P}$. We will use the Galerkin method to find P_{cr}. Let

$$w = \sum_{n=1}^{3} a_n x^{n+1}$$

Since the functions $x^{n+1} (n = 1, 2, 3)$ satisfy only the kinematic boundary conditions (at $x = 0$) and not the natural boundary conditions (at $x = L$), we will use the modified Galerkin method. Substitution of the above expression for $w(x)$ into Eq. (84) yields the following system of three homogeneous alebraic equations in a_n, $n = 1, 2, 3$. Wherever $\overline{P}$ appears, we must use $-\overline{P}$ because of the sign convention.

$$\int_0^L \left[EI24a_3 + \overline{P}(2a_1 + 6a_2 x + 12a_3 x^2) \right] \begin{Bmatrix} x^2 \\ x^3 \\ x^4 \end{Bmatrix} dx + \left[-EI(6a_2 + 24a_3 L) \right.$$

$$\left. \overline{P}(2a_1 L + 3a_2 L^2 + 4a_3 L^3) \right] \begin{Bmatrix} L^2 \\ L^3 \\ L^4 \end{Bmatrix} + EI(2a_1 + 6a_2 L + 12a_3 L^2) \begin{Bmatrix} 2L \\ 3L^2 \\ 4L^3 \end{Bmatrix} = 0 \tag{85}$$

If we perform the indicated operations, divide through by EI, and introduce the load parameter $\lambda = \overline{P}L^2/EI$, we obtain the following three equations

$$4\left(1 - \frac{1}{3}\lambda\right)a_1 + 3L\left(2 - \frac{1}{2}\lambda\right)a_2 + 8L^2\left(1 - \frac{1}{5}\lambda\right)a_3 = 0$$

$$3\left(2 - \frac{1}{2}\lambda\right)a_1 + 3L\left(4 - \frac{3}{5}\lambda\right)a_2 + 2L^2(9 - \lambda)a_3 = 0 \tag{86}$$

$$8\left(1 - \frac{1}{5}\lambda\right)a_1 + 2L(9 - \lambda)a_2 + 16L^2\left(\frac{9}{5} - \frac{1}{7}\lambda\right)a_3 = 0$$

These equations are similar to Eqs. (30) obtained by the Rayleigh-Timoshenko or Rayleigh-Ritz method, and the solution is identical to the one obtained, above, namely

$$\text{one-term solution} \quad \lambda_{cr} = 3$$
$$\text{two-term solution} \quad \lambda_{cr} = 2.4860$$
$$\text{three-term solution} \quad \lambda_{cr} = 2.4677$$

5.6   SOME COMMENTS ON KOITER'S THEORY

5.6.1   CRITICAL LOAD AND LOAD-CARRYING CAPACITY

As mentioned in Chapter 1, the interest in the stability of simple structural elements or overall structural configurations under external causes lies in the fact that the stability limit (critical load) in many cases forms the basic criterion for design. Because of safety reasons in designing structural configurations, the level of the external causes is usually kept at such a value that the load in the structural configurations is smaller than the critical load or condition. This line of thinking might suggest that there is no reason to concern ourselves with studies of how the structural configuration behaves past this critical condition, because the critical condition is directly associated with the load-carrying capacity of the structural configuration. It has been known since the last century, though, that certain structural configurations (the rectangular, simply supported plate under uniform edge compression, the simply supported or clamped plate under uniform radial compression around its circumference, and others) can carry loads higher than the first buckling load (and still behave elastically in many cases). This fact has been verified experimentally. In a number of other cases it has been demonstrated experimentally (thin spherical shells under uniform compression and thin circular cylindrical shells under uniform compression) that the buckling load is only a small fraction of the critical load predicted by the mathematical model, based on either the equilibrium approach or the energy criterion. Many attempts have been made to explain this discrepancy, and it is beyond the scope of this text to go into such explanations. What is important, though, is the observation that critical loads, derived on the basis that the primary path becomes unstable when a bifurcation point exists, cannot be directly associated with the load-carrying capacity of such structural configurations. Note that, at a point of bifurcation, the branch that characterizes the adjacent equilibrium positions can be stable or unstable (see Figs. 1.4 and 1.5). Therefore, in the interest of designing a safe structure, we must know how the structural configuration behaves past the critical load or condition. The first person to systematically develop a stability theory that deals with the question of post-buckling behavior of continuous elastic systems is Koiter (1945) in his famous Ph.D. Thesis.

His theory is an initial post-buckling analysis, and therefore it cannot possibly provide all the required answers in relating the critical load to the load-carrying capacity of the structural configuration. Although it has this limitation, it is an important first step toward achieving the true solution. These points will be discussed in more detail with general qualitative demonstrations in Section 5.6.2.

The initial post-buckling behavior of elastic systems has received the deserved attention only in the past fifteen years. An excellent review article on the subject was presented by Hutchinson and Koiter (1970).

5.6.2   CONCLUSIONS BASED ON KOITER'S THEORY

Koiter's theory, as mentioned previously, is primarily an initial post-buckling theory, and it is applicable to problems exhibiting bifurcational buckling only. In addition, the theory as presented in Koiter (1945) is limited to linearly elastic behavior. It first concerns itself with the investigation of the equilibrium position in the neighborhood of the buckling load. The most important conclusion of this investigation is that the stability or instability of these equilibrium positions is governed by the stability of equilibrium at the bifurcation point. If the equilibrium is stable at the critical load (bifurcation point), neighboring positions of equilibrium can exist only for loads greater than the critical load, and these positions are stable (see Fig. 2.12a). If the equilibrium position at the critical load is unstable, neighboring equilibrium positions do exist at loads smaller than the critical load and they are unstable (see Figs. 2.12b and 2.12c). It is true that, in this particular case, it is possible for stable equilibrium positions also to exist at loads greater than the critical load (see Fig. 2.12c), but these positions can be reached only by passing through the unstable critical position, and therefore their practical significance is, to say the least, doubtful.

Another important ingredient of Koiter's theory is the investigation of the influence of small imperfections in the actual structure in comparison to the idealized perfect model. The most important conclusion of this part of Koiter's work is that, if the equilibrium position at the critical load of the perfect model is unstable (Figs. 2.12b and 2.12c), the critical load of the structural configuration may be considerably smaller than that of the idealized perfect model because of the presence of small imperfections (see Fig. 2.11).

Stein (1968), in a review paper, suggests that the Koiter theory can serve to assess the imperfection sensitivity of a structural configuration through the slope of the post-buckling curve in the neighborhood of the critical load. In a plot of normalized load versus normalized characteristic displacement, Fig. 5.5, the pre-buckling curve for linear theory is a 45° straight line. If we introduce an angle φ between the horizontal and the tangent to the post-buckling curve (positive in a counterclockwise direction), then it is clear that if $0 < \varphi < \pi/4$, the structure is imperfection insensitive, while if $-3\pi/4 < \varphi < 0$, the structure is imperfection sensitive. Two questions arise for this latter case. First, how large must the negative angle φ be to distinguish the cases of small versus large effects of imperfection sensitivity? Second, by considering the tangent to the post-buckling curve at the critical load, do we have assurance that the effect of imperfection sensitivity is considerably large?

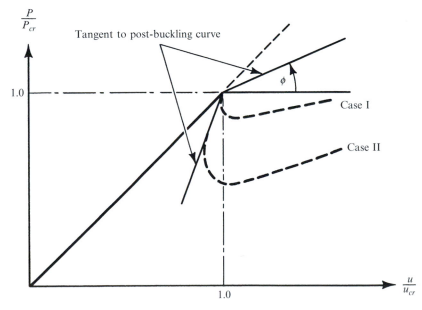

FIGURE 5.5   Initial post-buckling behavior.

Suppose that there is a possibility that two structural configurations are characterized by the same tangent at the critical load but their behavior differs considerably as we move away from the critical load (see Fig. 5.5, cases I and II). These questions and others with the assessment of the effect of imperfection sensitivity are the subject of much research. Again the reader is strongly advised to read Hutchinson and Koiter (1970).

PROBLEMS

In all of these problems, use one of the approximate methods discussed in this chapter.

1. Find P_{cr} for a column which is fixed at one end and simply supported at the other.
2. A continuous column of constant flexural stiffness and total length $3L$ is supported as shown. Find P_{cr}.

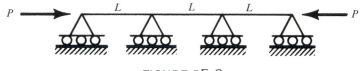

FIGURE P5.2

3. A column of constant flexural stiffness, EI, and length, L, is fixed at one end and supported elastically against translation (linear spring) at the other. If the column is loaded axially by P, find P_{cr}.
4. A cantilever column of constant flexural stiffness, EI, and length, L, is in an upright position with the fixed end at the lower part. If the direction of gravity is in line with the column, determine the critical weight if the column is to buckle under its own weight.
5. A simply supported column of length $3L$ is under the action of a compressive load. Find P_{cr} if the flexural stiffness varies according to

$$EI(x) = \begin{cases} EI_0 & 0 \leq x \leq L \\ 2EI_0 & L \leq x \leq 2L \\ EI_0 & 2L \leq x \leq 3L \end{cases}$$

6. A cantilever column of length $2L$ is fixed at $x = 0$ and loaded at the free end by a compressive load P. Find P_{cr} if the flexural stiffness varies according to

$$EI(x) = \begin{cases} 2EI_0 & 0 \leq x \leq L \\ EI_0 & L \leq x \leq 2L \end{cases}$$

7. A cantilever column, with the clamped end at $x = L$ is under a compressive load P (at the free end, $x = 0$). Find P_{cr} if the stiffness varies according to $EI(x) = EI_0(1 + I_1 x / I_0)$.

REFERENCES

Finlayson, B. A. and Scriven L. E. (1966). The method of weighted residuals—a review, *Appl. Mech. Rev.*, Vol. 19, No. 9, pp. 735–748.

Galerkin, B. G. (1941). "Sterzhnei i plastiny. Ryady V Nekotorykh Voprosakh Uprogogo Ravnoresiya Sterzhnei i plastiny," (Rods and Plates. Series Occuring in Some Problems of Elastic Equilibrium of Rods and Plates), *Vesin. Inzhen. i Tekh. Petrograd*, Vol. 19, pp. 897–908, 1915; English translation 63–18924 Clearinghouse Fed. Sci. Tech. Info. See also "On the Seventieth Anniversary of the Birth of B. G. Galerkin," *PMM*, Vol. 5, pp. 337–341.

Hutchinson, J. W. and Koiter, W. T. (1970). Postbuckling theory, *Appl. Mech. Rev.*, Vol. 13, pp. 1353–1366.

Koiter, W. T. (1945). "Elastic Stability and Postbuckling Behavior" in *Non-linear Problems*, edited by R. E. Langer, University of Wisconsin Press, Madison, 1963. Also "The Stability of Elastic Equilibrium," Thesis Delft (English translation, NASA TT-F-10833, 1967, and AFFDL TR-70-25, 1970).

Prager, W. and Taylor, J. E. (1968). Problem of optimal structural design, *J. Appl. Mech.*, Vol. 35, pp. 102–106.

Rayleigh, J. W. S. (1945). *Theory of Sound*. Dover Publications, New York.

Sagan, H. (1969). *Introduction to the Calculus of Variations*. McGraw-Hill Book Co., New York.

Simitses, G. J., Kamat, M. P. and Smith, C. V. Jr. (1972). "The Strongest Column by the Finite Element Displacement Method." AIAA Paper No. 72–141.

Singer, J. (1962). On the equivalence of the galerkin and Rayleigh-Ritz methods, *Journal of Royal Aeronautical Society*, Vol. 66, pp. 592–597.

Stein, M. (1968). "Recent Advances in Shell Buckling." AIAA Paper No. 68–103.

Tadjbakhsh, I. and Keller, J. B. 1962. Strongest columns and isoperimetric inequalities for Eigen values, *J. Appl. Mech.*, Vol. 29, pp. 159–164.

Timoshenko, S. P. and Gere, J. M. (1956). *Theory of Elastic Stability*. McGraw-Hill Book Co., New York.

Trefitz, E. (1933). "Zur Theorie der Stabilität des Elastichen Gleichogewihts." *Z. Angew. Math. Mech.*, Vol. 13, pp. 160–165.

6

COLUMNS ON ELASTIC
FOUNDATIONS

6.1   BASIC CONSIDERATIONS

Beams and columns supported elastically along their lengths are widely found in structural configurations. In some cases, the elastic support, called the elastic foundation, is provided by a medium which is indeed the foundation supporting the beams or columns such as in railroad tracks, in underground piping for different uses, and in footings for large-scale structures. In other cases, the elastic support is provided by adjacent elastic structural elements such as in stiffened plate and shell configurations. Regardless of the particular application, the mathematical model consists of a column supported in some manner at its ends and with a continuous distribution of springs of stiffness, $\bar{\beta}$, called the modulus of the foundation (see Fig. 6.1). The units of $\bar{\beta}$ are pounds per inch (force per length squared) and may be a constant or at most a function of position along the length of the column for linear spring behavior. In general, the spring behavior may be taken as nonlinear.

This chapter will present the analysis of some simple models and provide insight into the behavior of such columns under destabilizing compressive loads. An excellent and comprehensive treatment of the subject may be found in the text by Hetényi (1946).

To derive the buckling equations for a column on an elastic foundation, we refer to Chapter 3 and modify the expression for the total potential to include the energy stored into the foundation, U_f. For linear spring behavior

$$U_f = \frac{1}{2} \int_0^L \bar{\beta} w^2 dx \tag{1}$$

Therefore, if we use the principle of the stationary value of the total potential

$$\delta_\varepsilon U_T = \delta_\varepsilon U_i + \delta_\varepsilon U_f + \delta_\varepsilon U_p = 0 \tag{2}$$

we can derive the equilibrium equations for this configuration. Since $\bar{P} = $ constant, the buckling equation is

---

173

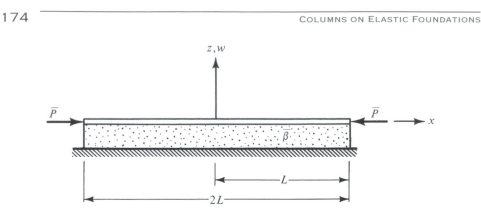

FIGURE 6.1   Column resting on an elastic foundation.

$$(EIw_{,xx})_{,xx} + \overline{P}w_{,xx} + \overline{\beta}w = 0 \tag{3}$$

The boundary conditions are not affected by the presence of the foundation.

6.2   THE PINNED-PINNED COLUMN

Consider a column of length, L, and constant flexural stiffness, EI, pinned at both ends and resting on an elastic foundation (see Fig. 6.2). Let the modulus of the foundation, $\overline{\beta}$, be a constant. The mathematical formulation of the problem is given by

$$\text{D.E.} \quad w_{,xxxx} + k^2 w_{,xx} + \left(\frac{\pi}{L}\right)^4 \beta w = 0 \tag{4a}$$

$$\begin{array}{lll} \text{B.C.'s} & w(0) = 0 & w(L) = 0 \\ & w_{,xx}(0) = 0 & w_{,xx}(L) = 0 \end{array} \tag{4b}$$

where

$$k^2 = \frac{\overline{P}}{EI} \quad \text{and} \quad \beta = \frac{\overline{\beta}L^4}{\pi^4 EI}$$

Thus, the problem has been reduced to an eigen-boundary-value problem and we are seeking the smallest value of k (P_{cr}) for which a nontrivial solution exists provided that β is fixed.

Since $\sin m\pi x/L$ satisfies all boundary conditions for all m, the solution to the buckling equation is taken in the form

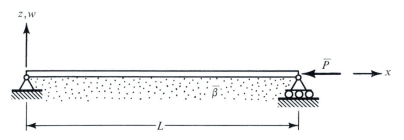

FIGURE 6.2   The pinned-pinned column.

$$w(x) = \sum_{m=1}^{\infty} A_m \sin\frac{m\pi x}{L} \tag{5}$$

where it is noted that m odd corresponds to symmetric modes while m even corresponds to antisymmetric ones.

Substitution into the differential equations results in the following characteristic equation:

$$\left(\frac{m\pi}{L}\right)^4 - k_m^2\left(\frac{m\pi}{L}\right)^2 + \beta\left(\frac{\pi}{L}\right)^4 = 0 \tag{6}$$

where k_m denote the eigenvalues.

From Eq. (6) we obtain

$$k_m^2\left(\frac{L}{\pi}\right)^2 = \frac{P_m}{P_E} = m^2 + \frac{\beta}{m^2} \tag{7}$$

where

$$P_E = \frac{\pi^2 EI}{L^2}$$

The critical load for a fixed value of the modulus of the foundation, β, is the smallest of P_m. We see from Eq. (7) that $\overline{P}_{cr}$ and the corresponding deformation mode depend on the value of β. Thus, $\overline{P}_{cr}$ is obtained from a plot of P_m/P_E versus β (see Fig. 6.3). As shown in the plot, $\overline{P}_{cr}$ is denoted by the solid piecewise linear curve and

$$\overline{P}_{cr} = P_1 \qquad 0 \le \beta \le 4$$
$$\overline{P}_{cr} = P_2 \qquad 4 \le \beta \le 36$$
$$\text{etc.}$$

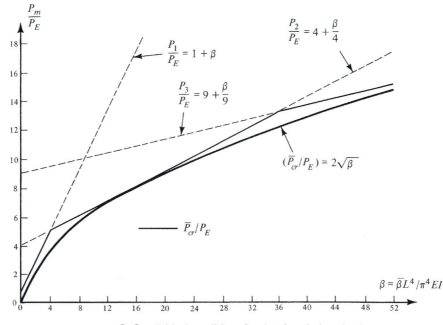

FIGURE 6.3 Critical conditions for the pinned-pinned column.

In general, the value of β at which the deformation mode changes from k half-sine waves to $(k + 1)$, and the critical load from P_k to P_{k+1}, is given by

$$\beta = [k(k + 1)]^2 \tag{8}$$

Finally, if we consider m^2 to be a continuous variable, minimization of P_m with respect to m^2 yields

$$m_{cr}^2 = \sqrt{\beta} \tag{9}$$

Substitution of this expression into Eq. (7) gives us the expression for the critical load:

$$\overline{P}_{cr} = 2P_E\sqrt{\beta} = 2\sqrt{\beta EI} \tag{10}$$

This expression is also plotted in Fig. 6-3. We notice that this form of the solution is approximate, and the approximation becomes more accurate as β increases.

Note that Eq. (10) is a solution for infinitely long columns, so that the boundary conditions do not matter.

6.3   RAYLEIGH-RITZ SOLUTION

Using Timoshenko's arguments about energy and work, we can derive a Rayleigh quotient for this problem:

$$\overline{P} = \frac{\int_0^L EIw_{,xx}^2 dx + \int_0^L \overline{\beta}w^2 dx}{\int_0^L w_{,x}^2 dx} \tag{11}$$

The solution may be expressed in terms of the following series for the pinned-pinned column:

$$w(x) = \sum_{m=1}^{N} A_m \sin\frac{m\pi x}{L} \tag{12}$$

When this expression is used in Eq. (11), the numerator and denominator become functions of A_m:

$$\overline{P} = \frac{f_1(A_m)}{f_2(A_m)} \tag{13}$$

Next, we adjust the coefficient A_m such that P_m is a minimum. This leads to the equation

$$\frac{\partial f_1}{\partial A_m} - \overline{P}\frac{\partial f_2}{\partial A_m} = 0 \quad m = 1, 2, \ldots, N \tag{14}$$

which becomes

$$L\left[EI\left(\frac{m\pi}{L}\right)^4 + \overline{\beta} - \overline{P}\left(\frac{m\pi}{L}\right)^2\right]A_m = 0 \quad m = 1, 2, \ldots, N \tag{15}$$

Thus, the introduction of P_E and β yields Eq. (7), and from this point on the arguments are the same as in Section 6.2.

6.4   THE GENERAL CASE

In this section we shall outline a procedure which may, in general, be applied to columns on an elastic foundation regardless of the boundary conditions. Consider a column of constant flexural stiffness and length $2L$ as shown in Fig. 6.1. Place the origin of the coordinate system used at the midpoint of the column.

The mathematical formulation of the general problem is given by

D.E.
$$w,_{xxxx} + \frac{\overline{P}}{EI} w,_{xx} + \frac{\overline{\beta}}{EI} w = 0$$

B.C.s   *Either* (Kinematic)   *Or* (Natural)

$$w = 0 \text{ at } x = \pm L \qquad w,_{xxx} + \frac{\overline{P}}{EI} w,_x = 0 \text{ at } x = \pm L$$

$$w,_x = 0 \text{ at } x = \pm L \qquad w,_{xx} = 0 \text{ at } x = \pm L$$

(16)

Note that the clamped-clamped problem is characterized by the satisfaction of the kinematic boundary conditions, whereas the free-free problem (floating ends) is characterized by the satisfaction of the natural boundary conditions, etc.

The general procedure is as follows. Let the solution to Eq. (16) be of the form e^{sx}. Then, substitution into Eq. (16) yields

$$s^4 + \frac{\overline{P}}{EI} s^2 + \frac{\overline{\beta}}{EI} = 0 \tag{17}$$

The solution for s is given by

$$s^2 = \frac{1}{2} \left[-\frac{\overline{P}}{EI} \pm \sqrt{\left(\frac{\overline{P}}{EI}\right)^2 - 4\frac{\overline{\beta}}{EI}} \right]$$

$$= \left(\frac{\overline{\beta}}{EI}\right)^{\frac{1}{2}} \left[-\frac{\overline{P}}{2\sqrt{\overline{\beta}EI}} \pm \sqrt{\left(\frac{\overline{P}}{2\sqrt{\overline{\beta}EI}}\right)^2 - 1} \right]$$

Denoting the expression $2\sqrt{\overline{\beta}EI}$ by P_1, which represents $\overline{P}_{cr}$ for large values of β (not $\overline{\beta}$) or extremely long columns with simply supported boundaries, then

$$s^2 = \left(\frac{\overline{\beta}}{EI}\right)^{\frac{1}{2}} \left[-\frac{\overline{P}}{P_1} \pm \sqrt{\left(\frac{\overline{P}}{P_1}\right)^2 - 1} \right] \tag{18}$$

Furthermore, if we let $\overline{P}/P_1 = \gamma$, the four roots of Eq. (17) are

$$s_1 = i\left(\frac{\overline{\beta}}{EI}\right)^{\frac{1}{4}} \left(\gamma - \sqrt{\gamma^2 - 1}\right)^{\frac{1}{2}}$$

$$s_2 = -s_1$$

$$s_3 = i\left(\frac{\overline{\beta}}{EI}\right)^{\frac{1}{4}} \left(\gamma + \sqrt{\gamma^2 - 1}\right)^{\frac{1}{2}}$$

$$s_4 = -s_3$$

(19)

Regardless of the boundary conditions, the following three cases must be considered.

Case I:

$$\gamma > 1$$

For this case, let us first introduce the real and positive parameters k_1 and k_2:

$$k_1 = \left(\frac{\bar{\beta}}{EI}\right)^{\frac{1}{4}} \left(\gamma - \sqrt{\gamma^2 - 1}\right)^{\frac{1}{2}}$$

$$k_2 = \left(\frac{\bar{\beta}}{EI}\right)^{\frac{1}{4}} \left(\gamma + \sqrt{\gamma^2 - 1}\right)^{\frac{1}{2}} \tag{20}$$

Then the four roots and the solution to Eq. (16) become

$$s_1 = ik_1 \quad s_2 = -ik_1 \quad s_3 = ik_2 \quad s_4 = -ik_2$$

and

$$w(x) = C_{11} \cos k_1 x + C_{12} \sin k_1 x + C_{13} \cos k_2 x + C_{14} \sin k_2 x \tag{21}$$

Case II:

$$\gamma = 1$$

For this case, we first introduce the real positive parameter k_3:

$$k_3 = \left(\frac{\bar{\beta}}{EI}\right)^{\frac{1}{4}} \tag{22}$$

Then, the four roots and the solution to Eq. (16) become

$$s_1 = ik_3 \quad s_2 = -ik_3 \quad s_3 = ik_3 \quad s_4 = -ik_3$$

and

$$w(x) = C_{21} \cos k_3 x + C_{22} \sin k_3 x + C_{23} x \cos k_3 x + C_{24} x \sin k_3 x \tag{23}$$

Note that we have two pairs of double roots for this case.

Case III:

$$\gamma < 1$$

Since γ is smaller than 1, then the four roots, Eqs. (19), are

$$s_1 = i\left(\frac{\bar{\beta}}{EI}\right)^{\frac{1}{4}} \left(\gamma - i\sqrt{1 - \gamma^2}\right)^{\frac{1}{2}}$$

$$s_2 = -s_1$$

$$s_3 = i\left(\frac{\bar{\beta}}{EI}\right)^{\frac{1}{4}} \left(\gamma + i\sqrt{1 - \gamma^2}\right)^{\frac{1}{2}}$$

$$s_4 = -s_3$$

If we take the square root of the complex number and introduce the real positive quantities

$$\rho = \left(\frac{\bar{\beta}}{EI}\right)^{\frac{1}{4}} \sqrt{\frac{1 - \gamma}{2}}$$

$$r = \left(\frac{\bar{\beta}}{EI}\right)^{\frac{1}{4}} \sqrt{\frac{1 + \gamma}{2}} \tag{24a}$$

the four roots become

$$s_1 = \rho + ir = \eta$$
$$s_2 = -\rho - ir = -\eta$$
$$s_3 = -\rho + ir = \omega \qquad \text{(24b)}$$
$$s_4 = \rho - ir = -\omega$$

Note that we have two complex conjugate pairs $s_4 = \bar{s}_1$ and $s_2 = \bar{s}_3$.

The solution to Eq. (16) for this case is

$$w(x) = A_1 e^{(\rho+ir)x} + A_2 e^{-(\rho+ir)x} + A_3 e^{-(-\rho+ir)x} + A_4 e^{(\rho-ir)x}$$

or

$$w(x) = C_{31} \cosh \eta x + C_{32} \cosh \omega x + C_{33} \sinh \eta x + C_{34} \sinh \omega x \qquad \text{(25)}$$

6.4.1 THE CLAMPED-CLAMPED COLUMN

To find $\bar{P}_{cr}$ for this particular problem, we must investigate all three cases by using the proper boundary conditions with the three corresponding solutions, Eqs. (21), (23), and (25). Note that in all three cases, when the boundary conditions are used, we end up with a system of four linear homogeneous algebraic equations in the constants C_{ij} $(i = 1, 2, 3; j = 1, 2, 3, 4)$. The first subscript, i, is associated with the particular case (I, II, and III) and the second subscript, j, with the four roots.

The boundary conditions for the clamped-clamped case are

$$w(-L) = w(L) = 0$$
$$w_{,x}(-L) = w_{,x}(L) = 0 \qquad \text{(26)}$$

Case III:

$$\gamma < 1$$

Employing the boundary conditions, Eqs. (26), and the expression for $w(x)$, Eq. (25), we obtain the following system of equations:

$$C_{31} \cosh \eta L + C_{32} \cosh \omega L \pm C_{33} \sinh \eta L \pm C_{34} \sinh \omega L = 0$$
$$\pm C_{31} \eta \sinh \eta L \pm C_{32} \omega \sinh \omega L + C_{33} \eta \cosh \eta L + C_{34} \omega \cosh \omega L = 0 \qquad \text{(27)}$$

If we add and subtract the first two equations and the last two equations, we have an equivalent system of four linear homogeneous algebraic equations in C_{3j} $(j = 1, 2, 3, \text{ and } 4)$:

$$C_{31} \cosh \eta L + C_{32} \cosh \omega L = 0$$
$$C_{33} \sinh \eta L + C_{34} \sinh \omega L = 0$$
$$C_{31} \eta \sinh \eta L + C_{32} \omega \sinh \omega L = 0$$
$$C_{33} \eta \cosh \eta L + C_{34} \omega \cosh \omega L = 0$$

Thus, the equations have decomposed into two systems of equations:

$$\left. \begin{array}{l} \textit{Either} \quad C_{31} \cosh \eta L + C_{32} \cosh \omega L = 0 \\ \qquad\quad C_{31} \eta \sinh \eta L + C_{32} \omega \sinh \omega L = 0 \end{array} \right\} \qquad \text{(28)}$$

$$\left.\begin{array}{l} \text{or} \qquad C_{33} \sinh \eta L + C_{34} \sinh \omega L = 0 \\ \qquad C_{33} \eta \cosh \eta L + C_{34} \omega \cosh \omega L = 0 \end{array}\right\} \tag{29}$$

The first system implies that $C_{31} \neq 0$, $C_{32} \neq 0$, and $C_{33} = C_{34} = 0$, which corresponds to a symmetric mode of deformations [see Eq. (25)]. The second system corresponds to an antisymmetric mode of deformation. Both systems must be used for finding $\overline{P}_{cr}$.

For the symmetric case, a nontrivial solution exists if

$$\begin{vmatrix} \cosh \eta L & \cosh \omega L \\ \eta \sinh \eta L & \omega \sinh \omega L \end{vmatrix} = 0 \tag{30}$$

The expansion of the determinant yields

$$\omega \cosh \eta L \ \sinh \omega L - \eta \sinh \eta L \cosh \omega L = 0 \tag{31}$$

This equation may now be written as

$$\omega \left(e^{\eta L} + e^{-\eta L} \right) \left(e^{\omega L} - e^{-\omega L} \right) - \eta \left(e^{\eta L} - e^{-\eta L} \right) \left(e^{\omega L} + e^{-\omega L} \right) = 0$$

If we substitute the expressions for η and ω from Eqs. (24b), we obtain

$$4i(\rho \sin 2rL + r \sinh 2\rho L) = 0 \tag{32a}$$

or

$$\frac{\sin 2rL}{2rL} + \frac{\sinh 2\rho L}{2\rho L} = 0 \tag{32b}$$

This equation has no solution, therefore $C_{31} = C_{32} = 0$ (trivial solution for the system). Similarly, for the antisymmetric case, the characteristic equation requires

$$\frac{\sin 2rL}{2rL} - \frac{\sinh 2\rho L}{2\rho L} = 0 \tag{33}$$

This equation has no solution for γ; therefore, for this case $(\gamma < 1)$, there is no bifurcation point and the only solution is $w(x) \equiv 0$ (straight configuration).

Case II:

$$\gamma = 1$$

If the steps outlined for case III are repeated for this case using Eq. (23) for the displacement, we obtain the following characteristic equations for symmetric buckling $(C_{21} \neq 0, C_{24} \neq 0, C_{22} = C_{23} = 0)$:

$$\sin 2k_3 L = -2k_3 L \tag{34a}$$

and for antisymmetric buckling $(C_{21} = C_{24} = 0, C_{22} \neq 0, C_{23} \neq 0)$:

$$\sin 2k_3 L = 2k_3 L \tag{34b}$$

There is no solution to Eqs. (34); therefore, there is no bifurcation for $\gamma = 1$.

Case I:

Substitution of Eq. (21) into the boundary conditions, Eqs. (26), yields

$$\begin{array}{c} C_{11} \cos k_1 L \pm C_{12} \sin k_1 L + C_{13} \cos k_2 L \pm C_{14} \sin k_2 L = 0 \\ \mp C_{11} k_1 \sin k_1 L + C_{12} k_1 \cos k_1 L \mp C_{13} k_2 \sin k_2 L + C_{14} k \cos k_2 L = 0 \end{array} \tag{35}$$

As in case III, we first obtain an equivalent system of equations through subtraction and addition, which separates the problem into symmetric and antisymmetric buckling.

For symmetric buckling ($C_{11} \neq 0$, $C_{13} \neq 0$, $C_{12} = C_{14} = 0$):

$$C_{11} \cos k_1 L + C_{13} \cos k_2 L = 0$$
$$C_{11} k_1 \sin k_1 L + C_{13} k_2 \sin k_2 L = 0 \tag{36}$$

This leads to the characteristic equation

$$k_2 \sin k_2 L \cos k_1 L = k_1 \sin k_1 L \cos k_2 L$$

or

$$(k_1 L) \tan (k_1 L) = (k_2 L) \tan (k_2 L) \tag{37}$$

For antisymmetric buckling ($C_{11} = C_{13} = 0$, $C_{12} \neq 0$, $C_{14} \neq 0$):

$$C_{12} \sin k_1 L + C_{14} \sin k_2 L = 0$$
$$C_{12} k_1 \cos k_1 L + C_{14} k_2 \cos k_2 L = 0 \tag{38}$$

The characteristic equation is

$$(k_1 L) \cot(k_1 L) = (k_2 L) \cot(k_2 L) \tag{39}$$

To find γ_{cr}, we must solve both Eq. (37) and Eq. (39) for fixed values of $\overline{\beta}$. Heteńyi (1946) presents graphically the solution to the two characteristic equations in a plot of γ versus $4L^2 \sqrt{\beta/EI}$. In the same reference, plots for the pinned-pinned and free-free columns are presented with the same coordinates. For the clamped-clamped and pinned-pinned columns, as $\overline{\beta}$ is increased from zero, the buckling mode changes from symmetric to antisymmetric back to symmetric, etc. For the free-free column, as $\overline{\beta}$ is increased from zero, the buckling mode changes from antisymmetric to symmetric, etc.

6.4.2   THE FREE-FREE COLUMN

The boundary conditions for this particular problem are:

$$w_{,xx}(-L) = w_{,xx}(L) = 0$$
$$w_{,xxx}(-L) + \frac{\overline{P}}{EI} w_{,x}(-L) = w_{,xxx}(L) + \frac{\overline{P}}{EI} w_{,x}(L) = 0 \tag{40}$$

To find $\overline{P}_{cr}$, we must again consider all three cases. It can be shown that no critical load exists for cases I and II ($\gamma > 1$, $\gamma = 1$; see Problem 1 at the end of this chapter). Therefore, if there is a $\overline{P}_{cr}$, the characteristic equation must be found from case III ($\gamma < 1$). Substitution of Eq. (25) into the boundary conditions, Eqs. (40), results in:

$$C_{31} \eta^2 \cosh \eta L + C_{32} \omega^2 \cosh \omega L \mp C_{33} \eta^2 \sinh \eta L \mp C_{34} \omega^2 \sinh \omega L = 0$$

$$\left[\mp C_{31} \eta^3 \sinh \eta L \mp C_{32} \omega^3 \sinh \omega L + C_{33} \eta^3 \cosh \eta L + C_{34} \omega^3 \cosh \omega L \right]$$
$$+ \frac{\overline{P}}{EI} \left[\mp C_{31} \eta \sinh \eta L \mp C_{32} \omega \sinh \omega L + C_{33} \eta \cosh \eta L + C_{34} \omega \cosh \omega L \right] = 0 \tag{41}$$

First, we observe that $\overline{P}/EI = -(\eta^2 + \omega^2)$. This can easily be verified through Eqs. (24) and the expressions for γ and P_1:

$$\eta^2 + \omega^2 = (\rho + ir)^2 + (-\rho + ir)^2$$
$$= 2(\rho^2 - r^2)$$
$$= 2\left(\frac{\overline{\beta}}{EI}\right)^{\frac{1}{2}}\left(\frac{1-\gamma}{2} - \frac{1+\gamma}{2}\right)$$
$$= -2\left(\frac{\overline{\beta}}{EI}\right)^{\frac{1}{2}}\gamma = -2\left(\frac{\overline{\beta}}{EI}\right)^{\frac{1}{2}}\frac{\overline{P}}{2\sqrt{\overline{\beta}EI}}$$
$$= -\frac{\overline{P}}{EI}$$

Next, if we add and subtract the first two and last two of Eqs. (41), we obtain the following two systems of equations:

Symmetric Buckling:

$$C_{31}\eta^2 \cosh \eta L + C_{32}\omega^2 \cosh \omega L = 0$$
$$C_{31}\omega \sinh \eta L + C_{32}\eta \sinh \omega L = 0 \tag{42}$$

Antisymmetric Buckling:

$$C_{33}\eta^2 \sinh \eta L + C_{34}\omega^2 \sinh \omega L = 0$$
$$C_{33}\omega \cosh \eta L + C_{34}\eta \cosh \omega L = 0 \tag{43}$$

The characteristic equations for both cases must be derived and solved for $\overline{P}_{cr}$.

First, for the symmetric buckling case, the characteristic equation is obtained by requiring the system of Eqs. (42) to have a nontrivial solution:

$$\eta^3 \sinh \omega L \cosh \eta L - \omega^3 \sinh \eta L \cosh \omega L = 0 \tag{44}$$

Use of Eqs. (24) gives

$$(\rho + ir)^3 \left[\left(e^{2irL} - e^{-2irL}\right) - \left(e^{2\rho L} - e^{-2\rho L}\right)\right]$$
$$- (\rho - ir)^3 \left[-\left(e^{2irL} - e^{-2irL}\right) - \left(e^{2\rho L} - e^{-2\rho L}\right)\right] = 0 \tag{45}$$

Since the second term of the above equation is the complex conjugate of the first term, and since the difference of two complex conjugate pairs is the imaginary part multiplied by $2i$, the characteristic equation is given by

$$\text{Im}\left[(\rho + ir)^3 \left(e^{2irL} - e^{-2irL} + e^{-2\rho L} - e^{2\rho L}\right)\right] = 0$$

or

$$\text{Im}\left\{\left[\rho^3 - 3\rho r^2 + i\left(3\rho^2 r - r^3\right)\right]\left(-2 \sinh 2\rho L + 2i \sin 2rL\right)\right\} = 0$$

This characteristic equation assumes the following final form

$$\left(3\rho^2 r - r^3\right) \sinh 2\rho L = \left(\rho^3 - 3\rho r^2\right) \sin 2rL \tag{46}$$

Similarly, for the antisymmetric buckling case, the characteristic equation is

$$\left(3\rho^2 r - r^3\right) \sinh 2\rho L = -\left(\rho^3 - 3\rho r^2\right) \sin 2rL \tag{47}$$

Solutions to these equations are presented graphically in Heteńyi (1946).

We see from Eqs. (46) and (47) that, as the length increases, the right-hand side of both equations remains finite. Since $\sinh 2\rho L$ increases indefinitely, the quantity $3\rho^2 r - r^3$ must approach zero. Therefore, as the length approaches infinity

$$r\left(3\rho^2 - r^2\right) = 0$$

Substitution of the expressions for r and ρ from Eq. (24a) gives

$$\gamma = \frac{1}{2}$$

and

$$\bar{P}_{cr} = \sqrt{\beta EI} \tag{48}$$

An important application of columns on an elastic foundation is in the prediction of wrinkling of the facings in a sandwich construction where the core acts as an elastic foundation (Goodier and Hsu, 1954). The mathematical model (buckling equations and boundary conditions) is similar to that used for axisymmetric buckling for thin, circular, cylindrical shells.

PROBLEMS

1. Show that there is no critical load for $\bar{P} \geq 2\sqrt{\beta EI}$ when the ends of the column are free (floating ends). Find the bifurcation loads.
2. Analyze the pinned-pinned column, resting on an elastic foundation, of length $2L$ and uniform flexural stiffness by using the approach outlined in Section 6.4.
3. Find $\bar{P}_{cr}$ for the clamped-clamped column, resting on an elastic foundation, of length L and uniform flexural stiffness, by employing a Rayleigh-Ritz technique and for $0 \leq \beta \leq 64$.
4. Find $\bar{P}_{cr}$ for the clamped-free column, resting on an elastic foundation, of length L and uniform flexural stiffness, by employing the Rayleigh-Ritz technique and for low values of β.
5. Find $\bar{P}_{cr}$ for a pinned-pinned column, resting on an elastic foundation, of length L and flexural stiffness $EI(x) = EI_1 \sin \pi x/L$, by employing some approximate technique and for low values of β.

REFERENCES

Goodier, J. N. and Hsu, C. S. (1954). Nonsinusoidal buckling modes of sandwich plates, *Journal of Aerospace Science*, pp. 525–532, August.
Heteńyi, M. (1946). *Beams on Elastic Foundation*. The University of Michigan Press, Ann Arbor.

7

BUCKLING OF RINGS
AND ARCHES

Thin rings and arches (high or low) are often used as structural elements and, when loaded in their plane and in a normal direction, are subject to instability. In this chapter, the analyses of thin circular rings, high circular arches, and low arches are presented. In addition, the analysis of a low half-sine arch, loaded by a half-sine distributed load, resting on an elastic foundation, and pinned at both ends is presented. This is an interesting model because, depending on the values of the rise parameter and the modulus of the foundation, it exhibits the possibilities of limit point stability (top-of-the-knee buckling), snapthrough buckling through unstable bifurcation, and classical stable bifurcation buckling. For all cases, it is assumed that the behavior of the material is linearly elastic.

7.1   THE THIN CIRCULAR RING

Buckling of thin circular rings was first investigated by Bresse in 1866. Timoshenko and Gere (1961) present the solution to this problem and a complete historical sketch of other investigations. In addition to the references cited in the text by Timoshenko and Gere, the investigations of Boresi (1955), Wasserman (1961), Wempner and Kesti (1962), and Smith and Simitses (1969) are important contributions to the solution of this problem.

7.1.1 KINEMATIC RELATIONS

The geometry and sign convention are given on Fig. 7.1. Let the deformation components of the neutral surface material points be denoted by $w(s)$ and $v(s)$. The strain at any material point z units from the neutral surface is given by

$$\varepsilon = \varepsilon^0 + z\kappa \qquad (1)$$

This equation is based on the assumptions that planes remain plane after deformation, the normals to the neutral axis are inextensional, and the ring is thin

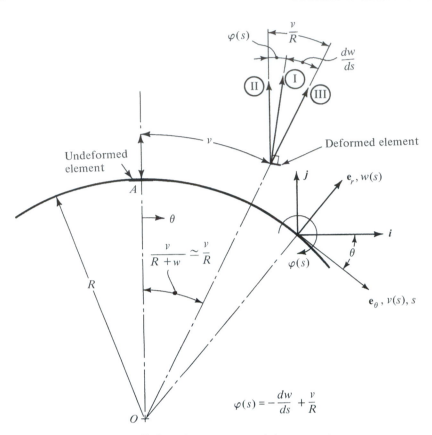

$$\varphi(s) = -\frac{dw}{ds} + \frac{v}{R}$$

FIGURE 7.1   Ring geometry and sign convention.

(thickness much smaller than the radius). In Eq. (1), ε^0 denotes the extensional strain of material points on the neutral surface and κ denotes the change in curvature. We present the development of the expressions of ε^0 and κ in terms of the deformation components and their gradients separately.

First, the expression for ε^0 is developed. The position vector from the origin O to any material point on the reference axis is given by

$$\mathbf{r}(s) = Re_r(s) \tag{2}$$

where e_r is a unit vector in the radial direction. Let $\mathbf{d}(s)$ represent the deformation vector; then

$$\mathbf{d}(s) = w(s)\mathbf{e}_r + v(s)\mathbf{e}_\theta \tag{3}$$

The position vector to a material point in the deformed state r^* is given by

$$\mathbf{r}^*(s) = \mathbf{r}(s) + \mathbf{d}(s) \tag{4}$$

or

$$\mathbf{r}^*(s) = R\left(1 + \frac{w}{R}\right)\mathbf{e}_r + v\mathbf{e}_\theta \tag{5}$$

Now consider a line segment which is tangent to the s-coordinate line (reference axis) in the undeformed state (of length ds). This segment is given by the vector

$$\mathbf{dr} = \frac{\mathbf{dr}}{ds}ds = \mathbf{e}_\theta ds \qquad (6)$$

After deformation, the segment is represented by the vector $\mathbf{dr}^*$, or

$$\mathbf{dr}^* = \frac{\mathbf{dr}^*}{ds}ds = \left[\left(\frac{dw}{ds} - \frac{v}{R}\right)\mathbf{e}_r + \left(1 + \frac{w}{R} + \frac{dv}{ds}\right)\mathbf{e}_\theta\right]ds \qquad (7)$$

The length of this segment is given by

$$|\mathbf{dr}^*| = \left[\left(\frac{dw}{ds} - \frac{v}{R}\right)^2 + \left(1 + \frac{w}{R} + \frac{dv}{ds}\right)^2\right]^{\frac{1}{2}}ds \qquad (8)$$

Now there are two possible definitions of extensional strain denoted by ε_E (engineering definition) and ε_L.

$$\varepsilon_E = \frac{|\mathbf{dr}^*| - |\mathbf{dr}|}{|\mathbf{dr}|} \qquad (9)$$

$$\varepsilon_L = \frac{1}{2}\frac{|\mathbf{dr}^*|^2 - |\mathbf{dr}|^2}{|\mathbf{dr}|^2} \qquad (10)$$

From Eq. (9)

$$\frac{|\mathbf{dr}^*|}{|\mathbf{dr}|} = \varepsilon_E + 1 \qquad (11)$$

Substitution of Eq. (11) into Eq. (10) results in

$$\varepsilon_L = \varepsilon_E + \frac{1}{2}\varepsilon_E^2 \qquad (12)$$

It is clear from Eq. (12) that, for small engineering extensional strains, both definitions give the same results. Therefore, in developing the strain-deformation relations for the thin ring, we will use Eq. (10) to obtain

$$\varepsilon^0 = \left(\frac{w}{R} + \frac{dv}{ds}\right) + \frac{1}{2}\left(\frac{dw}{ds} - \frac{v}{R}\right)^2 + \frac{1}{2}\left(\frac{w}{R} + \frac{dv}{ds}\right)^2 \qquad (13)$$

Note that the last term is negligibly small by comparison to the first term (in parenthesis). Thus,

$$\varepsilon^0 = \frac{w}{R} + \frac{dv}{ds} + \frac{1}{2}\left(\frac{dw}{ds} - \frac{v}{R}\right)^2 \qquad (14)$$

For small strains and moderately small rotations, the change in curvature can be accurately approximated by (for details see Smith and Simitses, 1969 and Sanders, 1963):

$$\kappa = \frac{d\varphi}{ds} \qquad (15)$$

where φ is the rotation of the element from its undeformed state to its deformed state, taken positive as shown in Fig. 7.1.

It is seen from Fig. 7.1, that

$$\varphi = -\frac{dw}{ds} + \frac{v}{R} \tag{16}$$

Therefore

$$\kappa = -\frac{d^2 w}{ds^2} + \frac{dv}{R\,ds} \tag{17}$$

Finally, if we use the variable θ instead of $s = R\theta$, then

$$\varepsilon^0 = \frac{w}{R} + \frac{1}{R}\frac{dv}{d\theta} + \frac{1}{2R^2}\left(\frac{dw}{d\theta} - v\right)^2 \tag{18a}$$

$$\kappa = -\frac{1}{R^2}\left(\frac{d^2 w}{d\theta^2} - \frac{dv}{d\theta}\right) \tag{18b}$$

and

$$\varepsilon = \frac{w}{R} + \frac{1}{R}\frac{dv}{d\theta} + \frac{1}{2R^2}\left(\frac{dw}{d\theta} - v\right)^2 - \frac{z}{R^2}\left(\frac{d^2 w}{d\theta^2} - \frac{dv}{d\theta}\right) \tag{19}$$

7.1.2   EQUILIBRIUM EQUATIONS

Consider the thin circular ring to be loaded by a uniformly distributed load around its circumference with components p_r and p_θ in the radial and tangential directions, respectively. The equilibrium equations for such a configuration are derived using the principle of the stationary value of the total potential

$$\delta U_T = \int_0^{2\pi R} \int_A E\left(\varepsilon^0 + z\kappa\right)\left(\delta\varepsilon^0 + z\delta\kappa\right)\,dA\,ds$$
$$- \int_0^{2\pi R} (p_r \delta w + p_\theta \delta v)ds = 0 \tag{20}$$

where A is the cross-sectional area of the thin ring. Note that linear elastic behavior is assumed.

Let N and M denote the hoop load and bending moment, respectively,

$$N = \int_A E\varepsilon^0 dA = EA\varepsilon^0$$
$$M = -\int_A Ez^2\kappa dA = -EI\kappa \tag{21}$$

where

$$A = \int_A dA \text{ and } I = \int_A z^2 dA$$

Substitution of Eqs. (21) into Eq. (20) yields

$$\int_0^{2\pi R} \left(N\delta\varepsilon^0 - M\delta\kappa - p_r\delta w - p_\theta\delta v\right)ds = 0 \tag{22}$$

From Eqs. (18), we find the expression for $\delta\varepsilon^0$ and $\delta\kappa$:

$$\delta\varepsilon^0 = \frac{\delta w}{R} + \frac{1}{R}\frac{d\delta v}{d\theta} + \frac{1}{R^2}\left(\frac{dw}{d\theta} - v\right)\left(\frac{d\delta w}{d\theta} - \delta v\right) \tag{23a}$$

$$\delta\kappa = -\frac{1}{R^2}\left(\frac{d^2\delta w}{d\theta^2} - \frac{d\delta v}{d\theta}\right) \tag{23b}$$

Substitution of Eqs. (23) into Eq. (22), integration by parts, and requiring continuity at any point of the reference axis leads to the following equilibrium equations.

$$-\frac{N}{R} + \frac{d}{ds}\left[N\left(\frac{dw}{ds} - \frac{v}{R}\right)\right] - \frac{d^2M}{ds^2} + p_r = 0$$

$$\frac{dN}{ds} + \frac{N}{R}\left(\frac{dw}{ds} - \frac{v}{R}\right) - \frac{1}{R}\frac{dM}{ds} + p_\theta = 0 \tag{24}$$

or

$$-NR + \frac{d}{d\theta}\left[N\left(\frac{dw}{d\theta} - v\right)\right] - \frac{d^2M}{d\theta^2} + p_r R^2 = 0$$

$$R\frac{dN}{d\theta} + N\left(\frac{dw}{d\theta} - v\right) - \frac{dM}{d\theta} + p_\theta R^2 = 0 \tag{25}$$

If we assume that the loading is a uniform pressure loading, p, then $p_r = p$, $p_\theta = 0$, and the primary path (prebuckling solution) is characterized by a uniform radial expansion (or contraction). The complete solution of Eqs. (25) for the primary path is given by

$$w^p = \frac{pR^2}{EA}$$
$$v^p = 0 \qquad p_r^p = p$$
$$N^p = pR \tag{26}$$
$$M^p = 0 \qquad p_\theta^p = 0$$
$$\varphi^p = 0$$

7.1.3 BUCKLING EQUATIONS

According to the bifurcation approach, close to the critical load

$$w = w^p + w^* = \frac{p_{cr}R^2}{EA} + w^*$$
$$v = v^p + v^* = v^*$$
$$N = N^p + N^* = p_{cr}R + N^* \tag{27}$$
$$M = M^p + M^* = M^*$$
$$p_r = p_{cr} + p_r^*, \quad p_\theta = p_\theta^*$$

Substitution of Eqs. (27) into Eqs. (25) yields

$$-N^*R + \frac{d}{d\theta}\left[(p_{cr}R + N^*)\left(\frac{dw^*}{d\theta} - v^*\right)\right] - \frac{d^2M^*}{d\theta^2} + p_r^* R^2 = 0$$

$$R\frac{dN^*}{d\theta} + (p_{cr}R + N^*)\left(\frac{dw^*}{d\theta} - v^*\right) - \frac{dM^*}{d\theta} + p_\theta^* R^2 = 0 \tag{28}$$

Since the starred quantities denote the increments which take the system from the unbuckled state to the adjacent buckled equilibrium state, they can be taken as small (but nonzero) as we wish. Therefore by neglecting N^* as small by comparison to $p_{cr}R$, we have

$$-N^*R + p_{cr}R\left(\frac{d^2w^*}{d\theta^2} - \frac{dv^*}{d\theta}\right) - \frac{d^2M^*}{d\theta^2} + p_r^*R^2 = 0$$

$$R\frac{dN^*}{d\theta} + p_{cr}R\left(\frac{dw^*}{d\theta} - v^*\right) - \frac{dM^*}{d\theta} + p_\theta^*R^2 = 0 \qquad (29)$$

Again, because the adjacent state can be taken as close to the primary state as desired, we may use the linearized version of the kinematic relation for the starred quantities:

$$N^* = EA\left(\frac{w^*}{R} + \frac{1}{R}\frac{dv^*}{d\theta}\right)$$

$$M^* = \frac{EI}{R^2}\left(\frac{d^2w^*}{d\theta^2} - \frac{dv^*}{d\theta}\right) \qquad (30)$$

Substitution of Eqs. (30) into Eqs. (29) results in the buckling equations

$$-EA\left(w^* + \frac{dv^*}{d\theta}\right) + p_{cr}R\left(\frac{d^2w^*}{d\theta^2} - \frac{dv^*}{d\theta}\right) - \frac{EI}{R^2}\left(\frac{d^4w^*}{d\theta^4} - \frac{d^3v^*}{d\theta^3}\right) + p_r^*R^2 = 0$$

$$EA\left(\frac{dw^*}{d\theta} + \frac{d^2w^*}{d\theta^2}\right) + p_{cr}R\left(\frac{dw^*}{d\theta} - v^*\right) - \frac{EI}{R^2}\left(\frac{d^3w^*}{d\theta^3} - \frac{d^2v^*}{d\theta^2}\right) + p_\theta^*R^2 = 0 \qquad (31)$$

Before we solve Eqs. (31), we must consider the behavior of the load during the buckling process. Since p_r^* and p_θ^* denote the incremental components of the pressure load in the buckled state, the following distinction must be made. There are three possibilities concerning the behavior of the load (cases I, II, and III). In case I, it is assumed that the load remains normal to the deflected reference axis. In Fig. 7.1 the pressure load is in direction I, and its components along the original radial and tangential directions are $p_{cr}\cos\varphi^*$ and $p_{cr}\sin\varphi^*$, respectively. Since φ^* is taken to be small, $p_r^* = 0$ and $p_\theta^* = p_{cr}\varphi^*$. In case II, it is assumed that the load remains parallel to its original direction (direction II in Fig. 7.1). For this case $p_r^* = p_\theta^* = 0$. Finally, in case III, it is assumed that the load is directed toward the initial center of curvature. For this case, $p_r^* = 0$ and $p_\theta^* = p_{cr}(\varphi^* + dw^*/ds)$. In summary

Case I:

$$p_r^* = 0 \quad p_\theta^* = -\frac{p_{cr}}{R}\left(\frac{dw^*}{d\theta} - v^*\right)$$

Case II:

$$p_r^* = 0 \quad p_\theta^* = 0 \qquad (32)$$

Case III:

$$p_r^* = 0 \quad p_\theta^* = p_{cr}\frac{v^*}{R}$$

If we now substitute. Eqs. (32) into Eqs. (31), we obtain the buckling equation for the three cases of load behavior

$$-EA\left(w^* + \frac{dv^*}{d\theta}\right) + p_{cr}R\left(\frac{d^2w^*}{d\theta^2} - \frac{dv^*}{d\theta}\right) - \frac{EI}{R^2}\left(\frac{d^4w^*}{d\theta^4} - \frac{d^3v^*}{d\theta^3}\right) = 0$$

$$EA\left(\frac{dw^*}{d\theta} + \frac{d^2v^*}{d\theta^2}\right) + p_{cr}R\left(\frac{dw^*}{d\theta} - v^*\right) - \frac{EI}{R^2}\left(\frac{d^3w^*}{d\theta^3} - \frac{d^2v^*}{d\theta^2}\right) - p_{cr}R\begin{bmatrix}\frac{dw^*}{d\theta} - v^* \\ 0 \\ -v^*\end{bmatrix} = 0$$

Combining like terms in the second equation, we obtain

$$-EA\left(w^* + \frac{dv^*}{d\theta}\right) + p_{cr}R\left(\frac{d^2w^*}{d\theta^2} - \frac{dv^*}{d\theta}\right) - \frac{EI}{R^2}\left(\frac{d^4w^*}{d\theta^4} - \frac{d^3v^*}{d\theta^3}\right) = 0$$

$$EA\left(\frac{dw^*}{d\theta} + \frac{d^2v^*}{d\theta^2}\right) - \frac{EI}{R^2}\left(\frac{d^3w^*}{d\theta^3} - \frac{d^2v^*}{d\theta^2}\right) + p_{cr}R\begin{bmatrix}0 \\ \frac{dw^*}{d\theta} - v^* \\ \frac{dw^*}{d\theta}\end{bmatrix} = 0 \qquad (33)$$

We clearly see from Eqs. (33) that the problem has been reduced to an eigenvalue problem in which we seek the smallest value for p_{cr} for which a nontrivial solution exists.

7.1.4 SOLUTION

Before obtaining and discussing the solution, let us multiply both equations by R^2/EI. Then, let $\lambda = p_{cr}R^3/EI$, and $\rho^2 = I/A$, where ρ is the radius of gyration of the cross-sectional geometry

$$-\left(\frac{R}{\rho}\right)^2\left(w^* + \frac{dv^*}{d\theta}\right) + \lambda\left(\frac{d^2w^*}{d\theta^2} - \frac{dv^*}{d\theta}\right) - \left(\frac{d^4w^*}{d\theta^4} - \frac{d^3v^*}{d\theta^3}\right) = 0$$

$$\left(\frac{R}{\rho}\right)^2\left(\frac{dw^*}{d\theta} + \frac{d^2v^*}{d\theta^2}\right) - \left(\frac{d^3w^*}{d\theta^3} - \frac{d^2v^*}{d\theta^2}\right) + \lambda\begin{bmatrix}0 \\ \frac{dw^*}{d\theta} - v^* \\ \frac{dw^*}{d\theta}\end{bmatrix} = 0 \qquad (34)$$

Assume solutions of the form

$$w^* = B_n \cos n\theta, \quad v^* = C_n \sin n\theta$$

or

$$w^* = B_n \sin n\theta, \quad v^* = C_n \cos n\theta \qquad (35)$$

which satisfy the continuity requirements. Note that $n = 1$ for the first equation and $n = 0$ for the second yield rigid-body modes. Substitution of the first set leads to the following system of homogeneous linear algebraic equations in B_n and C_n:

$$-\left[\left(\frac{R}{\rho}\right)^2 + \lambda n^2 + n^4\right]B_n - \left[\left(\frac{R}{\rho}\right)^2 n + \lambda n + n^3\right]C_n = 0$$

$$-\left[\left(\frac{R}{\rho}\right)^2 n + n^3 + \lambda n\begin{pmatrix}0 \\ 1 \\ 1\end{pmatrix}\right]B_n - \left[\left(\frac{R}{\rho}\right)^2 n^2 + n^2 + \lambda\begin{pmatrix}0 \\ 1 \\ 0\end{pmatrix}\right]C_n = 0 \qquad (36)$$

For a nontrivial solution to exist, the determinant of the coefficients of B_n and C_n must vanish.

The expansion of the determinant yields

$$\lambda^2 n^2 \begin{pmatrix} 0 \\ 0 \\ 1 \end{pmatrix} + \lambda \left(\frac{R}{\rho} \right)^2 \left[\begin{array}{c} n^2(n^2-1) \\ (n^2-1)^2 \\ n^2(n^2-2) - n^4\left(\frac{\rho}{R}\right)^2 \end{array} \right] + \left(\frac{R}{\rho} \right)^2 n^2(n^2-1)^2 = 0 \qquad (37)$$

The solutions for λ are:

Case I:

$$\lambda = -(n^2-1)$$

Case II:

$$\lambda = -n^2 \qquad (38)$$

Case III:

$$\lambda = -\frac{(n^2-1)^2}{(n^2-2)}$$

To obtain the solution for case III, it is necessary to assume that $(\rho/R)^2 \ll 1$.

The critical value is obtained by minimizing λ with respect to integer values of n. Since $n = 1$ corresponds to rigid body motion (not of interest in this buckling analysis), the critical condition corresponds to $n = 2$.

Case I:

$$\lambda_{cr} = -3, \qquad p_{cr} = -3\frac{EI}{R^3}$$

Case II:

$$\lambda_{cr} = -4, \qquad p_{cr} = -4\frac{EI}{R^3} \qquad (39)$$

Case III:

$$\lambda_{cr} = -4.5, \qquad p_{cr} = -4.5\frac{EI}{R^3}$$

Note that for all three cases, when we assume that $(\rho/R)^2 \ll 1$ and $n = 1$, from the first of Eqs. (36) we have $B_1 = -C_1$.

Next, if we introduce an orthogonal set of unit vectors, $\mathbf{i}$ and $\mathbf{j}$ (see Fig. 7.1) in the tangential and radial directions, respectively, when $\theta = 0$, we have

$$\mathbf{e}_\theta = \cos\theta\mathbf{i} - \sin\theta\mathbf{j}$$
$$\mathbf{e}_r = \sin\theta\mathbf{i} + \cos\theta\mathbf{j} \qquad (40)$$

The deformation vector, $\mathbf{d}$, of any material point on the reference axis for $n = 1$ (see Eqs. 35) is given by

$$\mathbf{d} = (B_1\cos\theta)\,\mathbf{e}_r - (B_1\sin\theta)\mathbf{e}_\theta \qquad (41)$$

Use of Eqs. (40) and Eq. (41) yields

$$\mathbf{d} = B_1\mathbf{j} \qquad (42)$$

Equation (42) shows that, when $n = 1$, we have a rigid-body translation.

For all three cases we have found the bifurcation load (classical buckling approach), and no attention is paid to the stability or instability of the system as a rigid body. The reason for considering the three different cases is because all those have been used as models for the real load case, which is pressure. The pressure behavior is best represented by case I. It is difficult to conceive of a true application which is represented by case II. Case III can serve as a mathematical model for the following problem: Consider a thin ring which is loaded by a very large number of closely spaced radial cables pulled together through a stiff, small, rigid ring at the center of the thin ring. Singer and Babcock (1970) have shown that for case II the thin ring is unstable as a rigid body and will rotate under arbitrarily small pressure.

7.2   HIGH CIRCULAR ARCHES UNDER PRESSURE

The buckling of a high circular arch under uniform pressure has been presented in Timoshenko (1961). As discussed in Timoshenko (1961) the solution was first obtained by Hurlbrink and the problem was also investigated by Timoshenko and Nicolai. The solution presented herein is based on the use of Eqs. (34). In order to apply these equations, it is assumed that first the arch is uniformly contracted (see Fig. 7.2). On this basis a primary state exists, which is identical to that of a complete circular ring. Then at the instant of buckling, the supports become immovable and the arch buckles as shown in Fig. 7.2.

If the arch is simply supported at both ends, the boundary conditions at $\theta = \pm\alpha$, for the incremental quantities, are given by

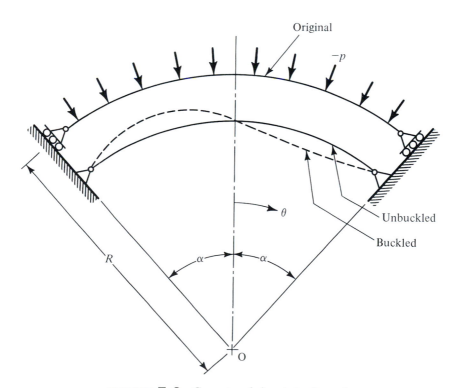

FIGURE 7.2   Geometry of pinned circular arch.

$$w^* = 0$$

$$M^* = \frac{EI}{R^2}\left(\frac{d^2 w^*}{d\theta^2} - \frac{dv^*}{d\theta}\right) = 0 \tag{43}$$

$$N^* = \frac{EA}{R}\left(w^* + \frac{dv^*}{d\theta}\right) = 0$$

These boundary conditions are satisfied by the following assumed form of solution:

$$w^* = B_n \sin\frac{n\pi\theta}{\alpha} \tag{44}$$

$$v^* = C_n \cos\frac{n\pi\theta}{\alpha}$$

Substitution of Eqs. (44) into Eqs. (43) gives

$$-\left[\left(\frac{R}{\rho}\right)^2 + \lambda\left(\frac{n\pi}{\alpha}\right)^2 + \left(\frac{n\pi}{\alpha}\right)^4\right]B_n$$

$$+\left[\left(\frac{R}{\rho}\right)^2\left(\frac{n\pi}{\alpha}\right) + \lambda\left(\frac{n\pi}{\alpha}\right) + \left(\frac{n\pi}{\alpha}\right)^3\right]C_n = 0$$

$$\left[\left(\frac{R}{\rho}\right)^2\left(\frac{n\pi}{\alpha}\right) + \left(\frac{n\pi}{\alpha}\right)^3 + \lambda\left(\frac{n\pi}{\alpha}\right)\begin{pmatrix}0\\1\\1\end{pmatrix}\right]B_n \tag{45}$$

$$-\left[\left(\frac{R}{\rho}\right)^2\left(\frac{n\pi}{\alpha}\right)^2 + \left(\frac{n\pi}{\alpha}\right)^2 + \lambda\begin{pmatrix}0\\1\\0\end{pmatrix}\right]C_n = 0$$

For a nontrivial solution to exist, the determinant of the coefficients of B_n and C_n must vanish. The expansion of the determinant yields an expression similar to Eq. (37), except that whenever n appears in Eq. (37), we must put $(n\pi/\alpha)$. The solutions for λ are:

Case I:

$$\lambda = -\left[\left(\frac{n\pi}{\alpha}\right)^2 - 1\right]$$

Case II:

$$\lambda = -\left(\frac{n\pi}{\alpha}\right)^2 \tag{46}$$

Case III:

$$\lambda = -\frac{\left[(n\pi/\alpha)^2 - 1\right]^2}{\left[(n\pi/\alpha)^2 - 2\right]}$$

The critical condition corresponds to the smallest n, which is one for this case. Therefore,

Case I:

$$p_{cr} = -\frac{EI}{R^3}\left[\left(\frac{\pi}{\alpha}\right)^2 - 1\right]$$

Case II:

$$p_{cr} = -\frac{EI}{R^3}\left(\frac{\pi}{\alpha}\right)^2 \tag{47}$$

Case III:

$$p_{cr} = -\frac{EI}{R^3}\frac{\left[(\pi/\alpha)^2-1\right]^2}{(\pi/\alpha)^2-2}$$

The solution for case I is the same as the one reported in Timoshenko (1961). For this case we note that, when $\alpha = \pi$, we have a complete ring and $p_{cr} = 0$. The reason for this undesirable result is that we have a complete ring with a hinge, and it is free to rotate as a rigid body about this hinge for arbitrarily small pressure. The continuous complete ring corresponds to $\alpha = \pi/2$. When $\alpha = \pi/2$, Eqs. (47) are identical to Eqs. (39).

These results, which are derived by assuming the buckling mode of Fig. 7.2, are not applicable to shallow arches. The low arch is treated in a later section.

The solution to the clamped arch is presented in Timoshenko (1961). This solution is due to E. L. Nicolai:

$$p_{cr} = -\frac{EI}{R^3}\left(k^2 - 1\right) \tag{48}$$

where k is the solution of the following transcendental equation

$$k\tan\alpha\cot k\alpha = 1 \tag{49}$$

7.3 ALTERNATE APPROACH FOR RINGS AND ARCHES

An alternate approach for solving the ring and high arch problem is to eliminate v^* from Eqs. (34) and obtain a single higher-order buckling equation in w^* alone. This single equation is then solved, subject to the appropriate boundary conditions (when applicable). This approach is used in this discussion for load case I. Equations (34) may be written as

$$\begin{aligned} L_1 w^* + L_2 v^* &= 0 \\ L_3 w^* + L_4 v^* &= 0 \end{aligned} \tag{50}$$

where L_i ($i = 1, 2, 3,$ and 4) are the following differential operators

$$\begin{aligned} L_1 &= -\left(\frac{R}{\rho}\right)^2 + \lambda\frac{d^2}{d\theta^2} - \frac{d^4}{d\theta^4} \\ L_2 &= -\left[\lambda + \left(\frac{R}{\rho}\right)^2\right]\frac{d}{d\theta} + \frac{d^3}{d\theta^3} \\ L_3 &= \left(\frac{R}{\rho}\right)^2\frac{d}{d\theta} - \frac{d^3}{d\theta^3} \\ L_4 &= \left[1 + \left(\frac{R}{\rho}\right)^2\right]\frac{d^2}{d\theta^2} \end{aligned} \tag{51}$$

Since these operators are linear, they are commutative

$$L_1 L_4 = L_4 L_1, \quad L_2 L_4 = L_4 L_2, \quad \text{etc.}$$

By operating with L_4 on the first of Eqs. (50), with L_2 on the second of Eqs. (50), and by subtracting the two resulting equations, we have

$$(L_1 L_4 - L_3 L_2) w^* = 0 \tag{52}$$

Substitution of the expressions, Eqs. (51), for the operators yields the following single buckling equation:

$$\frac{d^6 w^*}{d\theta^6} + (2 - \lambda) \frac{d^4 w^*}{d\theta^4} + (1 - \lambda) \frac{d^2 w^*}{d\theta^2} = 0 \tag{53}$$

If we let the solution to Eq. (53) be of the form $e^{\gamma\theta}$, we obtain

$$\gamma^2 \left[\gamma^4 - (\lambda - 2)\gamma^2 - (\lambda - 1) \right] = 0 \tag{54}$$

and the six roots are

$$
\begin{aligned}
\gamma_1 &= +i\sqrt{1-\lambda} = ik \\
\gamma_2 &= -i\sqrt{1-\lambda} = -ik \\
\gamma_3 &= +i \\
\gamma_4 &= -i \\
\gamma_5 &= \gamma_6 = 0
\end{aligned}
\tag{55}
$$

where

$$k = \sqrt{1-\lambda} \quad \text{and} \quad \lambda = 1 - k^2$$

From this, the general solution to Eq. (53) is

$$w^* = A_1 \sin k\theta + A_2 \cos k\theta + A_3 \sin\theta + A_4 \cos\theta + A_5\theta + A_6 \tag{56}$$

Note that λ is a negative number because buckling of the ring is possible only when the uniform radial pressure is compressive and k is the positive square root. Similarly, if we eliminate w^* from Eqs. (50), we obtain a single higher-order equation in v^* (the same as Eq. 53). The solution for v^* is

$$v^* = B_1 \sin k\theta + B_2 \cos k\theta + B_3 \sin\theta + B_4 \cos\theta + B_5\theta + A_7 \tag{57}$$

If we substitute Eqs. (56) and (57) into Eqs. (50), we have

$$
\begin{aligned}
B_1 &= -\frac{A_2}{k} \frac{k^2 + (R/\rho)^2}{1 + (R/\rho)^2}, \quad B_2 = \frac{A_1}{k} \frac{k^2 + (R/\rho)^2}{1 + (R/\rho)^2} \\
B_3 &= -A_4, \quad B_4 = A_3, \quad A_5 \equiv 0 \\
B_5 &= -\frac{(R/\rho)^2}{1 - k^2 + (R/\rho)^2} A_6
\end{aligned}
\tag{58}
$$

Next, if we make use of the thin ring assumption, $(\rho/R)^2 \ll 1$, and if we substitute Eqs. (58) into Eq. (57), we have

$$v^* = \frac{A_1}{k} \cos k\theta - \frac{A_2}{k} \sin k\theta + A_3 \cos\theta - A_4 \sin\theta - A_6\theta + A_7 \tag{59}$$

Substitution of Eqs. (56) and (59) into Eqs. (30) yields the following results for the incremental hoop load, N^*, and the incremental bending moment, M^*,

$$N^* \equiv 0$$

$$M^* = \frac{EI}{R^2}\left[(1-k^2)(A_1 \sin k\theta + A_2 \cos k\theta) + A_6\right] \tag{60}$$

We obtain the expression for the rotation at any point by substituting Eqs. (56) and (59) into Eq. (16):

$$\varphi^* = -\frac{1}{R}\left[\frac{k^2-1}{k}(A_1 \cos k\theta - A_2 \sin k\theta) + A_6\theta - A_7\right] \tag{61}$$

Finally, the expression for the radial shear, Q_r^*, is given by

$$Q_r^* = -\frac{1}{R}\frac{dM^*}{d\theta} = \frac{EI}{R^3}k(k^2-1)(A_1 \cos k\theta - A_2 \sin k\theta) \tag{62}$$

7.3.1 THE CIRCULAR RING

For this particular case, the characteristic equation is obtained from requiring continuity in w^*, v^*, φ^*, M^*, and Q_r^* at the ring reference axis. The continuity equations are:

$$\begin{aligned} w^*(0) &= w^*(2\pi) \\ v^*(0) &= v^*(2\pi) \\ \varphi^*(0) &= \varphi^*(2\pi) \\ M^*(0) &= M^*(2\pi) \\ Q_r^*(0) &= Q_r^*(2\pi) \end{aligned} \tag{63}$$

Substitution of Eqs. (56), and Eqs. (59) through (62) into Eqs. (63) gives

$$A_1 \sin 2k\pi + A_2(\cos 2k\pi - 1) = 0 \tag{64}$$

$$A_1 \frac{1}{k}(\cos 2k\pi - 1) - A_2 \frac{1}{k}\sin 2k\pi - A_6(2\pi) = 0 \tag{65}$$

$$A_1(1-k^2)\sin 2k\pi + A_2(1-k^2)(\cos 2k\pi - 1) = 0 \tag{66}$$

$$\frac{k^2-1}{k}\left[A_1(\cos 2k\pi - 1) - A_2 \sin 2k\pi\right] + A_6(2\pi) = 0 \tag{67}$$

and

$$A_1(\cos 2k\pi - 1) - A_2 \sin 2k\pi = 0 \tag{68}$$

Note that Eqs. (64) and (66) are dependent. Moreover, Eq. (68) can be obtained from a linear combination of Eqs. (65) and (67). Thus, Eqs. (65), (66), and (67) comprise a system of three linear homogeneous algebraic equations in A_1, A_2, and A_6. For a nontrivial solution to exist, the determinant of the coefficients must vanish.

$$\begin{vmatrix} \frac{1}{k}(\cos 2k\pi - 1) & -\frac{1}{k}\sin 2k\pi & 2\pi \\ (1-k^2)\sin 2k\pi & (1-k^2)(\cos 2k\pi - 1) & 0 \\ \frac{k^2-1}{k}(\cos 2k\pi - 1) & -\frac{k^2-1}{k}\sin 2k\pi & 2\pi \end{vmatrix} = 0 \tag{69}$$

The expansion of this determinant yields

$$\cos 2k\pi = 1 \tag{70}$$

Equation (70) is the characteristic equation, and the solution is

$$2k\pi = 2n\pi, \qquad n = 0, 1, 2, \ldots \tag{71}$$

From Eq. (71), the critical load parameter, λ, corresponds to $n = 2$ and $\lambda_{cr} = -3$. Since $k_{cr} = n = 2$, then from Eqs. (64) through (67),

$$A_6 = A_7 = 0 \tag{72}$$

In addition, for this value of k, Eq. (68) is satisfied and thus continuity in shear Q_r^* does exist. Note that A_3 and A_4 do not appear in any of the continuity equations. This is not surprising because the A_3 and A_4 terms, in the expressions for w^* and v^*, denote rigid body translation.

7.3.2   THE PINNED CIRCULAR ARCH

For this particular case, we assume that the ring supports are on rollers and a membrane state exists (see Fig. 7.2). At the instant of buckling, the pin supports become immovable. Thus, the boundary conditions for the buckling equations, Eqs. (34), are

$$
\begin{aligned}
w^*(-\alpha) &= w^*(\alpha) = 0 \\
M^*(-\alpha) &= M^*(\alpha) = 0 \\
v^*(-\alpha) &= v^*(\alpha) = 0
\end{aligned}
\tag{73}
$$

Using Eqs. (56), (59), and (60) in Eqs. (73), we obtain

$$\mp A_1 \sin k\alpha + A_2 \cos k\alpha \mp A_3 \sin \alpha + A_4 \cos \alpha + A_6 = 0 \tag{74}$$

$$\left(1 - k^2\right)\left(\mp A_1 \sin k\alpha + A_2 \cos k\alpha\right) + A_6 = 0 \tag{75}$$

$$A_1 \frac{1}{k} \cos k\alpha \pm A_2 \frac{1}{k} \sin k\alpha + A_3 \cos \alpha \pm A_4 \sin \alpha \pm A_6 \alpha + A_7 = 0 \tag{76}$$

Addition and subtraction of each pair of equations, Eqs. (74), (75), and (76), yield the following two sets of linear homogeneous algebraic equations in A_1, A_3, A_7, and A_2, A_4, and A_6.

$$
\begin{aligned}
A_1 \sin k\alpha + A_3 \sin \alpha &= 0 \\
A_1 \left(1 - k^2\right) \sin k\alpha &= 0 \\
A_1 \frac{1}{k} \cos k\alpha + A_3 \cos \alpha + A_7 &= 0
\end{aligned}
\tag{77}
$$

$$
\begin{aligned}
A_2 \cos k\alpha + A_4 \cos \alpha + A_6 &= 0 \\
A_2 \left(1 - k^2\right) \cos k\alpha + A_6 &= 0 \\
A_2 \frac{1}{k} \sin k\alpha + A_4 \sin \alpha + A_6 \alpha &= 0
\end{aligned}
\tag{78}
$$

Equations (77) correspond to an antisymmetric mode of deformation, w^*, while Eqs. (78) correspond to a symmetric mode, with respect to $\theta = 0$.

Antisymmetric Buckling. From Eqs. (77), we obtain the characteristic equation

$$\begin{vmatrix} \sin k\alpha & \sin \alpha & 0 \\ (1 - k^2) \sin k\alpha & 0 & 0 \\ \dfrac{1}{k} \cos k\alpha & \cos \alpha & 1 \end{vmatrix} = 0 \tag{79}$$

The expansion of the determinant gives

$$\sin k\alpha = 0 \tag{80}$$

and

$$k\alpha = n\pi, \quad n = 1, 2, 3, \ldots \tag{81}$$

Thus $k_{cr} = \pi/\alpha$ and $\lambda_{cr} = -(\pi/\alpha)^2 + 1$, as expected. Note that $A_3 = A_7 = 0$ for this case.

Symmetric Buckling. The characteristic equation for symmetric buckling is obtained from Eqs. (78):

$$\begin{vmatrix} \cos k\alpha & \cos \alpha & 1 \\ (1 - k^2) \cos k\alpha & 0 & 1 \\ \dfrac{1}{k} \sin k\alpha & \sin \alpha & \alpha \end{vmatrix} = 0 \tag{82}$$

The expansion of the determinant yields

$$\tan k\alpha = (k\alpha)^3 \frac{\tan \alpha - \alpha}{\alpha^3} + (k\alpha) \tag{83}$$

The corresponding expression for w^* is given by

$$w^* = A_2 \left[\cos k\theta - k^2 \frac{\cos k\alpha}{\cos \alpha} \cos \theta + (k^2 - 1) \cos k\alpha \right] \tag{84}$$

Evaluation of the roots of Eq. (83) reveals that $|\lambda_{cr}|$ for symmetric buckling is always greater than $|\lambda_{cr}|$ for antisymmetric buckling. For α less than $\pi/2$ it is greater by a factor of 2.25.

7.3.3 THE CLAMPED ARCH

Following the same line of thinking as in the pinned arch case, we find that the boundary conditions for this case are

$$\begin{aligned} w^*(-\alpha) &= w^*(\alpha) = 0 \\ \varphi^*(-\alpha) &= \varphi^*(\alpha) = 0 \\ v^*(-\alpha) &= v^*(\alpha) = 0 \end{aligned} \tag{85}$$

Use of Eqs. (56), (59), and (61) in Eqs. (85) yields

$$\mp A_1 \sin k\alpha + A_2 \cos k\alpha \mp A_3 \sin \alpha + A_4 \cos \alpha + A_6 = 0 \tag{86}$$

$$\frac{k^2 - 1}{k} (A_1 \cos k\alpha \mp A_2 \sin k\alpha) \pm A_6 \alpha - A_7 = 0 \tag{87}$$

$$A_1 \frac{1}{k} \cos k\alpha \pm A_2 \frac{1}{k} \sin k\alpha + A_3 \cos \alpha \pm A_4 \sin \alpha \pm A_6 \alpha + A_7 = 0 \tag{88}$$

Following the same steps as in the case of the pinned arch we have:

Antisymmetric Buckling.

$$A_1 \sin k\alpha + A_3 \sin \alpha = 0$$

$$A_1 \frac{k^2 - 1}{k} \cos k\alpha - A_7 = 0 \qquad (89)$$

$$A_1 \frac{1}{k} \cos k\alpha + A_3 \cos \alpha + A_7 = 0$$

From Eqs. (89) the characteristic equation is

$$\begin{vmatrix} \sin k\alpha & \sin \alpha & 0 \\ \dfrac{k^2 - 1}{k} \cos k\alpha & 0 & -1 \\ \dfrac{1}{k} \cos k\alpha & \cos \alpha & 1 \end{vmatrix} = 0 \qquad (90)$$

The expansion of the determinants yields

$$k \tan \alpha \cot k\alpha = 1 \qquad (91)$$

This equation is identical with Eq. (49).

Symmetric Buckling.

$$A_2 \cos k\alpha + A_4 \cos \alpha + A_6 = 0$$

$$A_2 \frac{k^2 - 1}{k} \sin k\alpha - A_6 \alpha = 0 \qquad (92)$$

$$A_2 \frac{1}{k} \sin k\alpha + A_4 \sin \alpha + A_6 \alpha = 0$$

The characteristic equation is

$$\begin{vmatrix} \cos k\alpha & \cos \alpha & 1 \\ \dfrac{k^2 - 1}{k} \sin k\alpha & 0 & -\alpha \\ \dfrac{1}{k} \sin k\alpha & \sin \alpha & \alpha \end{vmatrix} = 0 \qquad (93)$$

which yields

$$\cot k\alpha = \frac{k\alpha}{\alpha} \left(\frac{1}{\tan \alpha} - \frac{1}{\alpha} \right) + \frac{1}{k\alpha} \qquad (94)$$

Note that, when $\alpha = \pi$, both Eqs. (91) and (94) yield $k_{cr} = 2$ and therefore $\lambda_{cr} = -3$. When α is less than π, $|\lambda_{cr}|$ for symmetric buckling is always greater than $|\lambda_{cr}|$ for antisymmetric buckling. For $\alpha < \pi$, it is greater by a factor 1.6.

7.4   SHALLOW ARCHES

Shallow arches have been used widely as structural elements. One important response of such elements when loaded transversely (see Fig. 7.3) is snapthrough buckling or oil-canning. This buckling phenomenon is characterized by a visible and sudden jump from one equilibrium configuration to another for which displacements are distinctly larger than the first. The significance of snapthrough buckling, insofar

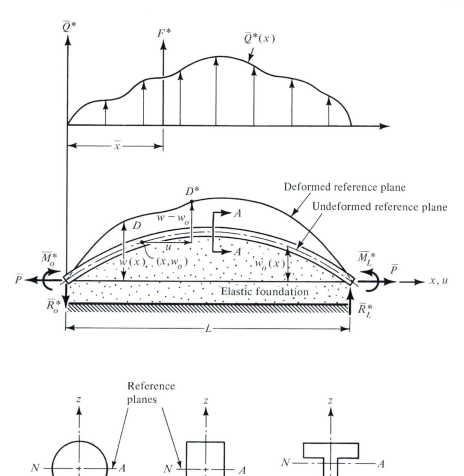

FIGURE 7.3 The shallow arch: geometry and sign convention.

as it illustrates certain important features in more complicated buckling problems of plates and shells, was pointed out by Marguerre (1938), who constructed a simplified mechanical model to demonstrate these features. Timoshenko (1935) obtained an approximate solution to the problem of a low arch under a uniformly distributed transverse load. Biezeno (1938) considered the problem of a low arch loaded transversely at the midpoint with a concentrated load. According to Timoshenko (1961), this problem was first discussed by Navier (1833). Fung and Kaplan (1952) investigated the problem of pinned low arches of various initial shapes (parabolic, half-sine, circular, etc.), and spatial distributions of the lateral load. They also considered the effect of prestress on the critical value of the load. About the same time, Hoff and Bruce (1954) presented results for the pinned half-sine low arch under a half-sine distributed load, as a special case of their dynamic analysis of the buckling of laterally loaded low arches. The results of these two analyses show that a very shallow arch snaps through symmetrically (limit point instability), whereas a higher shallow arch snaps through asymmetrically (unstable bifurcation).

Gjelsvik and Bodner (1962) obtained an approximate solution to the problem of a clamped low circular arch with a concentrated lateral load at the midpoint. Schreyer and Masur (1966) obtained an exact solution to this problem (and to the case of uniform pressure), and they showed that for the concentrated load case, the arch snaps-through symmetrically regardless of the value of the rise parameter. Masur and Lo (1972) presented a general discussion of the behavior of the shallow circular arch regarding buckling, post-buckling and imperfection sensitivity. The effects of inelastic material behavior have been considered by Franciosi, Augusti and Sparacio (1964), by Lee and Murphy (1968), and by Onat and Shu (1962). Experimental results have been reported in Fung and Kaplan (1952), and by Roorda (1965). Finally, snapping of low pinned arches resting on an elastic foundation has been investigated by Simitses (1973). This work is presented in a later section because it provides an interesting model for stability studies. This model exhibits all forms of experimentally observed buckling phenomena.

7.4.1 MATHEMATICAL FORMULATION

The equilibrium equations and proper boundary conditions will be derived first. Consider a slender arch of small initial curvature. We assume that all of the assumptions of slender beams are satisfied except that we now have initial shape, such that material points on the undeformed midline (midplane) are characterized by $w_0(x)$ (see Fig. 7.3). Let $u(x)$ and $w(x)$ denote the location of material points on the deformed midline. On the basis of these assumptions, the strain at any material point is given by

$$\varepsilon = \varepsilon^0 + z\kappa \qquad (95)$$

where ε^0 and κ denote the reference-plane extensional strain and change in curvature, respectively. The appropriate kinematic relations are derived by reference to Fig. 7.3. Let D and D^* denote the undeformed and deformed positions of a material point of the reference line [intersection of the plane of symmetry, x-z, with the reference plane (neutral surface)]. The coordinates of D and D^* are (x, w_0) and $[x + u(x), w]$, respectively. If ds and $(ds)^*$ denote the undeformed and deformed lengths of elements on the reference line, then for small strains, the reference plane extensional strain, ε^0, is given by

$$\varepsilon^0 = \frac{1}{2} \frac{(ds^*)^2 - (ds)^2}{(ds)^2} \qquad (96)$$

Since

$$(ds^*)^2 = (dx + du)^2 + (dw)^2$$
$$(ds)^2 = (dx)^2 + (dw_0)^2 \qquad (97)$$

then

$$\varepsilon^0 = \frac{du}{dx} + \frac{1}{2}\left(\frac{dw}{dx}\right)^2 - \frac{1}{2}\left(\frac{dw_0}{dx}\right)^2 \qquad (98)$$

The expression in Eq. (98) is based on the assumption that $(dw_0/dx)^2 \ll 1$ and $du/dx \ll 1$.

In addition, for small initial curvature and for $(dw/dx)^2 \ll 1$, the expression for the change in curvature, κ, is given by

$$\kappa = \frac{d\varphi}{dx} = -\left(\frac{d^2w}{dx^2} - \frac{d^2w_0}{dx^2}\right) \qquad (99)$$

Assuming that the behavior of the material is linearly elastic and denoting by P and M the axial force and bending moment, respectively, we have

$$P = EA\varepsilon^0$$
$$M^* = EI\kappa \tag{100}$$

where A is the cross-sectional area and I is the second moment of this area about the neutral axis.

We obtain the equilibrium equations by making use of the principle of the stationary value of the total potential. The total potential consists of the sum of the energy stored in the system and the potential of the external forces. The energy stored in the system is the sum of the stretching strain energy, bending strain energy, and energy stored into the foundation. If we let U_m^*, U_b^*, and U_f^* denote the linear stretching, bending, and foundation energy densities, we can write the following expressions

$$U_m^* = \frac{1}{2}EA\left(\varepsilon^0\right)^2 = \frac{P^2}{2EA}$$
$$U_b^* = \frac{1}{2}EI\kappa^2 = \frac{M^{*2}}{2EI} \tag{101}$$
$$U_f^* = \frac{1}{2}\overline{\beta}(w - w_0)^2$$

where $\overline{\beta}$ is the modulus of the foundation (same as in Chapter 6). The potential of the external forces, U_{PT}^*, includes the contributions of the distributed load $Q^*(x)$, the concentrated loads F^*, $\overline{R}^*$, and $\overline{P}$, and the couples $\overline{M}^*$.

$$U_{PT}^* = -\int_0^L [Q^* + F^*\delta^*(x - \overline{x})](w - w_0)dx$$
$$- \left[\overline{R}^*(w - w_0)\right]_0^L + \left[\overline{M}^*\varphi\right]_0^L - \left[\overline{P}u\right]_0^L \tag{102}$$

where $\delta^*(x - \overline{x})$ is the Dirac-delta function.

Since Eqs. (101) denote linear densities, the total potential is obtained through integration of these expressions, over the entire length, and added to U_{PT}^*, Eq. (102).

$$U_T^* = \int_0^L \left\{\frac{P^2}{2EA} + \frac{M^{*2}}{2EI} + \frac{1}{2}\overline{\beta}(w - w_0)^2 - [Q^* + F^*\delta^*(x - \overline{x})](w - w_0)\right\}dx$$
$$- \left[\overline{R}^*(w - w_0)\right]_0^L + \left[\overline{M}^*\varphi\right]_0^L - \left[\overline{P}u\right]_0^L \tag{103}$$

Before we proceed with the derivation of the equilibrium equations and the proper boundary conditions, it is convenient to express all of the parameters in a nondimensional form. We do this by introducing the following parameters:

$$\xi = \frac{\pi x}{L}, \quad \eta(\xi) = \frac{w(x)}{\rho}, \quad v(\xi) = \frac{u(x)}{\rho}, \quad \delta^*(x - \overline{x}) = \rho\delta\left(\xi - \overline{\xi}\right)$$
$$q(\xi) = \frac{\rho Q^*(x)}{P_E \overline{\varepsilon}_E}, \quad \beta = \frac{\overline{\beta}L^4}{\pi^4 EI}, \quad F = \frac{F^*}{P_E \overline{\varepsilon}_E} \tag{104}$$
$$\overline{R} = \frac{\overline{R}^*}{P_E \overline{\varepsilon}_E}, \quad p = \frac{P}{P_E}, \quad M = \frac{M^*}{\rho P_E}, \quad U_T = \frac{4U_T^*}{P_E \overline{\varepsilon}_E L}, \quad (\ )' = \frac{d}{d\xi}$$

where

$$\rho^2 = \frac{I}{A}, \quad P_E = \frac{\pi^2 EI}{L^2}, \text{ and } \overline{\varepsilon}_E = \left(\frac{\pi\rho}{L}\right)^2$$

Note that ρ is the radius of gyration for the cross-sectional area, P_E is the Euler load for a pinned column of length L, and $\bar{\varepsilon}_E$ is the corresponding Euler strain.

With these new nondimensionalized parameters, Eqs. (100), (101), and (102) become

$$P = \frac{P_E}{2}\left[2\frac{v'}{\sqrt{\bar{\varepsilon}_E}} + (\eta')^2 - (\eta_0')^2\right] \tag{105}$$

$$M^* = -\rho P_E(\eta'' - \eta_0'')$$

$$U_m^* = \frac{P_E\bar{\varepsilon}_E}{8}\left[2\frac{v'}{\sqrt{\bar{\varepsilon}_E}} + (\eta')^2 - (\eta_0')^2\right]$$

$$U_b^* = \frac{P_E\bar{\varepsilon}_E}{2}(\eta'' - \eta_0'')^2 \tag{106}$$

$$U_f^* = \frac{P_E\bar{\varepsilon}_E}{2}\beta(\eta - \eta_0)^2$$

$$U_{PT} = -\frac{1}{\pi}\int_0^\pi \left[4q + 4F\delta(\xi - \bar{\xi})\right](\eta - \eta_0)d\xi - \frac{1}{\pi}\left[\bar{R}(\eta - \eta_0)\right]_0^\pi\left(4\sqrt{\bar{\varepsilon}_E}\right)$$
$$-\frac{4}{\pi}\left[\bar{M}(\eta' - \eta_0')\right]_0^\pi - \frac{1}{\pi}\left(\frac{4}{\sqrt{\bar{\varepsilon}_E}}\right)\left[\bar{p}v\right]_0^\pi \tag{107}$$

Finally, the nondimensionalized total potential, Eq. (103), becomes

$$U_T = \frac{4U_T^*}{P_E\bar{\varepsilon}_E L} = \frac{1}{\pi}\int_0^\pi \left\{\frac{1}{2}\left[2\frac{v'}{\sqrt{\bar{\varepsilon}_E}} + (\eta')^2 - (\eta_0')^2\right]^2 + 2(\eta'' - \eta_0'')^2\right.$$
$$\left. + 2\beta(\eta - \eta_0)^2 - 4q(\eta - \eta_0) - 4F\delta(\xi - \bar{\xi})(\eta - \eta_0)\right\}d\xi \tag{108}$$
$$-\frac{1}{\pi}\left\{\left[\bar{R}(\eta - \eta_0)\right]_0^\pi\left(4\sqrt{\bar{\varepsilon}_E}\right) + 4\left[\bar{M}(\eta' - \eta_0')\right]_0^\pi + \frac{4}{\sqrt{\bar{\varepsilon}_E}}\left[\bar{p}v\right]_0^\pi\right\}$$

According to the principle of the stationary value of the total potential, the first variation of the total potential must be zero for equilibrium. To accomplish this, we express the functional U_T in terms of $v(\xi) + \varepsilon_1\zeta(\xi)$ and $\eta(\xi) + \varepsilon_2\gamma(\xi)$, where $\zeta(\xi)$ and $\gamma(\xi)$ are admissible functions of ξ, and ε_1 and ε_2 are small arbitrary constants. Thus $\varepsilon_1\zeta$ and $\varepsilon_2\gamma$ denote the variations in v and η, respectively (δv and $\delta\eta$).

$$U_T[v + \varepsilon_1\zeta,\ \eta + \varepsilon_2\gamma] = \frac{1}{\pi}\int_0^\pi\left\{\frac{1}{2}\left[2\frac{v' + \varepsilon_1\zeta'}{\sqrt{\bar{\varepsilon}_E}} + (\eta' + \varepsilon_2\gamma')^2 - (\eta_0')^2\right]^2\right.$$
$$+ 2(\eta'' + \varepsilon_2\gamma'' - \eta_0'')^2 + 2\beta(\eta + \varepsilon_2\gamma - \eta_0)^2 - 4q(\eta + \varepsilon_2\gamma - \eta_0)$$
$$\left. - 4F\delta(\xi - \bar{\xi})(\eta + \varepsilon_2\gamma - \eta_0)\right\}d\xi - \frac{1}{\pi}\left(4\sqrt{\bar{\varepsilon}_E}\right)\left[\bar{R}(\eta + \varepsilon_2\gamma - \eta_0)\right]_0^\pi$$
$$-\frac{4}{\pi}\left[\bar{M}(\eta' + \varepsilon_2\gamma' - \eta_0')\right]_0^\pi - \frac{1}{\pi}\left(\frac{4}{\sqrt{\bar{\varepsilon}_E}}\right)\left[\bar{p}(v + \varepsilon_1\zeta)\right]_0^\pi \tag{109}$$

If we perform the operations indicated in the integrand and group terms according to the powers in ε_1 and ε_2, we recognize that the terms that do not contain ε's denote $U_T[v, \eta]$ and thus

$$U_T[v + \varepsilon_1\zeta, \eta + \varepsilon_2\gamma] = U_T[v, \eta] + \frac{1}{\pi}\left\langle\!\!\left\langle \frac{\varepsilon_1}{\sqrt{\bar{\varepsilon}_E}} \int_0^\pi 2\left[2\frac{v'}{\sqrt{\bar{\varepsilon}_E}} + (\eta')^2 - (\eta_0')^2\right]\right\rangle\!\!\right\rangle\zeta'\,d\xi$$

$$-\frac{\varepsilon_1}{\sqrt{\bar{\varepsilon}_E}}[4\bar{p}\zeta]_0^\pi + \varepsilon_2\int_0^\pi \left\{2\left[\frac{2v'}{\sqrt{\bar{\varepsilon}_E}} + (\eta')^2 - (\eta_0')^2\right]\eta'\gamma' + 4(\eta'' - \eta_0'')\gamma''\right.$$

$$\left. + 4\beta(\eta - \eta_0)\gamma - 4q\gamma - 4F\delta(\xi - \bar{\xi})\gamma\right\}d\xi - \varepsilon_2\left[4\sqrt{\bar{\varepsilon}_E}\,\bar{R}\gamma\right]_0^\pi - \varepsilon_2\left[4\bar{M}\gamma'\right]_0^\pi$$

$$+\frac{1}{\pi}\left\{\varepsilon_1^2\int_0^\pi \frac{2}{\bar{\varepsilon}_E}(\zeta')^2\,d\xi + \varepsilon_1\varepsilon_2\int_0^\pi \frac{4}{\sqrt{\bar{\varepsilon}_E}}\eta'\gamma'\,d\xi\right.$$

$$\left. + \varepsilon_2^2\int_0^\pi \left\{2(\eta')^2(\gamma')^2 + \left[\frac{2v'}{\sqrt{\bar{\varepsilon}_E}} + (\eta')^2 - (\eta_0')^2\right](\gamma')^2 + 2(\gamma'')^2 + 2\beta\gamma^2\right\}d\xi\right\}$$

$$+\frac{1}{\pi}\left[\frac{\varepsilon_1\varepsilon_2^2}{\sqrt{\bar{\varepsilon}_E}}\int_0^\pi 2\zeta'(\gamma')^2\,d\xi + 2\varepsilon_2^3\int_0^\pi (\eta')(\gamma')^3\,d\xi\right] + \frac{\varepsilon_2^4}{2\pi}\int_0^\pi (\gamma')^4\,d\xi$$

$$\tag{110}$$

From Eqs. (105), the first variation becomes

$$\delta^1 U_T = \frac{1}{\pi}\left\langle\!\!\left\langle \frac{4\varepsilon_1}{\sqrt{\bar{\varepsilon}_E}}\left[\int_0^\pi p\zeta'\,d\xi - (\bar{p}\zeta)|_0^\pi\right]\right.\right.$$

$$+ 4\varepsilon_2\left\{\int_0^\pi \left[p\eta'\gamma' + (\eta'' - \eta_0'')\gamma'' + \beta(\eta - \eta_0)\gamma - q\gamma\right.\right. \tag{111}$$

$$\left.\left.\left.- F\delta(\xi - \bar{\xi})\gamma\right]d\xi - \left[\sqrt{\bar{\varepsilon}_E}\,\bar{R}\gamma\right]_0^\pi - \left[\bar{M}\gamma'\right]_0^\pi\right\}\right\rangle\!\!\right\rangle$$

By setting the first variation equal to zero, we obtain

$$\int_0^\pi p\zeta'\,d\xi - [\bar{p}\zeta]_0^\pi = 0$$

$$\int_0^\pi \left[p\eta'\gamma' + (\eta'' - \eta_0'')\gamma'' + \beta(\eta - \eta_0)\gamma - q\gamma - F\delta(\xi - \bar{\xi})\gamma\right]d\xi \tag{112}$$

$$- \left[\sqrt{\bar{\varepsilon}_E}\,\bar{R}\gamma\right]_0^\pi - \left[\bar{M}\gamma'\right]_0^\pi = 0$$

Integration by parts yields the following form for Eqs. (112)

$$-\int_0^\pi p'\zeta\,d\xi + [(p - \bar{p})\zeta]_0^\pi = 0$$

$$\int_0^\pi \left[-(p\eta')' + (\eta'' - \eta_0'')'' + \beta(\eta - \eta_0) - q - F\delta(\xi - \bar{\xi})\right]\gamma\,d\xi \tag{113}$$

$$+ \left\{\left[p\eta' - (\eta'' - \eta_0'')' - \sqrt{\bar{\varepsilon}_E}\,\bar{R}\right]\gamma\right\}\Big|_0^\pi + \left\{\left[(\eta'' - \eta_0'') - \bar{M}\right]\gamma'\right\}\Big|_0^\pi = 0$$

Through the fundamental lemma of the calculus of variations, we obtain from Eqs. (113) equilibrium equations and boundary conditions.

Equilibrium Equations.

$$p' = 0 \tag{114}$$

$$\left(\eta'' - \eta_0''\right)'' - \left(p\eta'\right)' + \beta(\eta - \eta_0) - q - F\delta\left(\xi - \bar{\xi}\right) = 0 \tag{115}$$

Boundary Conditions.

Either (*kinematic*)   *Or* (*Natural*)

$$\varepsilon_1\zeta = \delta v = 0 \qquad\qquad p = -\bar{p}$$

$$\varepsilon_2\gamma' = (\delta\eta)' = 0 \qquad\qquad \eta'' - \eta_0'' = \overline{M}$$

$$\varepsilon_2\gamma = \delta\eta = 0 \qquad\qquad -\left(\eta'' - \eta_0''\right)' + p\eta' = \sqrt{\varepsilon_E}\,\overline{R}$$

$$\tag{116}$$

Note that, if the supports are immovable, then $\delta v = 0$ or $v(0) = v(\pi) = 0$. Furthermore, if the immovable supports are pinned, $\delta\eta = 0$ or $\eta(0) = \eta(\pi) = 0$, $\eta''(0) = \eta_0''(0)$ and $\eta''(\pi) = \eta_0''(\pi)$. Finally, if the immovable supports are clamped $\eta(0) = \eta(\pi) = 0$, and $\eta'(0) = \eta_0'(0)$ and $\eta'(\pi) = \eta_0'(\pi)$.

When the supports are immovable, the expression for p, Eqs. (105), after an integration over the length, becomes

$$\int_0^\pi p\,d\xi = \frac{1}{2}\int_0^\pi \left[\frac{2v'}{\sqrt{\varepsilon_E}} + (\eta')^2 - (\eta_0')^2\right]d\xi$$

and

$$p = \frac{1}{2\pi}\int_0^\pi \left[(\eta')^2 - (\eta_0')^2\right]d\xi \tag{117}$$

Note that this expression uses the fact that $p(\xi) =$ constant, according to the first of the equilibrium equations, Eqs. (114). With this, we may express the total potential, Eq. (108), solely in terms of η and its space-dependent derivatives. The boundary terms vanish for supported ends (either pinned or clamped).

$$U_T = \frac{1}{\pi}\int_0^\pi \left\{\frac{1}{2\pi^2}\left[\int_0^\pi \left[(\eta')^2 - (\eta_0')^2\right]d\xi\right]^2 + 2(\eta'' - \eta_0'')^2 \right.$$

$$\left. + 2\beta(\eta - \eta_0)^2 - 4q(\eta - \eta_0) - F\delta\left(\xi - \bar{\xi}\right)(\eta - \eta_0)\right\}d\xi \tag{118}$$

7.5   THE SINUSOIDAL PINNED ARCH

The problem to be considered here is a low half-sine pinned arch loaded quasi-statically by a half-sine spatially distributed load. The initial shape is given by

$$\eta_0 = e\sin\xi \quad 0 \le \xi \le \pi \tag{119}$$

where e is the initial rise parameter. Since $(w_0)_{max} = \rho e$, then $e = (w_0)_{max}/\rho$, and if the cross section is rectangular of width l and thickness h, then $\rho = h/2\sqrt{3}$ and

$e = 2\sqrt{3}(w_0)_{max}/h$, which clearly shows that e is a measure of the ratio of the initial maximum rise to the thickness of the arch. The expression for the loading is given by

$$q(\xi) = q_1 \sin \xi \tag{120}$$

The deflection may be represented by an infinite sine series, each term of which satisfies the boundary conditions

$$\eta(\xi) = \eta_0(\xi) + \sum_{n=1}^{\infty} a_n \sin n\xi \tag{121}$$

Boundary conditions:

$$\eta(0) = \eta(\pi) = 0, \quad \eta''(0) = \eta''(\pi) = 0 \tag{122}$$

Substitution of Eq. (121) into the expression for the total potential, Eq. (118), yields

$$U_T = \frac{1}{8} \left(\sum_{n=1}^{\infty} n^2 a_n^2 + 2ea_1 \right)^2 + \sum_{n=1}^{\infty} n^4 a_n^2 - 2q_1 a_1 + \beta \sum_{n=1}^{\infty} a_n^2 \tag{123}$$

We are interested in finding, for the entire range of the free parameters β and e, the load at which instability (snapthrough or bifurcation buckling) is possible. This load is called critical load. We find it by first writing the equilibrium equations and then studying the character of these static equilibrium positions (stability in the small).

To find the static equilibrium positions, we use the principle of the stationary value of the total potential, or

$$\frac{\partial U_T}{\partial a_k} = 0 \quad k = 1, 2, 3, \ldots \tag{124}$$

This leads to

$$\frac{1}{4} \left(\sum_{n=1}^{\infty} n^2 a_n^2 + 2ea_1 \right)(a_1 + e) + a_1 + \beta a_1 = q_1$$

$$\frac{1}{4} \left(\sum_{n=1}^{\infty} n^2 a_n^2 + 2ea_1 \right) k^2 a_k + k^4 a_k + \beta a_k = 0, \quad k = 2, 3, 4, \ldots \tag{125}$$

There are three possible cases that result from Eqs. (125)

Case I: $a_1 \neq 0$ and $a_k \equiv 0$ for $k = 2, 3, 4, \ldots$

Case II: $a_1 \neq 0$, $a_m \neq 0$, and $a_k \equiv 0$ for $k = 2, 3, 4, \ldots$, except $m = k$.

Case III: When $\beta = m^2 j^2$, then it is possible that $a_1 \neq 0$, $a_m \neq 0$, $a_j \neq 0$, and $a_k \equiv 0$ for $k = 2, 3, 4, \ldots$, except $k = j$ and m.

Case III will be treated separately. For the first two cases, a more convenient form of the equilibrium equations may be obtained if we introduce the following new parameters:

$$r_1 = a_1 + e, \quad Q = q_1 + (1 + \beta)e \tag{126}$$

With these new parameters, the equilibrium equations are

$$\frac{1}{4} [r_1^2 - e^2 + k^2 a_k^2 + 4(1 + \beta)]r_1 = Q$$

$$\left[\frac{k^2}{4} (r_1^2 - e^2 + k^2 a_k^2) + (k^4 + \beta) \right] a_k = 0 \tag{127}$$

We see from the equilibrium equations, Eqs. (127), that there are two possibilities:
(1) $r_1 \neq 0$ and $a_k \equiv 0$, and (2) $r_1 \neq 0$ and $a_k \neq 0$. All the possible positions of static
equilibrium are shown in Fig. 7.4. Note that the starting (undeformed) position is A
(see Fig. 7.5), and the possibility of the existence of the a_k-mode is present for
$e^2 > 4(k^4 + \beta)/k^2$.

1. If $a_k \equiv 0$ then, from Eqs. (127), the equilibrium equation which also represents the
load-deflection curve is

$$r_1^3 - [e^2 - 4(1 + \beta)]r_1 = 4Q \tag{128}$$

We see from Eq. (128) that, for $e \leq 2\sqrt{1 + \beta}$, there is a one-to-one Q-to-r_1 dependence,
and there is no possibility of a snapthrough phenomenon. For $e > 2\sqrt{1 + \beta}$, since there
is a range of Q values for which there are three equilibrium positions for the same Q
value, the possibility for a snapthrough phenomenon exists (see Fig. 7.5). It will be
shown that for this case the near and far static equilibrium positions are stable and the

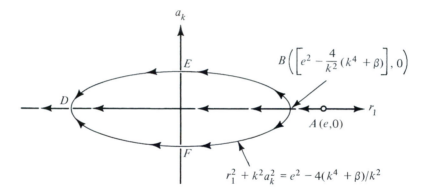

FIGURE 7.4   Positions of static equilibrium in the (r_1, a_k)-space.

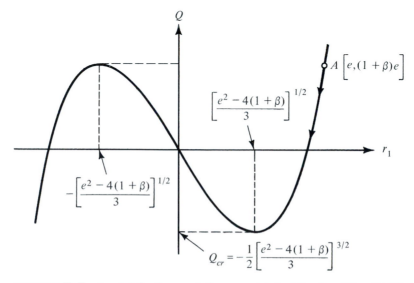

FIGURE 7.5   Load-deflection graph for symmetric buckling $[e > 2(1 + \beta)1/2]$.

intermediate one is unstable. When the near static equilibrium position becomes unstable in the small, snapping occurs and the corresponding load is a critical one.

2. If $e^2 > 4(k^4 + \beta)/k^2$, then it is possible for $a_k \neq 0$, and the load-displacement relation is given by the following set of equations in addition to Eq. (128):

$$r_1 \frac{k^2 - 1}{k^2}(\beta - k^2) = Q \tag{129}$$

$$a_k^2 = \frac{1}{k^2}\left[e^2 - \frac{4}{k^2}(k^4 + \beta) - \frac{Q^2 k^4}{(k-1)^2(\beta - k^2)^2}\right] \tag{130}$$

Thus we see from Eqs. (128), (129), and (130) that the load-displacement relation for the entire range of initial rise parameter values and all possible cases of its relation to the values of k and the modulus of the foundation may be represented by the six graphs of Fig. 7.6.

Critical load shall be defined as the smallest load for which the near static equilibrium position becomes unstable (in the small).

The necessary and sufficient condition for stability (in the small) of the static equilibrium positions given by the roots of Eqs. (128), (129), and (130) is that

$$\frac{\partial^2 U_T}{\partial r_1^2} > 0, \quad \frac{\partial^2 U_T}{\partial r_1^2}\cdot\frac{\partial^2 U_T}{\partial a_k^2} > \left(\frac{\partial^2 U_T}{\partial r_1 \partial a_k}\right)^2 \tag{131}$$

The expression for U_T obtained by substitution of expressions (126) into Eq. (123) is given by

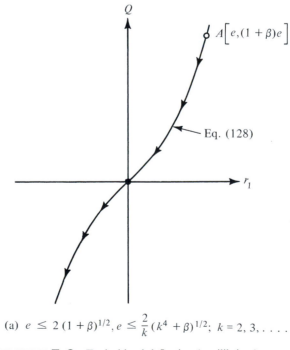

(a) $e \leq 2(1+\beta)^{1/2}, e \leq \frac{2}{k}(k^4 + \beta)^{1/2}; \ k = 2, 3, \dots$.

FIGURE 7.6 Typical load-deflection (equilibrium) curves.

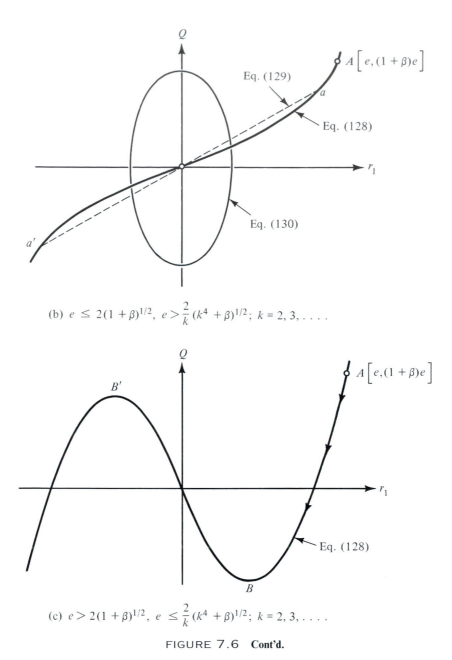

(b) $e \leq 2(1+\beta)^{1/2}$, $e > \dfrac{2}{k}(k^4+\beta)^{1/2}$; $k = 2, 3, \ldots$.

(c) $e > 2(1+\beta)^{1/2}$, $e \leq \dfrac{2}{k}(k^4+\beta)^{1/2}$; $k = 2, 3, \ldots$.

FIGURE 7.6   Cont'd.

$$U_T = \frac{1}{8}\left(r_1^2 - e^2 + k^2 a_k^2\right)^2 - (1+\beta)\left(e^2 - r_1^2\right) + 2Q(e - r_1) + \left(k^4 + \beta\right)a_k^2 \qquad (132)$$

From Eq. (132), we obtain the following expressions for the second derivatives

$$\frac{\partial^2 U_T}{\partial r_1^2} = \frac{1}{2}\left[3r_1^2 - e^2 + k^2 a_k^2 + 4(1+\beta)\right], \quad \frac{\partial^2 U_T}{\partial r_1 \partial a_k} = k^2 r_1 a_k$$

$$\frac{\partial^2 U_T}{\partial a_k^2} = \frac{1}{2}\left[r_1^2 - e^2 + 3k^2 a_k^2\right]k^2 + 2\left(k^4 + \beta\right) \qquad (133)$$

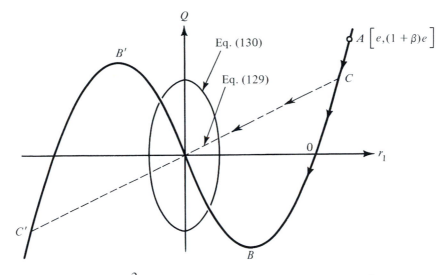

(d) $e > 2(1 + \beta)^{1/2} > \dfrac{2}{k}(k^4 + \beta)^{1/2}$; $k = 2, 3, \ldots$ (this implies $\beta > k^2$)

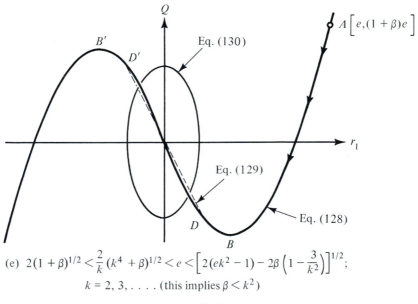

(e) $2(1 + \beta)^{1/2} < \dfrac{2}{k}(k^4 + \beta)^{1/2} < e < \left[2(ek^2 - 1) - 2\beta\left(1 - \dfrac{3}{k^2}\right)\right]^{1/2}$;

$k = 2, 3, \ldots$ (this implies $\beta < k^2$)

FIGURE 7.6 Cont'd.

First, we investigate the stability of the equilibrium positions that are characterized by the ellipse (see Fig. 7.4). These positions are shown in Figs. 7.6b (AA'), 7.6e (DD'), and 7.6f (EE'). Making use of the equilibrium equations, the necessary and sufficient conditions for stability, inequalities (131), become

$$r_1^2 + 2\frac{k^2 - 1}{k^2}(\beta - k^2) > 0, \quad k^4 a_k^2 \left[2\frac{k^2 - 1}{k^2}(\beta - k^2)\right] > 0 \qquad (134)$$

We see from these inequalities that, if $\beta > k^2$, the ellipse equilibrium positions are stable positions. The test fails only for the $a_k = 0$ positions of the ellipse, but it can

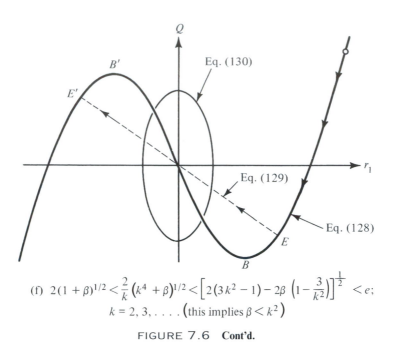

(f) $2(1+\beta)^{1/2} < \dfrac{2}{k}\left(k^4+\beta\right)^{1/2} < \left[2\left(3k^2-1\right)-2\beta\left(1-\dfrac{3}{k^2}\right)\right]^{\frac{1}{2}} < e;$

$k = 2, 3, \ldots$ $\left(\text{this implies } \beta < k^2\right)$

FIGURE 7.6   **Cont'd.**

easily be shown from the consideration of the third and fourth variations that these two positions are stable.

For the case of $\beta > k^2$, we see that at point A of Fig. 7.6b and point C of Fig. 7.6d there is a possibility of stable bifurcation, (classical buckling—paths AA', Fig. 7.6b, and CC', Fig. 7.6d). When this happens, the primary state equilibrium position becomes unstable past the bifurcation point (COB of Fig. 7.6d) because the second of Eqs. (131) is not satisfied.

Case III: The existence of a three-mode equilibrium shape.

A three-mode equilibrium position is possible only for $\beta = n^2k^2$ where n and k are differing integers greater than 2. For these distinct values of β, the equilibrium equations, Eq. (125), become

$$
\begin{aligned}
\left[\left(r_1^2 - e^2 + k^2 a_k^2 + n^2 a_n^2\right) + 4\left(1 + n^2 k^2\right)\right]r_1 &= 4Q \\
\left[k^2\left(r_1^2 - e^2 + k^2 a_k^2 + n^2 a_n^2\right) + 4\left(k^4 + n^2 k^2\right)\right]a_k &= 0 \\
\left[n^2\left(r_1^2 - e^2 + k^2 a_k^2 + n^2 a_n^2\right) + 4\left(n^4 + n^2 k^2\right)\right]a_n &= 0
\end{aligned}
\tag{135}
$$

All the possible static equilibrium positions are plotted in Fig. 7.7. Note that the starting point is characterized by $r_1 = e$, and the possibility of the existence of all three modes is realized when $r_1 = \sqrt{e^2 - 4(n^2 + k^2)}$ with the additional condition that $e > 2\sqrt{k^2 + n^2}$.

A typical load-deflection ($Q - r_1$) plot is shown in Fig. 7.8. The initial (unloaded) position is denoted by point A. As the system is loaded, position B (bifurcation point) is reached, and the system may follow either path BC (Eq. 128) or path BOB'.

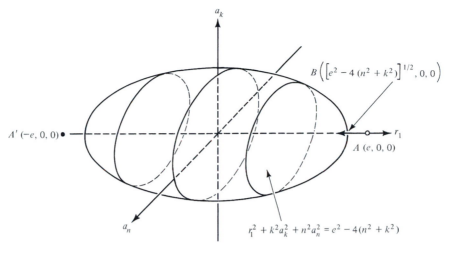

FIGURE 7.7

which is characterized by

$$Q = \left[\left(1 + n^2 k^2\right) - \left(n^2 + k^2\right)\right] r_1 \tag{136}$$

Note that, since $n^2 k^2 > n^2 + k^2$, the slope of the BOB' curve is positive as shown in Fig. 7.8.

It is shown next that the equilibrium positions on the ellipsoid (Fig. 7.7) are stable, and therefore points on BB' (Fig. 7.8) are stable equilibrium positions. Finally, since path BC becomes unstable, classical buckling takes place. Therefore, there is no possibility of a snapping phenomenon for this case.

The total potential for this case is

$$U_T = \frac{1}{8}\left(r_1^2 - e^2 + k^2 a_k^2 + n^2 a_n^2\right)^2 - \left(1 + n^2 k^2\right)\left(e^2 - r_1^2\right) \\ + 2Q(e - r_1) + \left(n^2 + k^2\right)\left(k^2 a_k^2 + n^2 a_n^2\right) \tag{137}$$

The necessary and sufficient condition for stability of the equilibrium positions on the ellipsoid is that the following determinant and all its principal minors (dashed lines) be positive definite

$$\begin{vmatrix} \dfrac{\partial^2 U_T}{\partial r_1^2} & \dfrac{\partial^2 U_T}{\partial r_1 \partial a_k} & \dfrac{\partial^2 U_T}{\partial r_1 \partial a_n} \\[2mm] \dfrac{\partial^2 U_T}{\partial r_1 \partial a_1} & \dfrac{\partial^2 U_T}{\partial a_k^2} & \dfrac{\partial^2 U_T}{\partial a_k \partial a_n} \\[2mm] \dfrac{\partial^2 U_T}{\partial r_1 \partial a_n} & \dfrac{\partial^2 U_T}{\partial a_k \partial a_2} & \dfrac{\partial^2 U_T}{\partial a_n^2} \end{vmatrix} > 0 \tag{138}$$

Use of the expression for the total potential and the equilibrium equations leads to the fact that the principal minors are positive definite, but the determinant is identically equal to zero; therefore the test fails.

Checking the higher variations, we can show that $\delta^3 U_T \equiv 0$ and $\delta^4 U_T$ (fourth variation) is positive definite because all of the fourth-order derivatives are zero except

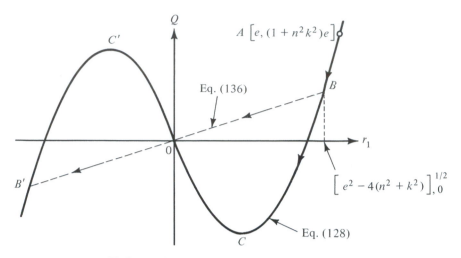

FIGURE 7.8   Typical load-deflection curves for the three-mode case.

$$\frac{\partial^4 U_T}{\partial r_1^4} = 3, \quad \frac{\partial^4 U_T}{\partial r_1^2 \partial a_k^2} = k^2, \quad \frac{\partial^4 U_T}{\partial r_1^2 \partial a_n^2} = n^2,$$

$$\frac{\partial^4 U_T}{\partial a_k^4} = 3k^4, \quad \frac{\partial^4 U_T}{\partial a_n^4} = 3n^2 \tag{139}$$

Because of this, the equilibrium positions on the ellipsoid are stable, and path BB' (Fig. 7.8) is a stable path. Finally, the primary path, BC, becomes unstable and the model exhibits classical buckling (adjacent equilibrium) at point B.

7.5.1   CRITICAL LOADS

Note from Eqs. (134) that if $\beta \geq k^2, k = 2, 3, \ldots$, there is no possibility of snapping but there is bifurcation buckling. Therefore, we must consider the following two ranges for β values separately.

Range $1, \beta < 4$. Snapping is possible and the following cases must be distinguished:

1. If $2\sqrt{1+\beta} < e < \sqrt{16 + \beta}$ (see Fig. 7.6c), then the system will reach point B and snap through in an a_1-only mode. The critical load for this case is obtained from Eq. (128) with $r_1 = \left[e^2 - 4(1 + \beta)\right]^{1/2} / 3^{1/2}$

$$Q_{cr} = -\frac{1}{2} \left[\frac{e^2 - 4(1 + \beta)}{3}\right]^{\frac{1}{2}}$$

$$q_{1cr} = -(1 + \beta)e - \frac{1}{2} \left[\frac{e^2 - 4(1 + \beta)}{3}\right]^{\frac{1}{2}} \tag{140}$$

2. If $2\sqrt{1+\beta} < \sqrt{16 + \beta} < e < \sqrt{22 - \beta/2}$, then, although an unstable bifurcation through mode a_2 is possible, the system will snap initially through an a_1-mode, because during the loading process point B will be reached before point D (see Fig. 7.6e). In this case, the critical load is still given by Eq. (140).

3. If $e > \sqrt{22 - \beta/2} > \sqrt{16 + \beta} > 2\sqrt{1 + \beta}$, then snapping will take place through an a_2-mode (point E of Fig. 7.6f), and the critical load is given by Eq. (129) with r_1 equal to the value corresponding at the bifurcation point or

$$q_{1_{cr}} = -(1+\beta)e - 3\left(1 - \frac{\beta}{4}\right)\left[e^2 - (16 + \beta)\right]^{\frac{1}{2}} \quad (141)$$

Range 2, $\beta \geq 4$. Stable bifurcational buckling takes place and the load at the bifurcation point is given by (the subscript cl.B. means classical buckling)

$$Q_{cl.B.} = \frac{k^2 - 1}{k^2}(\beta - k^2)\left[e^2 - \frac{4}{k^2}(k^4 + \beta)\right]^{\frac{1}{2}} \quad (142)$$

We see from this expression that the smallest bifurcation load and the corresponding mode of deformation depend on the value of the modulus of foundation, β.

For $4 \leq \beta \leq 36$, $k = 2$ and $Q_{cl.B.} = \frac{3}{4}(\beta - 4)\left[e^2 - (18 + \beta)\right]^{\frac{1}{2}}$

For $36 \leq \beta \leq 144$, $k = 3$ and $Q_{cl.B.} = \frac{8}{9}(\beta - 9)\left[e^2 - \frac{4}{9}(81 + \beta)\right]^{\frac{1}{2}}$

For $144 \leq \beta \leq 400$, $k = 4$ and $Q_{cl.B.} = \frac{15}{16}(\beta - 16)\left[e^2 - \frac{1}{4}(256 + \beta)\right]^{\frac{1}{2}}$

Note that at $\beta = k^2(k+1)^2$, bifurcation occurs either through an a_k-mode or through an a_{k+1}-mode or a combination of a_k- and a_{k+1}-modes (the three-mode case).

Numerical results are presented graphically in Figs. 7.9 and 7.10. For $\beta = 0$, the results reduce to those reported in Fung and Kaplan (1952) and Hoff and Bruce (1954). For $\beta = 2$, if $e \leq \sqrt{12}$, there is no possibility of snapthrough. If $\sqrt{12} < e < \sqrt{21}$, the critical load is given by

$$q_{1cr} = -3e - \frac{1}{2}\left(\frac{e^2 - 12}{3}\right)^{\frac{3}{2}}$$

and snapping occurs through a limit point instability (top-of-the-knee). If $e > \sqrt{21}$ the critical load is given by

$$q_{1cr} = -3e - \frac{9}{4}\left(e^2 - 18\right)^{\frac{1}{2}}$$

and snapping takes place through an unstable bifurcation. These results are shown in Fig. 7.9 as $(-q_{cr})$ versus the initial rise parameter.

For $\beta > 4$ the results are presented in Fig. 7.10 as classical buckling load versus the initial rise parameter. Note that when $\beta = 36$, the stable branch is characterized by the a_2-mode alone or a_3-mode alone, or a combined a_2-, a_3-mode. This phenomenon is similar to the pinned straight column on an elastic foundation. When $\beta = k^2(k^2 + 1)$, the column can buckle in either k or $k + 1$ half-sine waves (see Fig. 6.3).

7.6   THE LOW ARCH BY THE TREFFTZ CRITERION

According to the Trefftz criterion, we must set the first variation of the second variation of the total potential equal to zero at the critical condition. In order to have a convenient expression for the second variation, we shall use one of the conditions for equilibrium or $p = $ constant, Eq. (114). Through Eqs. (105), (114), and (117)

$$\frac{2v'}{\varepsilon_E^{1/2}} + (\eta')^2 - (\eta_0')^2 = \frac{1}{\pi}\int_0^\pi \left[(\eta')^2 - (\eta_0')^2\right]d\xi \quad (143)$$

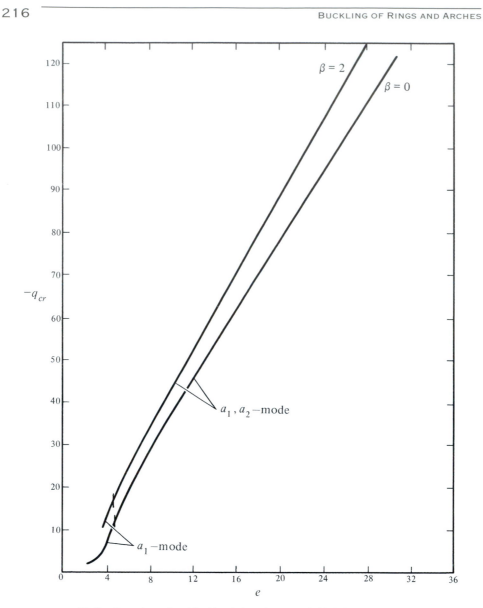

FIGURE 7.9   Snapthrough critical load, $(-q_{cr})$, versus initial rise parameter, $e(\beta < 4)$.

If we take the variations of both sides, we have

$$\pi\left[\frac{2v'}{\bar{\varepsilon}_E^{1/2}} + 2\varepsilon_1 \frac{\zeta'}{\bar{\varepsilon}_E^{1/2}} + (\eta')^2 - (\eta_0')^2 + 2\varepsilon_2\eta'\gamma' + \varepsilon_2^2(\gamma')^2\right]$$

$$= \int_0^\pi \left[(\eta')^2 - (\eta_0')^2\right]d\xi + \int_0^\pi \left[2\varepsilon_2\eta'\gamma' + \varepsilon_2^2(\gamma')^2\right]d\xi \qquad (144)$$

Now, making use of Eq. (143), we obtain

$$\pi\left[2\varepsilon_1 \frac{\zeta'}{\bar{\varepsilon}_E^{1/2}} + 2\varepsilon_2\eta'\gamma' + \varepsilon_2^2(\gamma')^2\right] = \int_0^\pi \left[2\varepsilon_2\eta'\gamma' + \varepsilon_2^2(\gamma')^2\right]\,d\xi \qquad (145)$$

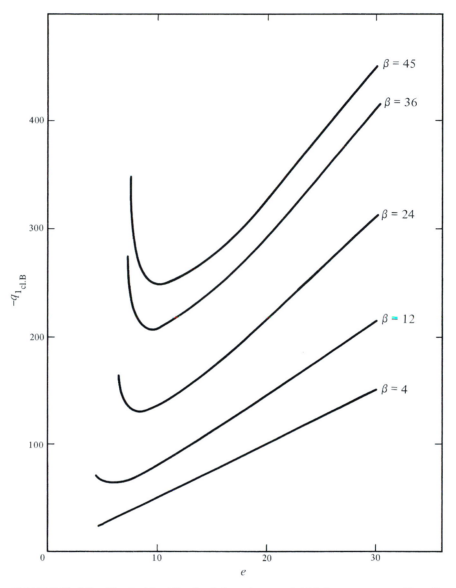

FIGURE 7.10   Classical buckling load, $(-q_{cl.B})$, versus initial rise parameter, e ($\beta \geq 4$).

Squaring both sides and integrating over π does not violate the validity of Eq. (145). Thus

$$\int_0^\pi \left[\varepsilon_1 \frac{\zeta'}{\sqrt{\varepsilon_E}} + \varepsilon_2 \eta' \gamma' + \frac{1}{2} \varepsilon_2^2 (\gamma')^2 \right]^2 d\xi = \frac{1}{4\pi} \left\{ \int_0^\pi \left[2\varepsilon_2 \eta' \gamma' + \varepsilon_2^2 (\gamma')^2 \right] d\xi \right\}^2 \quad (146)$$

Next we return to Eq. (110) which can be written as

$$\Delta U_T = \delta^1 U_T + \frac{1}{\pi} \int_0^\pi \left\{ 2 \left[\frac{\varepsilon_1^2}{\bar{\varepsilon}_E} (\zeta')^2 + 2 \frac{\varepsilon_1 \varepsilon_2}{\sqrt{\bar{\varepsilon}_E}} \eta' \zeta' \gamma' + \varepsilon_2^2 (\eta')^2 (\gamma')^2 \right. \right.$$
$$\left. \left. + \frac{\varepsilon_1 \varepsilon_2^2}{\sqrt{\bar{\varepsilon}_E}} (\gamma')^2 \zeta' + \varepsilon_2^3 \eta' (\gamma')^3 + \frac{1}{4} \varepsilon_2^4 (\gamma')^4 \right] + 2p\varepsilon_2^2 (\gamma')^2 + 2\varepsilon_2^2 (\gamma'')^2 + 2\beta \varepsilon_2^2 \gamma^2 \right\} d\xi$$

$$(147)$$

Rearranging the terms of the integrand on the right-hand side, we obtain

$$\Delta U_T = \delta^1 U_T + \frac{1}{\pi} \int_0^\pi \left\{ 2 \left[\varepsilon_1 \frac{\zeta'}{\sqrt{\bar{\varepsilon}_E}} + \varepsilon_2 \eta' \gamma' + \frac{1}{2} \varepsilon_2 (\gamma')^2 \right]^2 + 2p\varepsilon_2^2 (\gamma')^2 + 2\varepsilon_2^2 (\gamma'')^2 + 2\beta \varepsilon_2^2 \gamma^2 \right\} d\xi$$

$$(148)$$

Use of Eq. (146) for the first term in the integrand results in the following form for Eq. (148):

$$\Delta U_T = \delta^1 U_T + \frac{1}{2\pi^2} \left\{ \int_0^\pi \left[2\varepsilon_2 \eta' \gamma' + \varepsilon_2^2 (\gamma')^2 \right] d\xi \right\}^2$$
$$+ \frac{2\varepsilon_2^2}{\pi} \int_0^\pi \left[p(\gamma')^2 + (\gamma'')^2 + \beta \gamma^2 \right] d\xi$$

$$(149)$$

Performing the indicated operations and grouping terms on the right-hand side according to powers of ε_2, we have

$$\Delta U_T = \delta^1 U_T + \frac{2\varepsilon_2^2}{\pi} \int_0^\pi \left[\frac{1}{\pi} \left(\int_0^\pi \eta' \gamma' d\xi \right) \eta' \gamma' + p(\gamma')^2 + (\gamma'')^2 + \beta \gamma^2 \right] d\xi$$
$$+ \frac{2\varepsilon_2^3}{\pi} \int_0^\pi \eta' \gamma' d\xi \int_0^\pi (\gamma')^2 d\xi + \frac{\varepsilon_2^4}{2\pi} \left[\int_0^\pi (\gamma')^2 d\xi \right]^2$$

$$(150)$$

From Eq. (150), it is clear that

$$\delta^2 U_T = \frac{2\varepsilon_2^2}{\pi} \int_0^\pi \left[\frac{1}{\pi} \left(\int_0^\pi \eta' \gamma' d\xi \right) \eta' \gamma' + p(\gamma')^2 + (\gamma'')^2 + \beta(\gamma)^2 \right] d\xi \qquad (151)$$

Next, let $(\pi/2\varepsilon_2^2) \delta^2 U_T = V[\gamma]$ and find the first variation. Let $\theta(\xi)$ be a kinematically admissible function of ξ (same as γ) and ε_3 be a small constant; then

$$V[\gamma + \varepsilon_3 \theta] = \int_0^\pi \left[\left(\frac{1}{\pi} \int_0^\pi \eta' (\gamma' + \varepsilon_3 \theta') d\xi \right) \eta' (\gamma' + \varepsilon_3 \theta') \right.$$
$$\left. + p(\gamma' + \varepsilon_3 \theta')^2 + (\gamma'' + \varepsilon_3 \theta'')^2 + \beta(\gamma + \varepsilon_3 \theta)^2 \right] d\xi$$

$$(152)$$

Performing the indicated operations and grouping terms according to powers of ε_3, we have

$$V[\gamma + \varepsilon_3 \theta] = \int_0^\pi \left[\left(\frac{1}{\pi} \int_0^\pi \eta' \gamma' d\xi \right) \eta' \gamma' + p(\gamma')^2 + (\gamma'')^2 + \beta \gamma^2 \right] d\xi$$
$$+ \varepsilon_3 \int_0^\pi \left[\left(\frac{1}{\pi} \int_0^\pi \eta' \theta' d\xi \right) \eta' \gamma' + \left(\frac{1}{\pi} \int_0^\pi \eta' \gamma' d\xi \right) \eta' \theta' + 2p\gamma' \theta' + 2\gamma'' \theta'' + 2\beta \gamma \theta \right] d\xi$$
$$+ \varepsilon_3^2 \int_0^\pi \left[\left(\frac{1}{\pi} \int_0^\pi \eta' \theta' d\xi \right) \eta' \theta' + p(\theta')^2 + (\theta'')^2 + \beta \theta^2 \right] d\xi$$

$$(153)$$

It is clear from Eq. (153) that the first variation in V, which must vanish, is given by

$$\delta^1 V = 2\varepsilon_3 \int_0^\pi \left[\left(\frac{1}{\pi} \int_0^\pi \eta' \gamma' d\xi \right) \eta' \theta' + p\gamma' \theta' + \gamma'' \theta'' + \beta \gamma \theta \right] d\xi = 0 \qquad (154)$$

Integration by parts yields

$$\left[\left(-\gamma''' + p\gamma' + \left\{ \frac{1}{\pi} \int_0^\pi \eta' \gamma' d\xi \right\} \eta' \right) \theta \right]_0^\pi + \left[\gamma'' \theta' \right]_0^\pi$$
$$+ \int_0^\pi \left[\gamma'''' - p\gamma'' - \left(\frac{1}{\pi} \int_0^\pi \eta' \gamma' d\xi \right) \eta'' + \beta \gamma \right] \theta d\xi = 0 \qquad (155)$$

Since θ and γ are kinematically admissible, the first of the boundary terms is zero, Eq. (155), as long as the arch is supported, $\eta(0) = \eta(\pi) = 0$, and regardless of whether the support is pinned or clamped. Furthermore, the necessary condition for the vanishing of the first variation is the following differential equation in η and γ, and boundary conditions

$$\gamma'''' - p\gamma'' - \left(\frac{1}{\pi} \int_0^\pi \eta' \gamma' d\xi \right) \eta'' + \beta \gamma = 0 \qquad (156a)$$

$$\text{Either } \theta' = 0 \quad (\delta\gamma' = 0) \qquad\qquad \text{Or } \gamma'' = 0 \qquad (156b)$$

Since

$$\int_0^\pi \eta' \gamma' d\xi = -\int_0^\pi \eta'' \gamma d\xi + [\eta' \gamma]_0^\pi$$

Eq. (156) becomes

$$\gamma'''' - p\gamma'' + \left(\frac{1}{\pi} \int_0^\pi \eta'' \gamma d\xi \right) \eta'' + \beta \gamma = 0 \qquad (157)$$

In summary, we conclude that the response of the arch (primary path), $\eta(\xi)$, the critical load, q_{cr}, and the buckling mode, $\gamma(\xi)$, are established through the simultaneous solution of Eqs. (115) and (157) subject to the proper boundary conditions. This is demonstrated in the next section where we consider the pinned half-sine arch under a half-sine spatial distribution of the load.

7.6.1 THE SINUSOIDAL ARCH

Consider a half-sine arch under a half-sine load pinned at both ends. It has been demonstrated in Section 7.5 that Eq. (115) is satisfied (with $\beta = 0$) if

$$\frac{1}{4}\left(r_1^2 - e^2 + k^2 a_k^2 \right) r_1 + r_1 = Q$$
$$\left[\frac{1}{4}\left(r_1^2 - e^2 + k^2 a_k^2 \right) + k^2 \right] a_k = 0 \qquad (127)$$

where

$$\eta_0 = e \sin \xi, \quad q = q_1 \sin \xi$$
$$r_1 = a_1 + e, \quad Q = q_1 + e$$
$$\eta = (e + a_1) \sin \xi + a_k \sin k\xi \qquad (158)$$
$$p = \frac{1}{2\pi} \int_0^\pi \left[(\eta')^2 - (\eta_0')^2 \right] d\xi$$

Substitution of the needed expressions in Eqs. (158) into Eq. (157) yields

$$\gamma'''' - \frac{1}{4}\left(r_1^2 - e^2 + k^2 a_k^2\right)\gamma'' + \left(r_1 \sin \xi + k^2 a_k \sin k\xi\right)$$
$$\cdot \frac{1}{\pi}\int_0^\pi \left(r_1 \sin \xi + k^2 a_k \sin k\xi\right)\gamma d\xi = 0 \tag{159}$$

If we let $\gamma = \sum\limits_{m=1}^{\infty} A_m \sin m\xi$, we note that every term in the series for γ is kinematically admissible and satisfies the boundary conditions, $\gamma''(0) = \gamma''(\pi) = 0$. Substitution into Eq. (157), because of the linear independence of the functions $\sin m\xi$, yields

$$m = 1: \quad A_1\left[1 + \frac{1}{4}\left(r_1^2 - e^2 + k^2 a_k^2\right)\right] + \frac{1}{2}r_1\left(r_1 A_1 + k^2 a_k A_k\right) = 0 \tag{160}$$

$$m = k: \quad k^2 A_k\left[k^2 + \frac{1}{4}\left(r_1^2 - e^2 + k^2 a_k^2\right)\right] + \frac{k^2}{2}a_k\left(r_1 A_1 + k^2 a_k A_k\right) = 0 \tag{161}$$

$$m \neq 1, k: \quad m^2 A_m\left[m^2 + \frac{1}{4}\left(r_1^2 - e^2 + k^2 a_k^2\right)\right] = 0 \tag{162}$$

From Eq. (162) it is clear that, if one A_m is not zero, all other A_m must be zero. Furthermore, it can be shown that all A_m must be zero for a meaningful solution to exist (see Problems at the end of the chapter).

Now we can proceed to find the position of the bifurcation point $(r_{1_{cr}}, a_{k_{cr}}, Q_{cr})$ and the buckling mode (A_1, A_k). This is done by seeking the simultaneous satisfaction of Eqs. (127), (160), and (161).

$$\frac{1}{4}\left(r_{1_{cr}}^2 - e^2 + k^2 a_{k_{cr}}^2\right)r_{1_{cr}} + r_{1_{cr}} = Q_{cr}$$
$$\left[\frac{1}{4}\left(r_{1_{cr}}^2 - e^2 + k^2 a_{k_{cr}}^2\right) + k^2\right]a_{k_{cr}} = 0$$
$$A_1\left[1 + \frac{1}{4}\left(r_{1_{cr}}^2 - e^2 + k^2 a_{k_{cr}}^2\right)\right] + \frac{1}{2}a_{k_{cr}}\left(r_{1_{cr}} A_1 + k^2 a_{k_{cr}} A_k\right) = 0$$
$$A_k\left[k^2 + \frac{1}{4}\left(r_{1_{cr}}^2 - e^2 + k^2 a_{k_{cr}}^2\right)\right] + \frac{1}{2}a_{k_{cr}}\left(r_{1_{cr}} A_1 + k^2 a_{k_{cr}} A_k\right) = 0 \tag{163}$$

Equations (163) denote a system of four equations in five unknowns; $r_{1_{cr}}$ and $a_{k_{cr}}$ are the positions of the bifurcation point on the r_1-a_k equilibrium positions space (see Fig. 7.4), Q_{cr} is the corresponding critical load (limit point or bifurcation), and A_1, A_k are the amplitudes of the buckling mode (both cannot be determined uniquely).

Recognizing that, as the load is increased from zero, the primary path is associated with an r_1-only mode ($a_k \equiv 0$) and that through the Trefftz criterion we may obtain both the limit point as well as the bifurcation point, we must consider only the case of $r_1 \neq 0$ and $a_k \equiv 0$. There is no need to consider the case of $r_1 \neq 0$ and $a_k \neq 0$ which may also satisfy the equilibrium equations. (See cases I and II of Section 7.5.) With this, Eqs. (163) become

$$\frac{1}{4}\left(r_{1_{cr}}^2 - e^2\right)r_{1_{cr}} + r_{1_{cr}} = Q_{cr} \tag{164}$$

$$A_1\left[1 + \frac{1}{4}\left(r_{1_{cr}}^2 - e^2\right) + \frac{1}{2}r_{1_{cr}}^2\right] = 0 \tag{165}$$

$$A_k \left[k^2 + \frac{1}{4} \left(r_{1_{cr}}^2 - e^2 \right) \right] = 0 \tag{166}$$

Equations (164) through (166) suggest two possible solutions: (1) $A_1 \neq 0$, $A_k = 0$ and (2) $A_1 = 0$, $A_k \neq 0$. In the case for which $A_1 \neq 0$, and $A_k \equiv 0$, from Eq. (165) we have

$$r_{1cr} = \left(\frac{e^2 - 4}{3} \right)^{\frac{1}{2}} \tag{167}$$

Substitution of this expression for r_{1cr}, Eq. (167), into Eq. (164) gives

$$Q_{cr} = -\frac{1}{2} \left(\frac{e^2 - 4}{3} \right)^{\frac{3}{2}} \tag{168}$$

These are the results obtained in Section 7.5 for top-of-the-knee buckling (limit point stability). It is clear that buckling is possible for arches with $e > 2$.

In the case for which $A_1 = 0$ and $A_k \neq 0$, from Eq. (166) we obtain

$$r_{1cr} = \left(e^2 - 4k^2 \right)^{\frac{1}{2}} \tag{169}$$

Substitution into Eq. (164) yields

$$Q_{cr} = -\left(k^2 - 1 \right) \left(e^2 - 4k^2 \right) \tag{170}$$

Minimization of Q_{cr} with respect to integer values of k show that $k = 2$ and

$$\begin{aligned} r_{1cr} &= \left(e^2 - 16 \right)^{\frac{1}{2}} \\ Q_{cr} &= -3 \left(e^2 - 16 \right)^{\frac{1}{2}} \end{aligned} \tag{171}$$

These are identical to the results obtained in the case of bifurcation snapping in Section 7.5. Note that antisymmetric buckling is possible if $e > 4$. Note also that if the limit point is reached before the bifurcation point (see Fig. 7.6), Q_{cr} is still given by Eq. (168):

$$\left(\frac{e^2 - 4}{3} \right)^{\frac{1}{2}} > \left(e^2 - 16 \right)^{\frac{1}{2}}$$

$$e < \sqrt{22}$$

Thus, for $2 < e < \sqrt{22}$

$$Q_{cr} = -\frac{1}{2} \left(\frac{e^2 - 4}{3} \right)^{\frac{3}{2}}$$

and for $e > \sqrt{22}$

$$Q_{cr} = -3 \left(e^2 - 16 \right)^{\frac{1}{2}}$$

7.7 ENERGY FORMULATION BASED ON GEOMETRICALLY EXACT THEORY

In this section we specialize the geometrically exact theory for the case of stretching and bending of isotropic beams that have constant initial curvature. We consider the case in which both the undeformed beam and its deformed counterpart possess the same plane of symmetry. For the kinematics, the geometrically exact one-dimensional (1-D) measures of deformation are specialized for small strain. A 1-D constitutive law is used in which the magnitudes of initial radius of curvature and wavelength of deformation are assumed to be comparable, and both the strain and the ratio of cross-sectional diameter to initial radius of curvature (h/R) are taken to be small. In spite of a very simple final expression for the second variation of the total potential, it is shown that the only restriction on the validity of the buckling analysis is that the pre-buckling strain remains small.

Buckling of rings and high arches will be considered as examples. For illustrating the approach, we consider a pressure loading which is a constant force per unit deformed length and acting perpendicular to the reference line of the deformed beam. This is the closest representation of hydrostatic pressure. Although it is a follower force, we will prove that for many practical cases it is conservative, having a potential, in accordance with Berdichevsky (1983).[1]

7.7.1   1-D STRAIN ENERGY

To form the strain energy of a planar, constant-curvature beam, we develop the geometries of both undeformed and deformed states. The beam is symmetric about the plane in which it is initially curved, and its displacement field is symmetric about that plane. We then make use of a 1-D strain energy per unit length derived for initially curved beams by use of the variational-asymptotic method by Berdichevsky and Staroselskii (1983) and Hodges (1999). This function depends only on the geometrically exact stretching and bending measures, which we specialize for the case of small strain.

Undeformed State

Consider an initially curved beam with radius of curvature R in its undeformed state. The undeformed beam reference line (the line of area centroids will suffice in this case) is shown as the heavy, dark line in Fig. 7.11. The position vector from some fixed point to an arbitrary point p on the beam reference line is denoted by $\mathbf{r}(x_1)$, where $x_1 = R\phi$ is the arc-length coordinate along the undeformed beam reference line. The base vectors associated with the undeformed beam are $\mathbf{b}_1(x_1)$, $\mathbf{b}_2(x_1)$, and

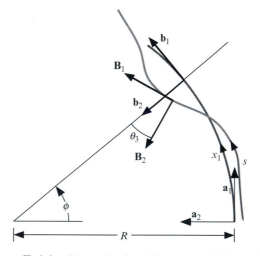

FIGURE 7.11   Schematic of undeformed and deformed beam.

[1]Portions of this material are based on Hodges (1999), used by permission.

$\mathbf{b}_3 = \mathbf{b}_1 \times \mathbf{b}_2 = \mathbf{a}_3$ ($\mathbf{b}_3$ is not a function of x_1). Spatially fixed base vectors are denoted by $\mathbf{a}_i$, for $i = 1$, 2, and 3, as shown in Fig. 7.11; note also that $\mathbf{a}_3 = \mathbf{a}_1 \times \mathbf{a}_2$. The relationship between these vectors is seen from the geometry to be

$$\left\{\begin{array}{c} \mathbf{b}_1 \\ \mathbf{b}_2 \\ \mathbf{b}_3 \end{array}\right\} = \left[\begin{array}{ccc} \cos\phi & \sin\phi & 0 \\ -\sin\phi & \cos\phi & 0 \\ 0 & 0 & 1 \end{array}\right] \left\{\begin{array}{c} \mathbf{a}_1 \\ \mathbf{a}_2 \\ \mathbf{a}_3 \end{array}\right\} \tag{172}$$

The unit vector tangent to the curve described by $\mathbf{r}(x_1)$ is

$$\frac{d\mathbf{r}}{dx_1} = \mathbf{r}' = \mathbf{b}_1 \tag{173}$$

where $(\ )' = d(\ )/dx_1$. The curvature vector for the undeformed state is defined as

$$\mathbf{k} = \frac{\mathbf{b}_3}{R} \tag{174}$$

so that

$$\mathbf{b}_i' = \mathbf{k} \times \mathbf{b}_i \tag{175}$$

The initial curvature then is exhibited, as expected, in

$$\mathbf{b}_1' = \frac{\mathbf{b}_2}{R} \qquad\qquad \mathbf{b}_2' = -\frac{\mathbf{b}_1}{R} \tag{176}$$

Deformed State

The deformed beam is shown as a heavy, gray line in Fig. 7.11. The mathematical description of the deformed state is a straightforward extension of the above. The position vector to the reference line of the deformed beam is

$$\mathbf{R}(x_1) = \mathbf{r}(x_1) + \mathbf{u}(x_1) \tag{177}$$

where $u_i(x_1) = \mathbf{u} \cdot \mathbf{b}_i$ is the displacement vector. The curvature vector for the deformed state is

$$\mathbf{K} = \left(\frac{1}{R} + \kappa_3\right)\mathbf{b}_3 \tag{178}$$

To derive a theory of the "classical" type, which neglects transverse shear deformation, we require the cross-sectional plane of the deformed beam to be normal to the tangent of the local deformed beam reference line, so that

$$\mathbf{R}' = (1 + \varepsilon)\mathbf{B}_1 \tag{179}$$

where $\varepsilon = s' - 1$, with s as the running arc-length of the deformed beam, and $\mathbf{B}_1$ is the unit vector tangent to the reference line of the deformed beam.

Strain Energy per Unit Length

Assuming the cross section to be doubly symmetric, the strain energy density is given by

$$\Psi = \frac{1}{2}\left[EA\varepsilon^2 + EI_3\kappa_3^2 - \frac{2(1+\nu)EI_3}{R}\varepsilon\kappa_3\right] + O\left(\frac{EAh^2\varepsilon^2}{R^2}\right) \tag{180}$$

where the $O(\ )$ means that all terms of order $(\ )$ are excluded. Based on this approximate 1-D energy, the corresponding 1-D constitutive law is then

$$
\begin{aligned}
F_1 &= \frac{\partial \Psi}{\partial \varepsilon} = EA\varepsilon - \frac{(1+\nu)EI_3}{R}\kappa_3 + O\left(\frac{EAh^2\varepsilon}{R^2}\right) \\
M_3 &= \frac{\partial \Psi}{\partial \kappa_3} = -\frac{(1+\nu)EI_3}{R}\varepsilon + EI_3\kappa_3 + O\left(\frac{EAh^3\varepsilon^2}{R^2}\right)
\end{aligned}
\tag{181}
$$

where F_1 and M_3 are the tangential force resultant and bending moment, respectively. The two underlined terms are $O(h/R)$ relative to the leading terms and represent a stretching-bending coupling indicative of a shift in the position of the neutral axis away from the area centroid. The only approximations in the dimensional reduction (i.e., the determination of the constitutive law in Eq. 181) are thus $\varepsilon \ll 1$ and $h^2/R^2 \ll 1$. Later it will be shown that these conditions dovetail into one condition for ring- and high-arch-buckling problems. The next approximation would produce terms in the 1-D energy which are $O(h^2/R^2)$ relative to the leading terms. These are associated with large initial curvature and transverse shear effects, not necessary in the present treatment. It should also be noted that the stretch-bending term does not affect the buckling loads for the idealized pre-buckling states considered here; however, general pre-buckling displacements, curvature, and bending moment of high arches are impossible to calculate accurately without this term, as shown by Hodges (1999).

1-D Strain-Displacement Relations

From the above, the unit tangent to the reference line of the deformed beam (see Fig. 7.1) is

$$
\frac{d\mathbf{R}}{ds} = \mathbf{B}_1
\tag{182}
$$

where $\mathbf{B}_1$ is a unit vector tangent to the reference line at P and s is the arc-length coordinate along the deformed beam. By choosing a specific set of displacement variables, one can find the relationship between s and x_1.

Let $\mathbf{u} = u_1\mathbf{b}_1 + u_2\mathbf{b}_2$. This way, u_1 is the "tangential" displacement and u_2 is the "radial" displacement. Using Eqs. (176) to express the derivatives of the base vectors, one finds

$$
\mathbf{B}_1 = \frac{d\mathbf{R}}{ds} = \frac{\mathbf{R}'}{s'} = \frac{\left(1+u_1'-\frac{u_2}{R}\right)\mathbf{b}_1 + \left(u_2'+\frac{u_1}{R}\right)\mathbf{b}_2}{s'}
\tag{183}
$$

However, it is also clear from the restriction to planar deformation that one can regard the unit vectors $\mathbf{B}_1$, $\mathbf{B}_2$, and $\mathbf{B}_3 = \mathbf{B}_1 \times \mathbf{B}_2 = \mathbf{b}_3 = \mathbf{a}_3$ in terms of a simple rotation by an angle, say, θ_3 such that

$$
\left\{ \begin{array}{c} \mathbf{B}_1 \\ \mathbf{B}_2 \\ \mathbf{B}_3 \end{array} \right\} =
\left[\begin{array}{ccc} \cos\theta_3 & \sin\theta_3 & 0 \\ -\sin\theta_3 & \cos\theta_3 & 0 \\ 0 & 0 & 1 \end{array} \right]
\left\{ \begin{array}{c} \mathbf{b}_1 \\ \mathbf{b}_2 \\ \mathbf{b}_3 \end{array} \right\}
\tag{184}
$$

Now, we can determine three things by comparing the expressions for $\mathbf{B}_1$ in Eqs. (183) and (184). The first two are that $\cos\theta_3$ and $\sin\theta_3$ can be found as

$$\cos\theta_3 = \frac{1 + u_1' - \frac{u_2}{R}}{s'}$$

$$\sin\theta_3 = \frac{u_2' + \frac{u_1}{R}}{s'} \tag{185}$$

and the third is that, since $\mathbf{B}_1$ must be a unit vector

$$s' = \sqrt{\left(1 + u_1' - \frac{u_2}{R}\right)^2 + \left(u_2' + \frac{u_1}{R}\right)^2} \tag{186}$$

Since $s' = 1 + \varepsilon$, for completeness we note that

$$\varepsilon = \sqrt{\left(1 + u_1' - \frac{u_2}{R}\right)^2 + \left(u_2' + \frac{u_1}{R}\right)^2} - 1 \tag{187}$$

Finally, the state of deformation of a beam is characterized by the bending energy, which depends on the curvature. Reissner (1972) makes use of a "moment strain" which is $\kappa_3 = \theta_3'$ rather than $d\theta_3/ds$. These only differ by a factor of $s' = 1 + \varepsilon$, and for a small strain analysis one can ignore ε compared to unity (i.e. set s' equal to unity) in the bending measure while retaining the stretching energy. Thus, we can develop $\kappa_3 = \theta_3'$ by differentiating the second of Eqs. (185), yielding

$$\theta_3' \cos\theta_3 = \frac{s'\left(u_2'' + \frac{u_1'}{R}\right) - s''\left(u_2' + \frac{u_1}{R}\right)^2}{s'^2} \tag{188}$$

Now using the first of Eqs. (185) and noting that

$$s'' = \frac{\left(1 + u_1' - \frac{u_2}{R}\right)\left(u_1'' - \frac{u_2'}{R}\right) + \left(u_2'' + \frac{u_1'}{R}\right)\left(u_2' + \frac{u_1}{R}\right)}{s'} \tag{189}$$

we find (after a remarkable series of cancellations!)

$$\theta_3' = \frac{\left(1 + u_1' - \frac{u_2}{R}\right)\left(u_2'' + \frac{u_1'}{R}\right) - \left(u_2' + \frac{u_1}{R}\right)\left(u_1'' - \frac{u_2'}{R}\right)}{s'^2} \tag{190}$$

which, when specialized for $R \to \infty$, is in agreement with Hodges (1984). We note that to be consistent with Hooke's law, one must restrict $\varepsilon = \max(\varepsilon, h\kappa_3)$ to be small compared to unity. Thus, for small strain we may regard $s' = 1$ in the denominator of Eq. (190), yielding a polynomial in the displacement functions and their derivatives for the moment strain, similar to the results of Epstein and Murray (1976):

$$\kappa_3 = \left(1 + u_1' - \frac{u_2}{R}\right)\left(u_2'' + \frac{u_1'}{R}\right) - \left(u_2' + \frac{u_1}{R}\right)\left(u_1'' - \frac{u_2'}{R}\right) \tag{191}$$

Other than small stretching strain $\varepsilon \ll 1$, we have made no approximations in the 1-D variables. The difference between this expression for curvature and that found in calculus texts is discussed fully in Hodges (1984). An alternative approximation that is discussed in Hodges (1984) for straight beams is to make use of the fact that the stretching strain is essentially zero in order to altogether eliminate u_1 from the moment strain (see also Hodges et al., 1980). However, this approach cannot be used to eliminate u_1 for an initially curved beam; it can be used only to write $u_1' - \frac{u_2}{R}$ in terms of $u_2' + \frac{u_1}{R}$ as done by Gellin (1980). This does not serve to eliminate a variable but instead introduces an unnecessary mathematical singularity into the formulation, along with a limit on the resulting rotation variable $|u_2' + \frac{u_1}{R}| < 1$. The magnitude of the difference between results based on Gellin's approach and the singularity-free one obtained from Eq. (191) is of the order of the strain compared to unity.

The 1-D strain energy is then the integral of Ψ over the length, namely

$$U = \frac{1}{2} \int_L \left[EA\varepsilon^2 - \frac{2EI_3(1+\nu)}{R} \varepsilon\kappa_3 + EI_3\kappa_3^2 \right] dx_1 \qquad (192)$$

where L is the total length of the beam. As one can see, the strain energy density becomes quite complicated when Eqs. (187) and (191) are substituted into Eq. (192). There are many problems for which the result does become tractable, however, and for this reason this approach is to be preferred over ad hoc approaches in which one cannot easily assess the error associated with particular approximations.

7.7.2   POTENTIAL ENERGY OF APPLIED PRESSURE LOADING

In anticipation of applying the above theory to inplane deformation and buckling, here we develop the potential energy by first finding the virtual work of the applied loading. Then we establish the criteria by which the virtual work can be represented as the variation of a functional, namely the negative of the potential energy.

Virtual Work of Pressure

We consider the case of a distributed follower force that is a constant per unit deformed beam length. This means that the local force on an element of the deformed beam is, say, $f_2\mathbf{B}_2 ds$ where f_2 is a constant. The work done by this force through a virtual displacement is then

$$\overline{\delta W} = \int_L f_2 s' \mathbf{B}_2 \cdot (\delta u_1 \mathbf{b}_1 + \delta u_2 \mathbf{b}_2) dx_1 \qquad (193)$$

where the $\overline{\delta W}$ is the virtual work and the bar over the symbol indicates that it is not necessarily equal to the variation of a functional W. We already know that $\mathbf{B}_2 = -\sin\theta_3 \mathbf{b}_1 + \cos\theta_3 \mathbf{b}_2$ so that, from Eqs. (185), we have

$$\overline{\delta W} = f_2 \int_L \left[\left(1 + u_1' - \frac{u_2}{R}\right)\delta u_2 - \left(u_2' + \frac{u_1}{R}\right)\delta u_1 \right] dx_1 \qquad (194)$$

Potential Energy Functional

For a beam of length ℓ, this can now be put into the form

$$\overline{\delta W} = f_2 \delta \int_L \left(u_2 - \frac{u_1^2}{2R} - \frac{u_2^2}{2R} - u_1 u_2' \right) dx_1 + f_2 u_1 \delta u_2 \Big|_0^\ell \qquad (195)$$

It is clear then that there are situations in which the trailing term vanishes which, in turn, allows the follower force to be derived from a potential function. Namely, this is the case if the ends of the beam are not allowed to displace, or if the beam is a closed ring, for which the ends are joined so that $u_1(\ell)\delta u_2(\ell) = u_1(0)\delta u_2(0)$; for a discussion of this type of "holonomicity" see Berdichevsky (1983), pp. 159–162. In these cases, the potential energy functional is

$$V = -f_2 \int_L \left(u_2 - \frac{u_1^2}{2R} - \frac{u_2^2}{2R} - u_1 u_2' \right) dx_1 \tag{196}$$

7.7.3 BUCKLING OF RINGS AND HIGH ARCHES

Inplane deformation and buckling of circular rings and high arches are considered as applications. A simple buckling analysis will be developed from the total potential energy, and the pre-buckling deflections will be determined for cases in which they are not trivial.

To facilitate these analyses, it is now helpful to nondimensionalize the equations. This we do by dividing through the total potential $U + V$ by EAR while simultaneously changing the meaning of certain symbols. We replace u_1 and u_2 with Ru_1 and Ru_2, respectively; we replace κ_3 with κ/R; and finally we let $(\ )'$ denote $d(\ )/d\phi$. We also introduce the new symbols $\rho^2 = I_3/AR^2$ and $\lambda = f_2 R^3/EI_3$. All these operations yield, for the nondimensional total potential $\Phi = (U + V)/EAR$

$$\Phi = \int_{-\alpha}^{\alpha} \left[\frac{\varepsilon^2}{2} - (1 + \nu)\rho^2 \varepsilon \kappa + \frac{\rho^2 \kappa^2}{2} - \lambda \rho^2 \left(u_2 - \frac{u_1^2}{2} - \frac{u_2^2}{2} - u_1 u_2' \right) \right] d\phi \tag{197}$$

where

$$\varepsilon = \sqrt{\left(1 + u_1' - u_2\right)^2 + \left(u_2' + u_1\right)^2} - 1 \tag{198}$$

and

$$\kappa = \left(1 + u_1' - u_2\right)\left(u_2'' + u_1'\right) - \left(u_2' + u_1\right)\left(u_1'' - u_2'\right) \tag{199}$$

Note that $h/R = O(\rho)$, so that $\rho^2 \ll 1$; for a ring $\alpha = \pi$.

It is helpful, before proceeding further, to rewrite κ^2 in a more compact way. To do so, we note that

$$\kappa^2 = \left(1 + u_1' - u_2\right)^2 \left(u_2'' + u_1'\right)^2$$
$$- 2\left(u_2' + u_1\right)\left(u_1'' - u_2'\right)\left(1 + u_1' - u_2\right)\left(u_2'' + u_1'\right) + \left(u_2' + u_1\right)^2 \left(u_1'' - u_2'\right)^2 \tag{200}$$

and that

$$(1 + \varepsilon)^2 = \left(1 + u_1' - u_2\right)^2 + \left(u_2' + u_1\right)^2 \tag{201}$$

Thus, Eq. (200) can be rearranged, making use of Eq. (201), to obtain

$$\kappa^2 = \left[(1 + \varepsilon)^2 - \left(u_2' + u_1\right)^2\right]\left(u_2'' + u_1'\right)^2 + \left[(1 + \varepsilon)^2 - \left(1 + u_1' - u_2\right)^2\right]\left(u_1'' - u_2'\right)^2$$
$$- 2\left(u_2' + u_1\right)\left(u_1'' - u_2'\right)\left(1 + u_1' - u_2\right)\left(u_2'' + u_1'\right) \tag{202}$$

which, in light of the fact that $\varepsilon' = s''$, given in Eq. (189), simplifies to

$$\kappa^2 = (1 + \varepsilon)^2 \left[\left(u_2'' + u_1'\right)^2 + \left(u_1'' - u_2'\right)^2 - \varepsilon'^2\right] \tag{203}$$

When $\rho^2 \kappa^2$ is compared to ε^2, the last term in Eq. (203) becomes negligible because $\rho^2 \ll 1$. For small strain κ^2 can finally be written as

$$\kappa^2 = \left(u_2'' + u_1'\right)^2 + \left(u_1'' - u_2'\right)^2 \tag{204}$$

For the first application we consider the buckling of rings and high arches. For the buckling analysis of high arches, we will follow the approach used earlier in this Chapter of assuming that the boundary conditions are such that the displacements in the pre-buckled state are the same as those for a ring with the same values of λ, ν, and ρ. This has the effect of simplifying the analysis of the pre-buckled state.

Pre-Buckled State

In the pre-buckled state, we note that the ring remains circular so that all derivatives with respect to ϕ vanish. Denoting the pre-buckled state variables with overbars and noting that $\bar{u}_2$ is the only nonzero displacement or rotation variable, we find that $\bar{\varepsilon} = -\bar{u}_2$ and $\bar{\kappa} = 0$, so that the functional reduces to

$$\overline{\Phi} = \int_{-\alpha}^{\alpha} \left[\frac{\bar{u}_2^2}{2} - \lambda \rho^2 \left(\bar{u}_2 - \frac{\bar{u}_2^2}{2} \right) \right] d\phi \tag{205}$$

from which we find, upon equating the variation to zero

$$\bar{u}_2 = \frac{\lambda \rho^2}{1 + \lambda \rho^2} \tag{206}$$

Here we make an important observation: the strain in the pre-buckled state

$$\bar{\varepsilon} = -\bar{u}_2 = -\frac{\lambda \rho^2}{1 + \lambda \rho^2} \tag{207}$$

is of the order of ρ^2. So, for a consistent small-strain analysis we need to ignore ρ^2 with respect to unity. To improve on this analysis we would not only need to keep ρ^2 compared to unity, we would also have to take transverse shear into account thereby improving on Eq. (192) so that ρ^2 is not neglected compared to unity in the dimensional reduction. This would be much more complicated. Furthermore, because ρ^2 is of the order of the strain, if terms of order ρ^2 are to be kept to be consistent we would be compelled to treat material non-linearities. Obviously, since the ring is slender and the pre-buckling strain is small compared to unity, these modifications are not necessary. This observation leads to a great simplification in the buckling analysis.

Buckling Analysis

To further simplify the total potential, we consider that the perturbations of the pre-buckled state at the onset of buckling can be regarded as arbitrarily small. We need to keep all terms of powers 1 and 2 in the perturbation quantities. Using the concept of the Taylor series to make certain all such terms are retained, we note that

$$\begin{aligned}
\varepsilon &= \bar{\varepsilon} + \hat{\varepsilon}_1 + \hat{\varepsilon}_2 \\
\kappa &= \hat{\kappa}_1 + \hat{\kappa}_2
\end{aligned} \tag{208}$$

The subscripts indicate the power of the perturbation displacements. Because of the nonzero value of $\bar{\varepsilon}$, we need both first- and second-order terms. For small strain, we find

$$\hat{\varepsilon}_1 = \hat{u}'_1 - \hat{u}_2$$

$$\hat{\varepsilon}_2 = \frac{1}{2(1 + \overline{\varepsilon})}\left(\hat{u}'_2 + \hat{u}_1\right)^2 = \frac{1}{2}\left(\hat{u}'_2 + \hat{u}_1\right)^2$$

$$\hat{\kappa}_1 = \hat{u}''_2 + \hat{u}'_1$$ $$\tag{209}$$

$$\hat{\kappa}_2 = \left(\hat{u}'_1 - \hat{u}_2\right)\left(\hat{u}''_2 + \hat{u}'_1\right) + \left(\hat{u}'_2 + \hat{u}_1\right)\left(\hat{u}''_1 - \hat{u}'_2\right)$$

Now we can write the perturbations of the energy. First, keeping only terms that are linear in the ($\wedge$) quantities, we obtain

$$\hat{\Phi}_1 = \int_{-\alpha}^{\alpha}\left[\overline{\varepsilon}\hat{\varepsilon}_1 - (1 + \nu)\rho^2\overline{\varepsilon}\hat{\kappa}_1 - \lambda\rho^2(1 + \overline{\varepsilon})\hat{u}_2\right]d\phi \tag{210}$$

the variation of which is identically zero, as expected. Equating to zero the variation with respect to $\hat{u}_1$, one obtains an identity; equating to zero the variation with respect to $\hat{u}_2$, one finds an equation that is satisfied given Eq. (206).

Now, let us consider the second-order terms (which amounts to a second variation):

$$\hat{\Phi}_2 = \frac{1}{2}\int_{-\alpha}^{\alpha}\left[2\overline{\varepsilon}\hat{\varepsilon}_2 + \hat{\varepsilon}_1^2 - 2(1 + \nu)\rho^2\overline{\varepsilon}\hat{\kappa}_2 - 2(1 + \nu)\rho^2\hat{\varepsilon}_1\hat{\kappa}_1 + \rho^2\left(\hat{u}''_2 + \hat{u}'_1\right)^2\right.$$
$$\left. + \rho^2\left(\hat{u}''_1 - \hat{u}'_2\right)^2 + \lambda\rho^2\left(\hat{u}_1^2 + \hat{u}_2^2 + 2\hat{u}_1\hat{u}'_2\right)\right]d\phi \tag{211}$$

When $\overline{\varepsilon} = -\lambda\rho^2$ is substituted into Eq. (211), the third term drops out, being $O(\rho^4)$ relative to the leading term. It should be clear that all the remaining terms in $\hat{\Phi}_2$ are proportional to ρ^2 except the $\hat{\varepsilon}_1^2$ term. Minimization of $\hat{\Phi}_2$ with respect to $\hat{u}_1$ shows that the leading term is essentially driven to zero and that

$$\hat{u}'_1 = \hat{u}_2 + \rho^2\nu\left(\hat{u}''_2 + \hat{u}_2\right) + \ldots \tag{212}$$

or alternatively

$$\hat{u}_2 = \hat{u}'_1 - \rho^2\nu\left(\hat{u}'''_1 + \hat{u}'_1\right) + \ldots \tag{213}$$

so that

$$\hat{\varepsilon}_1 = \rho^2\nu\left(\hat{u}''_2 + \hat{u}_2\right) + \ldots$$
$$= \rho^2\nu\left(\hat{u}'''_1 + \hat{u}'_1\right) + \ldots \tag{214}$$

Either $\hat{u}_1$ or $\hat{u}_2$ can be eliminated completely from the energy using these relations. Considering first the elimination of $\hat{u}_2$, substitution of Eq. (213) into Eq. (211), one obtains

$$\hat{\Phi}_2 = \frac{\rho^2}{2}\int_{-\alpha}^{\alpha}\left[\left(\hat{u}'''_1 + \hat{u}'_1\right)^2 - \lambda\left(\hat{u}''_1 + \hat{u}_1\right)^2 + \lambda\left(\hat{u}_1^2 + \hat{u}_1'^2 + 2\hat{u}_1\hat{u}''_1\right)\right]d\phi \tag{215}$$

which simplifies to

$$\hat{\Phi}_2 = \frac{\rho^2}{2}\int_{-\alpha}^{\alpha}\left[\left(\hat{u}'''_1 + \hat{u}'_1\right)^2 - \lambda\hat{u}''^2_1 + \lambda\hat{u}'^2_1\right]d\phi \tag{216}$$

The essential boundary conditions on $\hat{u}_2$ must be transferred over as essential boundary conditions on $\hat{u}_1'$ in order to make proper use of this energy functional.

Alternatively, the variable $\hat{u}_1$ can be eliminated but not without a somewhat unusual treatment of the boundary conditions. Integrating both sides of Eq. (212) for a ring, when specialized for $\rho^2 \ll 1$, one finds

$$\hat{u}_1 \big|_{-\alpha}^{\alpha} = 0 = \int_{-\alpha}^{\alpha} \hat{u}_2 d\phi \tag{217}$$

This equation is satisfied for the sinusoidal comparison functions used in predicting ring buckling (for which $\alpha = \pi$). However, for buckling of high arches one must be careful. Although functions that are antisymmetric about $\phi = 0$ automatically satisfy this condition for high arches, symmetric functions do not, in general. This condition is an essential (i.e., a displacement) boundary condition, and it is therefore mandatory that any admissible/comparison function satisfy it or else the results from Rayleigh's method, for example, will be wrong. Using Eq. (212), one can write the energy functional in terms of $\hat{u}_2$ only as

$$\hat{\Phi}_2 = \frac{\rho^2}{2} \int_{-\alpha}^{\alpha} \left[\left(\hat{u}_2'' + \hat{u}_2 \right)^2 - \lambda \left(\hat{u}_2'^2 - \hat{u}_2^2 \right) \right] d\phi \tag{218}$$

subject to

$$\int_{-\alpha}^{\alpha} \hat{u}_2 d\phi = 0 \tag{219}$$

These expressions for the second variation of the total potential provide very simple treatments relative to most published work. In spite of this simplicity, the only approximation employed is that $\bar{\varepsilon} \ll 1$, which, because of the pre-buckling state, is equivalent to $\rho^2 \ll 1$.

Now, using either Eq. (216) or Eqs. (218) and (219) one can derive an upper bound for the buckling load of a ring from Rayleigh's quotient. For example, using the latter

$$\lambda_{\text{cr}} = \frac{\int_{-\alpha}^{\alpha} \left(\hat{u}_2'' + \hat{u}_2 \right)^2 d\phi}{\int_{-\alpha}^{\alpha} \left(\hat{u}_2'^2 - \hat{u}_2^2 \right) d\phi} \tag{220}$$

and assuming that $\hat{u}_2 = \sin m\phi$, which satisfies Eq. (219), one finds that

$$\lambda_{\text{cr}} = m^2 - 1 \tag{221}$$

Since $m = 1$ is a rigid-body mode, as shown in Section 7.1, the critical load is then at $m = 2$ so that

$$\lambda_{\text{cr}} = 3 \tag{222}$$

in agreement with results therein.

Earlier in this chapter, high arches were treated approximately by allowing the boundaries to move in the pre-buckling problem, yielding a simplified pre-buckling state identical to that of the ring. For those cases described in Section 7.2, one can quite easily verify that Eq. (220), subject to Eq. (219), as well as its analog in terms of $\hat{u}_1$, provide upper bounds for the published symmetric or antisymmetric buckling loads when either symmetric or antisymmetric admissible or comparison functions are substituted therein.

7.8 ALTERNATIVE FORMULATION BASED ON ELASTICA THEORY

In this section we provide an alternative formulation based on the intrinsic equations augmented by appropriate kinematical and constitutive equations. For this problem the constitutive law can be written as

$$
\left\{ \begin{array}{c} F_1 \\ M_3 \end{array} \right\} = \left[\begin{array}{cc} EA & -\dfrac{EI(1+\nu)}{R} \\ -\dfrac{EI(1+\nu)}{R} & EI \end{array} \right] \left\{ \begin{array}{c} \varepsilon \\ \kappa_3 \end{array} \right\} \tag{223}
$$

The pre-buckling state can be described in terms of

$$
\begin{aligned}
F &= \overline{F}_1 e_1 \\
M &= \overline{M}_3 e_3 \\
K &= \frac{1}{R} e_3 \quad \text{or} \quad \overline{\kappa}_3 = 0
\end{aligned} \tag{224}
$$

where $\overline{F}_1$ and $\overline{M}_3$ are constants, $e_1 = \lfloor 1 \ \ 0 \ \ 0 \rfloor^T$, and $e_3 = \lfloor 0 \ \ 0 \ \ 1 \rfloor^T$. The pre-buckling state is brought about by a constant force per unit length

$$
f = \frac{\lambda EI}{R^3} e_2 \tag{225}
$$

where $e_2 = \lfloor 0 \ \ 1 \ \ 0 \rfloor^T$ and λ is a nondimensional pressure parameter. Thus, the only non-trivial pre-buckling equilibrium equation is

$$
\frac{\overline{F}_1}{R} + (1 + \overline{\varepsilon}) \frac{\lambda EI}{R^3} = 0 \tag{226}
$$

which can be solved yielding

$$
\overline{\varepsilon} = -\frac{\lambda \rho^2}{1 + \lambda \rho^2} \approx -\lambda \rho^2 \tag{227}
$$

where

$$
\rho^2 = \frac{I}{AR^2} \tag{228}
$$

in agreement with Eq. (207). We note that the pre-buckling strain is of the order of ρ^2, so that $\rho^2 \ll 1$.

The perturbations of the equilibrium equations are then

$$
\begin{aligned}
\hat{F}_1' - \frac{\hat{F}_2}{R} &= 0 \\
\hat{F}_2' + \overline{F}_1 \hat{\kappa}_3 + \frac{\hat{F}_1}{R} + \frac{\lambda EI \hat{\varepsilon}}{R^3} &= 0 \\
\hat{M}_3' + \hat{F}_2 &= 0
\end{aligned} \tag{229}
$$

and the perturbations of the constitutive equations can be written as

$$
\left\{ \begin{array}{c} \hat{F}_1 \\ \dfrac{\hat{M}_3}{R} \end{array} \right\} = EA \left[\begin{array}{cc} 1 & -(1+\nu)\rho^2 \\ -(1+\nu)\rho^2 & \rho^2 \end{array} \right] \left\{ \begin{array}{c} \hat{\varepsilon} \\ R\hat{\kappa}_3 \end{array} \right\} \tag{230}
$$

Given the smallness of ρ^2, we may write the inverse relation as

$$\begin{Bmatrix} \hat{\varepsilon} \\ R\hat{\kappa}_3 \end{Bmatrix} = \frac{1}{EA} \begin{bmatrix} 1 & 1+\nu \\ 1+\nu & \dfrac{1}{\rho^2} \end{bmatrix} \begin{Bmatrix} \hat{F}_1 \\ \dfrac{\hat{M}_3}{R} \end{Bmatrix} \tag{231}$$

Letting $(\ )'$ be replaced by $(\ )'/R = d(\ )/d\phi$ and dropping terms of order ρ^2 compared to unity, the perturbation equations in terms of force and moment quantities only become

$$\hat{F}_1' - \hat{F}_2 = 0$$

$$\hat{F}_2' - \lambda \frac{M_3}{R} + \hat{F}_1 = 0 \tag{232}$$

$$\frac{\hat{M}_3'}{R} + \hat{F}_2 = 0$$

These equations suggest that that $\hat{F}_1$, $\hat{F}_2$, and $\hat{M}_3/R$ are all of the same order and can be collapsed into a single equation, for example,

$$\hat{F}_2'' + (\lambda + 1)\hat{F}_2 = 0 \tag{233}$$

For a ring, we can assume that $\hat{F}_2 = F_2 \sin(m\phi)$ so that

$$F_2(\lambda - m^2 + 1) = 0 \tag{234}$$

or

$$\lambda = m^2 - 1 \tag{235}$$

Since $m = 0$ and $m = 1$ are rigid-body modes, the minimum value corresponds to $m = 2$ so that

$$\lambda_{\text{cr}} = 3 \tag{236}$$

in agreement with the treatment earlier in this Chapter and illustrating the simplicity of the approach based on the intrinsic equations.

PROBLEMS

1. Find the critical load for a thin ring under uniform pressure (load case I) when one section of the ring is fixed in space (say, at $\theta = 0$, $v = 0$, $\varphi = 0$, and $w = 0$).
2. Consider an arch as shown in Fig. 7.2 and labeled "original" and find p_{cr} (load case I). As boundary conditions, assume that the shear in the radial direction is zero instead of the displacement.
3. Consider a clamped arch on rollers (similar to Fig. 7.2 labeled "original") and find p_{cr} (load case I).
4. Using the alternate solution (Section 7.4.2), find p_{cr} for a pinned arch for load case II.
5. Using the alternate solution (Section 7.4.2) find p_{cr} for a pinned arch for load case III.
6. Show that there is no meaningful solution if $A_m \neq 0$ in Eq. (162).

7. Using the Trefftz criterion approach (Section 7.6.1), analyze a pinned half-sine arch under a half-sine loading and resting on an elastic foundation.

8. Using the energy formulation for Section 7.7, find approximate Raleigh quotient solutions for a clamped high arch under case I loading and comment on their accuracy.

9. Using the elastica formulation of Section 7.8, find the exact solution for a pinned high arch under case I loading.

REFERENCES

Berdichevsky, V. L. (1983). *Variational Principles of Continum Mechanics.* Nauka, Moscow.

Berdichevsky, V. L., and Starosel'skii, L. A. (1983). On the theory of curvilinear Tinoshenko-type rods. *Prikl. Matem. Mekham.* 47, 809–817.

Biezeno, C. B. (1938). "Das Durchschlagen eines Schwach Gekrummten Stabes." *Zeitschrift Ange. Math. und Mekh.,* Vol. 18, p. 21.

Boresi, A. (1955). A refinement of the theory of buckling of rings under uniform pressure, *J. Appl. Mech.,* Vol. 22, pp. 95–102.

Epstein, M. and Murray, D. W. (1976). Large deformation in-plane analysis of elastic beams. *Computers and Structures* 6; 1–9.

Franciosi, V., Augusti, G., and Sparacio, R. (1964). Collapse of arches under repeated loading, *J. Struct. Div.,* ASCE, Vol. 90, STI, p. 165.

Fung, Y. C. and Kaplan, A. (1952). "Buckling of Low Arches or Curved Beams of Small Curvature." NACA TN 2840.

Gellin, S. (1980). The plastic buckling of long cylindrical shells under pure bending. *International Journal of Solids and Structures* 16, 397–407.

Gjelsvik, A. and Bodner, S. R. (1962). The energy criterion and snap buckling of arches, *J. Eng. Mech. Div.,* ASCE, Vol. 88, EM5, p. 87.

Hodges, D. H. (1984). Proper definition of curvature in nonlinear beam kinematics. *AIAA Journal* 22, 1825–1827.

Hodges, D. H. (1999). Non-linear inplane deformation and buckling of rings and high arches. *International Journal of Non-Linear Mechanics* 34, 723–737.

Hodges, D. H., Ormiston, R. A., and Peters, D. A. (1980). On the nonlinear deformation geometry of Euler-Bernoulli beams. Technical Report TP-1566, NASA. Also AVRADCOM Technical Report 80-A-1.

Hoff, N. J. and Bruce, V. G. (1954). Dynamic analysis of the buckling of laterally loaded flat arches, *J. Math and Phys.,* Vol. XXXII. No. 4.

Lee, H. N. and Murphy, L. M. (1968). Inelastic buckling of shallow arches, *J. Eng. Mech. Div.,* ASCE, Vol. 94, EM1, p. 225.

Marguerre, K. (1938). "Die Durchschlagskraft eines Schwach Gekrummten Balken." *Sitz. Berlin Math. Gess.,* Vol. 37, p. 92.

Masur, E. F. and Lo, D. L. C. (1972). The shallow arch-general buckling, post-buckling, and imperfection analysis, *J. Struct. Mech.,* Vol. 1, No. 1, p. 91.

Navier, (1833). "Résumé des Leçons sur L'Application de la Méchanique." 2nd ed., p. 273, Paris.

Onat, E. T. and Shu, L. S. (1962). Finite deformation of a rigid perfectly plastic arch, *J. Appl. Mech.,* Vol. 29, No. 3, p. 549.

Reissner, E. (1972). On one-dimensional finite-strain beam theory: The plane problem. *Journal of Applied Mathematics and Physics (ZAMP)* 23, 795–804.

Roorda, J. (1965). Stability of structures with small imperfections, *J. Eng. Mech. Div.,* ASCE, Vol. 91, EMI, p. 87.

Sanders, J. L., Jr. (1963). Nonlinear theories for thin shells, *Quart. Appl. Math,* Vol. 21, pp. 21–36.

Schreyer, H. L. and Masur, E. F. (1966). Buckling of shallow arches, *J. Eng. Mech. Div.,* ASCE, Vol. 92, EM4, p. 1.

Simitses, G. J. (1973). Snapping of low pinned arches on an elastic foundation, *J. Appl. Mech.*, Vol. 40, No. 3, p. 741.

Singer, J. and Babcock, C. O. (1970). On the buckling of rings under constant directional and centrally directed pressure, *J. Appl. Mech.*, Vol. 37, No. 1, pp. 215–218.

Smith, C. V., Jr. and Simitses, G. J. (1969). "Effect of Shear and Load Behavior on Ring Stability." *Proc. ASCE*, EM3, pp. 559–569.

Timoshenko, S. P. (1935). Buckling of curved bars with small curvature, *J. Appl. Mech.*, Vol. 2, No. 1, p. 17.

Timoshenko, S. P. and Gere, J. (1961). *Theory of Elastic Stability*. McGraw-Hill Book Co., New York, pp. 287–293.

Wasserman, E. (1961). The effect of the behavior of the load on the frequency of free vibrations of a ring, NASA TT-F-52.

Wempner, G. and Kesti, N. (1962). On the buckling of circular arches and rings, *Proceedings, Fourth U.S. National Congress of Applied Mechanics*, ASME, Vol. 2, pp. 843–852.

8

BUCKLING OF SHAFTS

A shaft is defined here as a beam-like structural member that is designed to carry a large twisting moment. Shafts can buckle under such loads, and the buckling load can, in turn, be influenced by an axial force. In this Chapter we consider the stability analysis of flexible shafts subject to twisting moments and to combined axial force and twisting moments. This Chapter is in four parts, the first being the governing equations of the shaft interior, the second the strain energy of the shaft interior, the third the applied loads and other conditions at the boundaries, and the fourth the treatment of specific example problems.

8.1 PERTURBATION EQUATIONS GOVERNING BUCKLING

We will make use of elastica theory as specialized for a beam of circular cross section, so that the constitutive law reduces to

$$
\begin{Bmatrix} F_1 \\ M_1 \\ M_2 \\ M_3 \end{Bmatrix} =
\begin{bmatrix}
EA & 0 & 0 & 0 \\
0 & GJ & 0 & 0 \\
0 & 0 & EI & 0 \\
0 & 0 & 0 & EI
\end{bmatrix}
\begin{Bmatrix} \varepsilon \\ \kappa_1 \\ \kappa_2 \\ \kappa_3 \end{Bmatrix}
\tag{1}
$$

for a transversely isotropic shaft (where G and E are independent). When the shaft is subjected to an axial compressive force P and a twisting moment Q, one may write $F = -Pe_1 + \hat{F}$ and $M = Qe_1 + \hat{M}$ where $\hat{F}$ and $\hat{M}$ are perturbations of the cross-sectional force and moment and $e_1 = \lfloor 1 \quad 0 \quad 0 \rfloor^T$. With these changes of variable, one has

$$
\kappa = \frac{Q}{GJ}e_1 + \hat{\kappa}
\tag{2}
$$

and the equilibrium equations, Eqs. (3.115), reduce to

$$
\hat{F}' + \frac{Q}{GJ}\widetilde{e}_1\hat{F} + P\widetilde{e}_1\hat{\kappa} = 0
$$

$$
\hat{M}' + \frac{Q}{GJ}\widetilde{e}_1\hat{M} - Q\widetilde{e}_1\hat{\kappa} + \widetilde{e}_1\hat{F} = 0
\tag{3}
$$

where $(\tilde{\ })_{ij} = -e_{ijk}(\ )_k$ as in Eq. (3.116)

Using Eqs. (8.1), one may eliminate $\hat{M}$ in favor of $\hat{\kappa}$ to rewrite the moment equations as

$$\hat{\kappa}' + \left(\frac{Q}{GJ} - \frac{Q}{EI}\right)\tilde{e}_1\hat{\kappa} + \frac{1}{EI}\tilde{e}_1\hat{F} = 0 \tag{4}$$

One can relate $\hat{\kappa}$ to the rotation as follows. First, by definition for a prismatic beam

$$\tilde{\kappa} = -C'C^T \tag{5}$$

with

$$C = \left(I - \tilde{\hat{\theta}}\right)\overline{C} \tag{6}$$

where I is the 3×3 identity matrix and $\overline{C}$ is the matrix of direction cosines for the pre-buckled state, given by

$$\overline{C} = \begin{bmatrix} 1 & 0 & 0 \\ 0 & \cos\overline{\theta}_1 & \sin\overline{\theta}_1 \\ 0 & -\sin\overline{\theta}_1 & \cos\overline{\theta}_1 \end{bmatrix} \tag{7}$$

and $\overline{\theta}_1$ is the pre-buckling twist angle along the beam. The column matrix of rotation variables then becomes

$$\theta = \overline{\theta}_1 e_1 + \hat{\theta} \tag{8}$$

Note that

$$\tilde{\overline{\kappa}} = -\overline{C}'\overline{C}^T \tag{9}$$

or

$$\overline{C}' = -\tilde{\overline{\kappa}}\,\overline{C} \tag{10}$$

Substituting Eq. (6) into Eq. (5), making use of Eq. (9), and dropping all terms of second and higher degree in the $\hat{\ }$ quantities, one finds that

$$\tilde{\kappa} = \tilde{\overline{\kappa}} + \tilde{\hat{\theta}}' + \tilde{\overline{\kappa}}\tilde{\hat{\theta}} \tag{11}$$

which can be simplified to obtain

$$\overline{\kappa} = \overline{\theta}'_1 e_1$$
$$\hat{\kappa} = \hat{\theta}' + \tilde{\overline{\kappa}}\hat{\theta} \tag{12}$$
$$= \hat{\theta}' + \frac{Q}{GJ}\tilde{e}_1\hat{\theta}$$

The generalized strains of the reference line are here written as the column matrix γ in terms of the reference line displacement $u = \overline{u}_1 e_1 + \hat{u}$ so that

$$\gamma = C(e_1 + u') - e_1$$
$$= \left(I - \tilde{\hat{\theta}}\right)\overline{C}(e_1 + \overline{u}'_1 e_1 + \hat{u}') - e_1 \tag{13}$$
$$= \overline{u}'_1 e_1 + \overline{C}\hat{u}' + \tilde{e}_1\hat{\theta}$$

Constraining this result so that $\tilde{e}_1\gamma = 0$ gives the results that $\overline{\gamma} = \overline{\varepsilon}e_1 = \overline{u}'_1 e_1$ and

$$\hat{\theta} = \hat{\theta}_1 e_1 + \tilde{e}_1\overline{C}\hat{u}' \tag{14}$$

Making use of these results, we find that

$$\hat{\kappa} = \hat{\theta}_1' e_1 + \tilde{e}_1 \overline{C} \hat{u}'' \tag{15}$$

Taking the derivative and using Eqs. (10), one finds that

$$\hat{\kappa}' = \hat{\theta}_1'' e_1 - \frac{Q}{GJ} \tilde{e}_1 \tilde{e}_1 \overline{C} \hat{u}'' + \tilde{e}_1 \overline{C} \hat{u}''' \tag{16}$$

The perturbations of θ_1 are uncoupled from the bending equations and can be ignored. Therefore, by virtue of Eqs. (4), (15), and (16), one obtains

$$\tilde{e}_1 \hat{u}''' + \frac{Q}{EI} \hat{u}'' + \frac{1}{EI} \overline{C}^T \tilde{e}_1 \hat{F} = 0 \tag{17}$$

Solving the first of Eqs. (3) for $\hat{F}'$, and substituting the result into the derivative of Eq. (17), one eventually obtains two scalar equations for the bending deflections, given by

$$\begin{aligned} \hat{u}_2'''' + \frac{Q}{EI} \hat{u}_3''' + \frac{P}{EI} \hat{u}_2'' &= 0 \\ \hat{u}_3'''' - \frac{Q}{EI} \hat{u}_2''' + \frac{P}{EI} \hat{u}_3'' &= 0 \end{aligned} \tag{18}$$

It can be seen that all effects of GJ have disappeared from the governing equations. Of course, the boundary conditions may contain some effects of GJ in the pre-buckling torsional rotation (see Section 8.3).

8.2   ENERGY APPROACH

It can be shown that the above equations can also be derived from the energy approach. In order to do so, however, one must have all the terms in the total potential up through second degree in the perturbations. Because of the axial compressive force P and twisting moment Q, one must use geometrically nonlinear strain-displacement relations for ε and κ_1 that are valid up through second degree in the perturbations. Eqs. (3.137) can be simplified to obtain these as well as expressions for κ_2 and κ_3 linearized about the deformed state described by the twist angle $\overline{\theta}_1 = Qx_1/GJ$ and the compressive strain $\overline{\varepsilon} = -P/EA$. Thus,

$$\begin{aligned} \varepsilon &= -\frac{P}{EA} + \hat{u}_1' + \frac{1}{2}\left(\hat{u}_2'^2 + \hat{u}_3'^2\right) \\ \kappa_1 &= \frac{Q}{GJ} + \hat{\theta}_1' + \frac{1}{2}\left(\hat{u}_2'' \hat{u}_3' - \hat{u}_3'' \hat{u}_2'\right) \\ \kappa_2 &= \hat{u}_2'' \sin \overline{\theta}_1 - \hat{u}_3'' \cos \overline{\theta}_1 \\ \kappa_3 &= \hat{u}_2'' \cos \overline{\theta}_1 + \hat{u}_3'' \sin \overline{\theta}_1 \end{aligned} \tag{19}$$

where we have used the direction cosines for the twisted beam, given by Eq. (7). The strain energy for a shaft of length ℓ is given by

$$U = \frac{1}{2} \int_0^\ell \left[EA\varepsilon^2 + GJ\kappa_1^2 + EI\left(\kappa_2^2 + \kappa_3^2\right) \right] dx_1 \tag{20}$$

from which one finds a set of terms that are linear in the perturbation quantities and another that is quadratic. The linear terms lead to equations that are identically satisfied, while the quadratic terms are

$$U = \frac{1}{2} \int_0^\ell \left[EI\left(\hat{u}_2''^2 + \hat{u}_3''^2 \right) + Q\left(\hat{u}_2'' \hat{u}_3' - \hat{u}_2' \hat{u}_3'' \right) - P\left(\hat{u}_2'^2 + \hat{u}_3'^2 \right) \right] dx_1 \qquad (21)$$

Note that this expression does not include the potential of the applied loads from the ends of the shaft, if any, which are taken into account in Section 8.3.

8.3   APPLICATION OF FORCES AND MOMENTS—BOUNDARY CONDITIONS

The boundary conditions that we need to impose on our model of shafts are either displacement or natural. The former is on the displacement or rotation at an end of the shaft, and the latter relates to the system of forces and moments acting at an end of the shaft. Typically one or both ends of the shaft may have prescribed loads. In addition, some components of displacement and rotation may be prescribed at one or both ends. Bearings constrain the rotation caused by bending, and they are generally idealized in accordance with whether they are short or long. A very short bearing restrains displacement at the point where it is located, but the rotation induced by bending is allowed at that point. This makes it similar to a pin, but the kinematical details are slightly different. On the other hand, a long bearing will restrain both displacement and rotation, similar to the case of a clamped condition.

The shaft may also be subjected to twisting moments modeled in a variety of ways depending on the kinematics of the device that restrains or applies loads to the shaft. These include nonconservative axial or follower torques, as well as a variety of conservative torques such as those from quasi-tangential, semi-tangential, and Hooke joint mechanisms. Natural boundary conditions are generally expressed in the basis of the deformed beam, $\mathbf{B}_i$, when applying the intrinsic equations of the elastica theory. On the other hand, when applying the energy approach they are normally cast in terms of derivatives of the displacement variables.

It is well known that when taking into account the application of forces to nonlinear structural models, one must account for the possible reorientation of the force as the structure deforms. For example, when the orientation of the line of action for any force of constant magnitude remains fixed in space, we speak of the force as being conservative, because its virtual work can be represented as the variation of the work done by that force. In such cases the work is, of course, the negative of the potential energy. For example, consider a force given by $\mathbf{F} = F_{b_1}\mathbf{b}_1 + F_{b_2}\mathbf{b}_2 + F_{b_3}\mathbf{b}_3$ applied at a point with $\mathbf{b}_i$ being a set of unit vectors that does not change with the displacement and the values of F_{b_i} are constants. The virtual work is thus

$$\begin{aligned} \overline{\delta W} &= F_{b_1}\delta u_1 + F_{b_2}\delta u_2 + F_{b_3}\delta u_3 \\ &= \delta u^T F_b \end{aligned} \qquad (22)$$

where $u_i = \mathbf{u} \cdot \mathbf{b}_i$, $\mathbf{u}$ is the displacement vector of the point at which the load is applied, and δu and F_b are column matrices. Thus,

$$V = -F_{b_1}u_1 - F_{b_2}u_2 - F_{b_3}u_3 = -u^T F_b \qquad (23)$$

Forces of constant magnitude which have lines of action that change in orientation with the deformation of the structure do not in general possess a potential energy and are called follower forces. Consider a follower force given by $\mathbf{F} = F_1\mathbf{B}_1 + F_2\mathbf{B}_2 + F_3\mathbf{B}_3$ with $\mathbf{B}_i$ being the unit vectors fixed in the cross-sectional frame of the deformed beam at the point of application of the load, $C_{ij} = \mathbf{B}_i \cdot \mathbf{b}_j$, and F_i being constants. Thus,

$$\overline{\delta W} = \delta u^T C^T F \qquad (24)$$

which clearly cannot be expressed as the variation of a functional.

Unfortunately, the classification of structures loaded with torques is much more complex. For the purpose of our present discussion, let us apply the torque to a frame to which a structural system is rigidly attached. Thus, to apply a torque to, for example, an end of a beam, one actually applies the torque to a frame which is attached to an end cross section; the displacement and orientation measures at that end of the beam are constrained to be the same as those of the frame. First, one can show that a torque vector with constant components along axes that rotate with the frame is nonconservative. That is, given $\mathbf{M} = M_1\mathbf{B}_1 + M_2\mathbf{B}_2 + M_3\mathbf{B}_3$ and the virtual rotation vector

$$\overline{\delta\psi} = \overline{\delta\psi_1}\mathbf{B}_1 + \overline{\delta\psi_2}\mathbf{B}_2 + \overline{\delta\psi_3}\mathbf{B}_3 \qquad (25)$$

with the column matrix form being

$$\overline{\delta\psi} = R\delta\theta \qquad (26)$$

and $R(\theta)$ being the matrix in Eq. (3.121),

$$\overline{\delta W} = \overline{\delta\psi}^T M = \delta\theta^T R^T M \qquad (27)$$

There is no functional, the variation of which will give this virtual work. This is not at all unexpected, given the nature of follower forces. After all, such a torque can be constructed from a set of equal and opposite follower forces acting equidistant from, and on opposite sides of, the deformed beam reference line.

However, it is also true that a torque vector with constant components along space-fixed axes is, in general, nonconservative. That is, given $\mathbf{M} = M_{b_1}\mathbf{b}_1 + M_{b_2}\mathbf{b}_2 + M_{b_3}\mathbf{b}_3$,

$$\overline{\delta W} = \overline{\delta\psi}^T C M_b = \delta\theta^T R^T C M_b \qquad (28)$$

As before, one cannot find this virtual work by taking the variation of a functional. The only exception for such torques is a special case: when the attached frame to which the torque is applied is constrained to rotate about an axis fixed in space the resulting torque is clearly conservative.

On the other hand, we can specify conservative torques by constructing them from conservative forces. As alluded to above, the most elementary form of a conservative torque is applied about a rigid shaft that is fixed in space. One can then attach to that shaft mechanisms, such as universal joints or deforming structural elements, and thus transmit the conservative torque to other parts of the system. (This assumes that we may neglect any dissipative effects in these mechanisms, of course.) Another way to apply a conservative torque is with pulleys and taut cables. Any torque that can be applied as a system of conservative forces acting on taut cables connected to the structure by means of pulleys is conservative.

Fig. 8.1 shows a beam represented by a heavy line, the dashed portion of which is hidden from view by the frame attached at the end. This frame is represented by a circular disk of radius a. Cords are wrapped around the periphery of the disk, and equal and opposite conservative forces are applied by them, as shown, to obtain a moment that is about $\mathbf{b}_1$ in the pre-buckled state. Thus, at the onset of buckling, the applied moment has components in other directions that may be regarded as

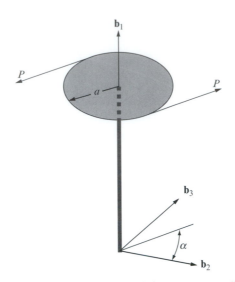

FIGURE 8.1   Schematic of a quasi-tangential twisting moment applied to the end of a shaft

infinitesimally small. For a moment of magnitude Q, $P = Q/(2a)$. As the beam deforms, the end of the beam rotates, and the direction associated with the tangent vector to the end of the beam ($\mathbf{B}_1$) becomes distinct from that of the moment vector. Such a conservative torque is called *quasi-tangential* and is similar to the torque applied by means of a Hooke joint, also referred to as a Cardan joint or universal joint (see below).

Another type of conservative torque may be formed by superposing an additional set of equal and opposite forces 90° out of phase with the two shown in Fig. 8.1 to achieve a system like that shown in Fig. 8.2. For a moment of magnitude Q, $P = Q/(4a)$. This is called a *semi-tangential* torque.

Other possibilities include the pseudo-tangential torque, achieved by applying the equal and opposite forces to the ends of a rod attached rigidly to the end of the beam. We will not consider this case herein. A simplified treatment for all of these torques is presented by Ziegler (1968) based on a collection of older work by Greenhill.

In addition to the quasi-tangential and semi-tangential cases, we also consider in this Chapter the important case of a Hooke joint. In the present Chapter we will analyze the buckling of shafts subject to applied twisting moments that are of these forms. In Chapter 9 the lateral-torsional buckling of deep beams caused by an applied bending moment at one end is considered. In that treatment the present expressions for the torques will be adapted to the different direction.

Here as in the remainder of the text, elastica theory will be used as needed for treatment of various elastic stability problems. For additional examples that involve its use as an alternative to the more usual approaches, the reader is referred to its use in Chapter 7. The methods for assessing stability for systems with follower forces and nonconservative moments are more involved, requiring application of the kinetic method. Hence we will postpone extensive discussion of nonconservative loading until Chapter 11 and treat mostly conservatively loaded structures for the present.

8.3.1   QUASI-TANGENTIAL TORQUE

In Fig. 8.1 we see that this applied moment depends on the direction of the applied forces in a plane parallel to $\mathbf{b}_2$ and $\mathbf{b}_3$, which is determined by the angle α

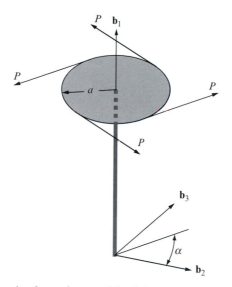

FIGURE 8.2   Schematic of a semi-tangential twisting moment applied to the end of a shaft

between the line of action of either force and $\mathbf{b}_2$. The expression for the torque is easily obtained by making use of the following observation: The position vector from the end of the deformed beam to either of the points on the periphery of the disk that lie on the line of action of the force is normal both to the force and to $\mathbf{B}_1$, the unit tangent to the deformed beam. The expression for the torque is lengthy in the general case of the geometrically-exact theory, but it may be greatly simplified for the case of small rotations of the disk caused by bending (with no restriction on the rotation of the disk caused by torsion). Thus, C_{12} and C_{13} are small compared to unity, and $C_{22} = C_{33} = \cos\bar{\theta}_1$ and $C_{23} = \sin\bar{\theta}_1 = -C_{32}$ with the pre-buckling twist angle $\bar{\theta}_1$ not restricted to be small. The simplified quasi-tangential torque becomes

$$
\begin{aligned}
\mathbf{M} &= Q\{\mathbf{B}_1 - (C_{12}\cos\alpha + C_{13}\sin\alpha)[\cos(\alpha - \bar{\theta}_1)\mathbf{B}_2 + \sin(\alpha - \bar{\theta}_1)\mathbf{B}_3]\} \\
&= Q\left[\mathbf{b}_1 + (C_{12}\sin^2\alpha - C_{13}\cos\alpha\sin\alpha)\mathbf{b}_2 + (-C_{12}\cos\alpha\sin\alpha + C_{13}\cos^2\alpha)\mathbf{b}_3\right]
\end{aligned}
$$
$$(29)$$

We note that quasi-tangential moments for the other two directions can be found by permutation of the indices of Eq. (29).

8.3.2   SEMI-TANGENTIAL TORQUE

Recalling Fig. 8.2, one can show that the full expression for this type of torque is even more lengthy than that of the quasi-tangential case. However, for the case of small rotations of the disk caused by bending, and again with no restriction on the rotation of the disk caused by torsion, the expression for the torque becomes independent of the angle α and actually simpler than Eq. (29). Again, C_{12} and C_{13} are small compared to unity, and $C_{22} = C_{33} = \cos\bar{\theta}_1$ and $C_{23} = \sin\bar{\theta}_1 = -C_{32}$ with the pre-buckling twist angle $\bar{\theta}_1$ not restricted to be small. The semi-tangential torque then becomes

$$\mathbf{M} = Q\left[\mathbf{B_1} - \frac{1}{2}(C_{12}C_{22} + C_{13}C_{23})\mathbf{B_2} - \frac{1}{2}(C_{12}C_{32} + C_{13}C_{33})\mathbf{B_3}\right]$$
$$= Q\left(\mathbf{b_1} + \frac{1}{2}C_{12}\mathbf{b_2} + \frac{1}{2}C_{13}\mathbf{b_3}\right) \tag{30}$$

As before, semi-tangential moments for the other two directions can be found by permutation of the indices of Eq. (30).

8.3.3   TORQUE APPLIED BY A HOOKE JOINT

Here we consider a beam loaded through a Hooke joint. Hooke joints have three parts: an interior part called the spider and two outer parts called yokes, all assumed here to be rigid. Consider a shaft to which a Hooke joint is attached and constrained as indicated in Fig. 8.3. As indicated here, the spider is free to rotate relative to the inner and outer yokes about bearings (axes of relative rotation) fixed in the yokes. The outer yoke is attached to the loading fixture and is constrained to rotate about a spatially fixed axis. The inner yoke is attached to the end cross-sectional frame of the beam. Thus, one axis of rotation for the spider is fixed in the end cross-sectional frame of the deformed beam, and the other is fixed in the loading fixture and in the outer rigid yoke.

Consider a shaft to which a Hooke joint is attached and constrained as indicated in Fig. 8.3. For the configuration under consideration, the loading fixture is constrained so that it can only rotate about $\mathbf{b_1}$; the angle of rotation is $\bar{\theta}_1$ and is caused by the application of a moment about $\mathbf{b_1}$ of magnitude Q. Thus, a tip twisting moment applied to the beam, and the pre-buckling deflection will only consist of pure twist. At buckling small perturbations of the beam deformation occur. The spider is free to rotate relative to the fixture about a line passing through the beam reference line and parallel to $\cos(\alpha + \bar{\theta}_1)\mathbf{b_2} + \sin(\alpha + \bar{\theta}_1)\mathbf{b_3}$, where α is a parameter that is specified according to the way the joint is attached to the beam; note that $\alpha = 0$ in the case illustrated. The other axis of rotation is fixed in the deformed beam cross-sectional frame at the beam end; assuming the cross members of the spider to be perpendicular to each other, the spider is thus free to rotate about a line parallel to

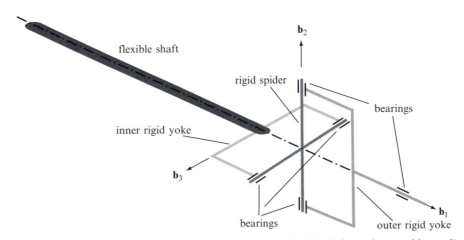

FIGURE 8.3   Schematic of Hooke joint attached to a shaft (undeformed state with $\alpha = 0$)

$-\sin\alpha\mathbf{B}_2 + \cos\alpha\mathbf{B}_3$. Given an applied moment so that $\mathbf{M}\cdot\mathbf{b}_1 = Q$ and $\mathbf{M}\cdot\left[\cos(\alpha+\bar{\theta}_1)\mathbf{b}_2 + \sin(\alpha+\bar{\theta}_1)\mathbf{b}_3\right] = \mathbf{M}\cdot(-\sin\alpha\mathbf{B}_2 + \cos\alpha\mathbf{B}_3) = 0$, one may deduce that $\mathbf{M}$ is given by

$$
\begin{aligned}
\mathbf{M} &= Q\{\mathbf{B}_1 - \left[C_{12}\cos(\alpha+\bar{\theta}_1) + C_{13}\sin(\alpha+\bar{\theta}_1)\right](\cos\alpha\mathbf{B}_2 + \sin\alpha\mathbf{B}_3)\} \\
&= Q\{\mathbf{b}_1 + \left[C_{12}\sin^2(\alpha+\bar{\theta}_1) - C_{13}\cos(\alpha+\bar{\theta}_1)\sin(\alpha+\bar{\theta}_1)\right]\mathbf{b}_2 \\
&\quad + \left[-C_{12}\cos(\alpha+\bar{\theta}_1)\sin(\alpha+\bar{\theta}_1) + C_{13}\cos^2(\alpha+\bar{\theta}_1)\right]\mathbf{b}_3\}
\end{aligned}
\tag{31}
$$

This expression thus is identical to that for the quasi-tangential torque with α replaced by $\alpha + \bar{\theta}_1$.

8.3.4 POTENTIAL ENERGY OF CONSERVATIVE TORQUES

Regarding the direction cosines $C_{12} \approx \hat{\theta}_3 = \hat{u}'_2$ and $C_{13} \approx -\hat{\theta}_2 = \hat{u}'_3$ as small perturbation quantities at the beam end, one can now find potential energy expressions for all three cases of conservative torques. In the potential energy, all terms of third and higher degree of the unknowns C_{12} and C_{13} may be dropped. To find the potential energy, one must first find the virtual work of the torques. Since the $\mathbf{b}_1$ component of the torque does not contain these unknowns, this component of the virtual rotation must be taken to a higher order. Thus, the virtual rotation can be written as

$$
\overline{\delta\boldsymbol{\psi}} = \left(\delta\bar{\theta}_1 + \frac{1}{2}C_{12}\delta C_{13} - \frac{1}{2}C_{13}\delta C_{12}\right)\mathbf{b}_1 - \delta C_{13}\mathbf{b}_2 + \delta C_{12}\mathbf{b}_3
\tag{32}
$$

For the quasi-tangential case the virtual work is simply the dot product of $\overline{\delta\boldsymbol{\psi}}$ and the torque, Eq. (29), given by

$$
\overline{\delta W} = Q\left[\delta\bar{\theta}_1 + \frac{\cos(2\alpha)(\delta C_{13}C_{12} + \delta C_{12}C_{13})}{2} + \cos\alpha\sin\alpha(C_{13}\delta C_{13} - C_{12}\delta C_{12})\right]
\tag{33}
$$

For the quasi-tangential case, one easily finds by integration of Eq. (33) a quantity V such that $\delta V = -\overline{\delta W}$ where

$$
V = -Q\bar{\theta}_1 + \frac{Q}{2}\left[\cos\alpha\sin\alpha\left(C_{12}^2 - C_{13}^2\right) - \cos(2\alpha)C_{12}C_{13}\right]
\tag{34}
$$

For the Hooke joint, one need only substitute $\alpha + \bar{\theta}_1$ for α, while for the semi-tangential torque,

$$
V = -Q\bar{\theta}_1
\tag{35}
$$

The simplicity of the potential energy of the applied load suggests that the energy approach may be a good one to follow for problems involving conservative torques. This will be illustrated later when specific problems are solved.

8.4   EXAMPLE PROBLEMS

Four examples will be presented in this Section. The first is a clamped-clamped shaft undergoing a twisting moment. The second is for a pinned-pinned shaft undergoing various types of twisting moments. The third brings in the effect of an axial compressive force, and the fourth considers rotation about the shaft axis.

8.4.1 CLAMPED-CLAMPED SHAFT UNDER TWISTING MOMENT

This case is among the simplest and is idealized for long bearings at each end. We will undertake the solution from the differential equations, which are Eqs. (8.18), with boundary conditions written in terms of displacements as $\hat{u}_2(0) = \hat{u}_2(\ell) = \hat{u}_3(0) = \hat{u}_3(\ell) = \hat{u}_2'(0) = \hat{u}_2'(\ell) = \hat{u}_3'(0) = \hat{u}_3'(\ell) = 0$. We presuppose a pre-buckling state with zero axial force such that one end of the shaft is rotated about $\mathbf{b}_1$. Thus, the torque along the shaft is uniform with value $Q\mathbf{b}_1$. Thus, Eqs. (8.18) become

$$\hat{u}_2'''' + \frac{Q}{EI}\hat{u}_3''' = 0$$
$$\hat{u}_3'''' - \frac{Q}{EI}\hat{u}_2''' = 0 \tag{36}$$

If we introduce a complex variable u such that $u = \hat{u}_2 + i\hat{u}_3$, then a single equation in u can be found by multiplying the second equation by i and adding the equations, so that

$$u'''' - iqu''' = 0 \tag{37}$$

where $q = Q\ell/EI$ and the nondimensional axial coordinate $x = x_1/\ell$ has been introduced, along with redefining $(\ )'$ as $d(\ )/dx$. The equation easily reduces to

$$u' - iqu = \frac{a}{2}\left(x^2 - x\right) \tag{38}$$

subject to $u(0) = u(1) = 0$ with a as an arbitrary constant. The solution is in terms of polynomial and trigonometric functions, and the characteristic equation can be reduced to

$$\tan\left(\frac{q}{2}\right) = \frac{q}{2} \tag{39}$$

the smallest root of which is $q_{\text{cr}} = \pm 8.98682 = 2.86059\pi$ so that $Q_{\text{cr}} = \pm 8.98682$ EI/ℓ, obtained by Greenhill in 1883 and Nicolai in 1926.

8.4.2 SHAFT WITH SHORT BEARINGS UNDER TWISTING MOMENT

For this case one must consider a variety of possibilities, depending on how the restraint at each end is constructed. The easiest case to deal with analytically is the tangential torque, for which $\hat{M}_\alpha(0) = \hat{M}_\alpha(\ell) = 0$ for $\alpha = 2$ and 3. Thus, $\hat{u}_\alpha''(0) = \hat{u}_\alpha''(\ell) = 0$ in addition to $\hat{u}_\alpha(0) = \hat{u}_\alpha(\ell) = 0$. Using the same complex variable u as before, one finds

$$u'' - iqu' = ax + b \tag{40}$$

where a and b are constants. The solution is again in terms of polynomial and trigonometric functions, and the characteristic equation is

$$\cos q = 1 \tag{41}$$

so that $q_{\text{cr}} = \pm 2\pi$.

Another simple case is the axial torque, for which $\mathbf{M} = Q\mathbf{b}_1 = Q(\mathbf{B} - \hat{\theta}_3\mathbf{B}_2 + \hat{\theta}_2\mathbf{B}_3)$ so that

$$\hat{M}_2(0) = -Q\hat{\theta}_3(0)$$
$$\hat{M}_2(\ell) = -Q\hat{\theta}_3(\ell)$$
$$\hat{M}_3(0) = Q\hat{\theta}_2(0)$$
$$\hat{M}_3(\ell) = Q\hat{\theta}_2(\ell) \tag{42}$$

which gives the governing equation as

$$u'' - iqu' = 0 \qquad (43)$$

with boundary condition $u(0) = u(1) = 0$. This yields the same critical load as the tangential case. We should not place too much stock in either of these answers, however, because both of these types of torques are nonconservative. Hence, one really needs to use the kinetic method in case there is a flutter instability at a value of q below 2π.

Next, let's consider the case of the semi-tangential torque. The most straightforward way to get the boundary conditions is to consider the variation of the total potential, recalling that $V = 0$ and U is given by Eq. (21). This yields the governing ordinary differential equation, Eq. (40), with boundary conditions

$$u(0) = u(1) = u''(0) - \frac{iq}{2}u'(0) = u''(1) - \frac{iq}{2}u'(1) = 0 \qquad (44)$$

The solution is again in the form of polynomial and trigonometric functions, with a characteristic equation given by

$$\frac{q}{6} + \tan\left(\frac{q}{2}\right) = 0 \qquad (45)$$

The critical torque is $q_{cr} = \pm 4.91129 = \pm 1.56331\pi$.

Finally, let's consider the case of a quasi-tangential torque with $\alpha = 0$ at the end where $x_1 = 0$ and varying α at the other end. (The closely related case of a Hooke joint with varying α and GJ/EI is left as an exercise for the reader.) Straightforward variation of the total potential yields the governing equations as

$$\begin{aligned} u_2'''' + qu_3''' &= 0 \\ u_3'''' - qu_2''' &= 0 \end{aligned} \qquad (46)$$

and boundary conditions

$$\begin{aligned} u_2(0) = u_3(0) = u_2(1) &= u_3(1) = u_2''(0) = u_3''(0) - qu_2'(0) = 0 \\ u_2''(1) + q\sin^2\alpha\, u_3'(1) + q\sin\alpha\cos\alpha\, u_2'(1) &= 0 \\ u_3''(1) - q\cos^2\alpha\, u_2'(1) - q\sin\alpha\cos\alpha\, u_3'(1) &= 0 \end{aligned} \qquad (47)$$

where $q = Q\ell/EI$, $u_2 = \hat{u}_2/\ell$, $u_3 = \hat{u}_3/\ell$, $x = x_1/\ell$, and $(\ )' = d(\ )/dx$. This system of equations can be integrated to yield

$$\begin{aligned} u_2'' + qu_3' &= a_1 x + a_2 \\ u_3'' - qu_2' &= a_3 x \end{aligned} \qquad (48)$$

the characteristic equation of which is

$$\begin{aligned} &(8 - 8q^2 + q^4)\cos q - (4 - 8q^2 + q^4)\cos(q - 2\alpha) \\ &+4\{(2 - q^2)\cos(2\alpha) - \cos(q + 2\alpha) + q(2 + q^2)\sin q - (2 - q^2)[1 + q\sin(q - 2\alpha)] \\ &-2q\sin(2\alpha)\} = 0 \end{aligned} \qquad (49)$$

The lowest root for $|q|$ is plotted against α in Fig. 8.4 for $-\pi/2 \le \alpha \le \pi/2$. At $\alpha = 2.31675$, the nondimensional critical load is 0.915531π whereas for $\alpha = 1.09012$ the nondimensional critical load is 1.29155π – a 41% fluctuation in the critical load depending on α. It is important to note that although these results (tangential/axial,

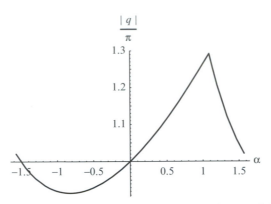

FIGURE 8.4   Buckling torque for a shaft loaded by a quasi-tangential twisting moment

at $\pm 2\pi$; semi-tangential, at $\pm 1.56331\pi$; and quasi-tangential, varying from $\pm 0.915531\pi$ to $\pm 1.29155\pi$) are nowhere near each other, the differences among these torques are infinitesimal!

It is interesting and instructive to verify that the boundary conditions for the moment are consistent with those on the second derivative of displacement as derived from the energy approach. The internal moment in the shaft can be expressed as

$$\mathbf{M} = (Q + GJ\hat{\kappa}_1)\mathbf{B}_1 + EI(\hat{\kappa}_2\mathbf{B}_2 + \hat{\kappa}_3\mathbf{B}_3)$$
$$\approx (Q + GJ\hat{\kappa}_1)\mathbf{b}_1 + \left(-EI\hat{u}_3'' + Q\hat{u}_2'\right)\mathbf{b}_2 + \left(EI\hat{u}_2'' + Q\hat{u}_3'\right)\mathbf{b}_3 \qquad (50)$$

so that

$$\hat{M}_{b_2} = -EI\hat{u}_3'' + Q\hat{u}_2'$$
$$\hat{M}_{b_3} = EI\hat{u}_2'' + Q\hat{u}_3' \qquad (51)$$

The use of the second of Eqs. (8.29) shows that at $x_1 = \ell$

$$\hat{M}_{b_2}(\ell) = Q\left[\hat{u}_2'(\ell)\sin^2\alpha - \hat{u}_3'(\ell)\cos\alpha\sin\alpha\right]$$
$$\hat{M}_{b_3}(\ell) = Q\left[-\hat{u}_2'(\ell)\cos\alpha\sin\alpha + \hat{u}_3'(\ell)\cos^2\alpha\right] \qquad (52)$$

which is consistent with the boundary conditions derived from the energy in Eqs. (8.47).

8.4.3   EFFECT OF AXIAL COMPRESSIVE FORCE

Next, we turn to the case of a shaft clamped at $x_1 = 0$ and undergoing axial compression and a semi-tangential torque at $x_1 = \ell$. The Euler equations can be expressed as

$$u'''' - iqu''' + pu'' = 0 \qquad (53)$$

where $p = P\ell^2/EI$. The essential boundary conditions are $u(0) = u'(0) = 0$ and the natural boundary conditions are

$$u''(1) - \frac{iq}{2}u'(1) = 0$$
$$u'''(1) - iqu''(1) + pu'(1) = 0 \qquad (54)$$

The general solution is given by

$$u = a_1 + a_2 x + a_3 \exp(i\alpha_1 x) + a_4 \exp(i\alpha_2 x) \tag{55}$$

where

$$\alpha_1 = \frac{q + \sqrt{q^2 + 4p}}{2} > 0$$
$$\alpha_2 = \frac{q - \sqrt{q^2 + 4p}}{2} < 0 \tag{56}$$

After imposing the boundary conditions, one can find (after a great deal of algebra) that

$$\cos\left(\frac{\sqrt{q^2 + 4p}}{2}\right) = 0 \tag{57}$$

or

$$q^2 + 4p = \pi^2 \tag{58}$$

The relationship between q and p is shown in Fig. 8.5, from which one can conclude that shafts undergoing compression buckle much more easily under a twisting moment, and shafts undergoing tension buckle less easily under a twisting moment. Similarly, shafts undergoing twisting moments buckle more easily under compressive forces.

8.4.4 CRITICAL SPEEDS OF LOADED, ROTATING SHAFTS

The inertial load has a quasi-static component when the shaft is rotating about its undeformed axis at a constant angular speed. The kinetic energy per unit length caused by this quasi-static component is

$$T = \frac{1}{2} m\Omega^2 \left(u_2^2 + u_3^2\right) \tag{59}$$

where $\Omega \mathbf{b}_1$ is the angular velocity of the shaft. Application of Hamilton's principle leads to a weak form of the equilibrium equations of the form

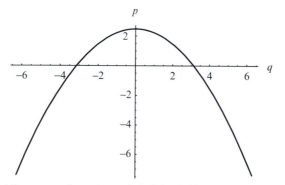

FIGURE 8.5   Buckling torque for a clamped shaft loaded by a semi-tangential twisting moment q and an axial force p

$$\delta \frac{1}{2} \int_0^1 \left[u_2''^2 + u_3''^2 - q(u_2'u_3'' - u_2''u_3') - p(u_2'^2 + u_3'^2) - \omega^2(u_2^2 + u_3^2) \right] dx = 0 \qquad (60)$$

with $u_2 = \hat{u}_2/\ell$, $u_3 = \hat{u}_3/\ell$, q and p defined as before, and $\omega^2 = m\ell^4\Omega^2/EI$. From setting the first variation equal to zero and integrating by parts, the resulting Euler-Lagrange equations can be put into the form

$$u'''' - iqu''' + pu'' - \omega^2 u = 0 \qquad (61)$$

with $u = u_2 + iu_3$ as before. The general solution to this equation is difficult to manipulate algebraically. However, one can observe a number of things. First, this equation with q set equal to zero is identical to that used to find the natural frequencies and mode shapes of an axially compressed beam. The critical angular speeds from this equation are the same as the natural frequencies of the loaded and nonrotating beam. As the beam is loaded with both p and q, the critical speeds decrease significantly. An approximate solution via the Rayleigh-Ritz method is left as an exercise for the reader.

Two important points must be made here. First, the static method can only tell us where the critical speeds are. It cannot tell us about the stability of the shaft when its angular speed exceeds the first critical speed but is less than the second, for example. Second, it is more typical for critical speeds to be calculated for the combination of a shaft and heavy disk. The shaft mass and disk flexibility are often neglected in such cases. This analysis requires the presence of gyroscopic terms in the differential equations, and hence the kinetic approach must be used. However, even the kinetic approach leaves unanswered the stability of the shaft when its angular speed is in between two critical speeds. This is because, fundamentally, the problem of the critical speed is a resonance problem, which can be treated by use of imperfections and obtaining the solution of the governing ordinary differential equations in the time domain. A detailed discussion of the problem of finding critical speeds can be found in Ziegler (1968).

PROBLEMS

1. Consider a shaft loaded only by a twisting moment which at the $x_1 = 0$ end is clamped, and at the $x_1 = \ell$ end has a long bearing that is free to move but that is not free to rotate as the shaft bends. Find the critical twisting moment.

   ans.: $Q_{cr} = 2\pi EI/\ell$

2. Find the critical torsional buckling load for a shaft that is clamped at one end and with zero shear/zero rotation condition at the other. Determine and plot the mode shape.

3. Show that the conditions stated in the text do in fact lead to Eqs. (31).

4. Show that the boundary conditions obtained from the energy approach are equivalent to those obtained using the intrinsic equations for a pinned-pinned shaft loaded by a quasi-tangential torques at both ends. Let $\alpha = 0$ at the end $x_1 = 0$ and α be a variable at the end $x_1 = \ell$.

5. Consider a shaft which is pinned at both ends and is loaded by quasi-tangential twisting moments at both ends. Taking $\alpha = 0$ at the end $x_1 = 0$ and letting α be a variable at the end $x_1 = \ell$, find the critical twisting moment for the case of $GJ/EI = 3/4$ for $\alpha = -1$ and $\alpha = \pi/2$.

   ans.: For $\alpha = -1$, $Q_{cr} = 0.920217\pi EI/\ell$, and for $\alpha = \pi/2$, $Q_{cr} = 1.02094\pi EI/\ell$

6. Consider a shaft which is pinned at both ends and is loaded by twisting moments applied by Hooke joints at both ends. Taking $\alpha = 0$ at the end $x_1 = 0$ and letting α be a variable at the end $x_1 = \ell$, find the critical twisting moment for the case of $GJ/EI = 3/4$ for $\alpha = -1$ and $\alpha = \pi/2$, taking into account the pre-buckling twist.

   ans.: For $\alpha = -1$, $Q_{cr} = 1.25831\pi\ EI/\ell$, and for $\alpha = \pi/2$, $Q_{cr} = 0.915904\pi\ EI/\ell$

7. Consider a loaded shaft with nonzero q and p, with the $x_1 = 0$ end clamped and the $x_1 = \ell$ end loaded with a semi-tangential torque. Using three polynomial admissible functions for u_2 and u_3, determine an approximate relationship among the critical speeds ω, q, and p. Check your result to make certain it gives good results for the Euler load, the critical torque, and the critical speeds of the unloaded beam (same as the natural frequencies of the nonrotating beam).

REFERENCES

Ziegler. H. (1968). *Principles of Structural Stability*. Blaisdell Publishing Company, Waltham, Massachusetts.

9

LATERAL-TORSIONAL BUCKLING OF DEEP BEAMS

Lateral-torsional buckling is the name commonly associated with an instability of the "plane form of bending" of deep beams. When such a beam is loaded in the plane of greatest flexural rigidity, small inplane deformation results. In this pre-buckled state, the beam typically undergoes deflection caused by inplane bending, extension, and possibly inplane shear. When the critical load is reached, the beam deflects out of that plane and undergoes combined deformation involving both out-of-plane bending and torsion. In this chapter we will consider two sets of examples of such buckling analysis, one very straightforward and the other quite the opposite. The straightforward examples involve pure, constant bending in the pre-buckled state. The complex examples are cantilevered beams, some of which have a more complicated pre-buckled state depending on the loading. The changes of the critical load associated with the Vlasov effect, initial curvature, load offset, and methods of applying moments are examined for some of the cases. The equations of Section 3.7 are used, augmented by the Vlasov correction of Section 3.8.1, where appropriate.

9.1   PINNED-PINNED BEAM

In this section we will consider the lateral-torsional buckling of a pinned-pinned beam subject to loads that produce constant planar bending deformation in the pre-buckled state. We first consider a straight beam that is much stiffer in one flexural plane than in the other and for which the Vlasov effect is not important. A thin strip-like beam would fulfill this condition. Next we consider the Vlasov correction. Beams with thin-walled, open cross sections would require this correction, such as I-beams. Finally, we introduce a small amount of initial curvature and determine its effect on the lateral-torsional buckling.

9.1.1   PRISMATIC STRIP-BEAM

In this example, the beam is attached to rigid frames at each end, as depicted in Fig. 9.1. The beam is free to rotate relative to the frames about $\mathbf{B}_2$ at each end. At the

FIGURE 9.1 Schematic of pinned-pinned strip-beam undergoing pure bending from torque applied at free end

left end where $x_1 = 0$, the frame is stationary. However, at the right end where $x_1 = \ell$, the frame is free to displace in any direction, but it is only free to rotate about the spatially fixed direction $\mathbf{b}_3$ and is subjected to a specified torque $Q\mathbf{b}_3$. The governing equations are Eqs. 3.118, the intrinsic force and moment equilibrium equations. Thus, the beam, in its pre-buckled state, is undergoing pure bending, i.e.,

$$\overline{F} = 0$$
$$\overline{\varepsilon} = 0$$
$$\overline{M} = Qe_3 \tag{1}$$
$$\overline{\kappa} = \frac{Q}{EI_3}e_3$$

where $e_3 = \lfloor 0 \quad 0 \quad 1 \rfloor^T$. One can now write linearized perturbation equations about this exact equilibrium state to examine its stability. These equations are written in terms of perturbation force, stretch, moment, and curvature variables ($\hat{F}$, $\hat{\varepsilon}$, $\hat{M}$, and $\hat{\kappa}$, respectively) so that

$$\hat{F}' + \frac{Q}{EI_3}\tilde{e}_3\hat{F} = 0 \tag{2}$$
$$\hat{M}' + \frac{Q}{EI_3}\tilde{e}_3\hat{M} - Q\tilde{e}_3\hat{\kappa} + \tilde{e}_1\hat{F} = 0$$

where $e_1 = \lfloor 1 \quad 0 \quad 0 \rfloor^T$. The force equations, along with boundary conditions, $\hat{F}(\ell) = 0$, show that $\hat{F}_3 \equiv 0$ and that there is an instability at a large value of torque which can be found from

$$\hat{F}_1'' + \left(\frac{Q}{EI_3}\right)^2 \hat{F}_1 = 0 \tag{3}$$

or

$$\hat{F}_2'' + \left(\frac{Q}{EI_3}\right)^2 \hat{F}_2 = 0 \tag{4}$$

as $Q_{cr} = \pi EI_3/\ell$. This instability is only of academic importance, however, owing to the much smaller critical load that can be found from the first two of the moment equations

$$\hat{M}_1' - \frac{Q}{EI_3}\hat{M}_2 + Q\hat{\kappa}_2 = 0 \tag{5}$$
$$\hat{M}_2' + \frac{Q}{EI_3}\hat{M}_1 - Q\hat{\kappa}_1 = 0$$

After noting that $\hat{\kappa}_1 = \hat{M}_1/GJ$ and $\hat{\kappa}_2 = \hat{M}_2/EI_2$, one can rewrite these equations as

$$\hat{M}_1' + Q\left(\frac{1}{EI_2} - \frac{1}{EI_3}\right)\hat{M}_2 = 0$$

$$\hat{M}_2' - Q\left(\frac{1}{GJ} - \frac{1}{EI_3}\right)\hat{M}_1 = 0$$

(6)

which can be combined into a single equation for $\hat{M}_2$, viz.,

$$\hat{M}_2'' + \beta^2\hat{M}_2 = 0 \tag{7}$$

where

$$\beta^2 = Q^2\left(\frac{1}{EI_2} - \frac{1}{EI_3}\right)\left(\frac{1}{GJ} - \frac{1}{EI_3}\right) > 0 \tag{8}$$

The solution is

$$\hat{M}_2 = a\sin(\beta x_1) + b\cos(\beta x_1) \tag{9}$$

Since $\hat{M}_2(0) = \hat{M}_2(\ell) = 0$, then $b = 0$ and $a\sin(\beta\ell) = 0$. Thus, the lowest critical load is such that $\beta\ell = \pi$ so that

$$Q_{\text{cr}} = \pm\frac{\pi}{\ell}\sqrt{\frac{GJ\,EI_2}{\left(1 - \frac{GJ}{EI_3}\right)\left(1 - \frac{EI_2}{EI_3}\right)}} \tag{10}$$

In the limit of infinitely deep beams, for which $GJ \ll EI_3$ and $EI_2 \ll EI_3$, the critical torque is found to be

$$Q_{\text{cr}} = \pm\frac{\pi}{\ell}\sqrt{GJ\,EI_2} \tag{11}$$

For deep beams, this value is close to but somewhat smaller than its more accurate counterpart in Eq. (10), showing that the effects of pre-buckling deformation, which are associated with EI_3, are secondary. Since they raise the predicted buckling load they are typically regarded as more of academic interest.

9.1.2 CORRECTION FOR THE VLASOV EFFECT

The Vlasov effect stems from the effect of warping rigidity, generally appropriate only for thin-walled beams with open cross sections. In Section 3.8.1 the strain energy per unit length and equilibrium equations for this effect are derived. They have the form

$$\Psi = \frac{1}{2}\left(E\Gamma\kappa_1'^2 + GJ\kappa_1^2 + \ldots\right) \tag{12}$$

This has the effect of changing only the torsional equation, so that the equations governing the buckling are

$$-E\Gamma\hat{\kappa}_1''' + \hat{M}_1' + Q\left(\frac{1}{EI_2} - \frac{1}{EI_3}\right)\hat{M}_2 = 0$$

$$\hat{M}_2' - Q\left(\frac{1}{GJ} - \frac{1}{EI_3}\right)\hat{M}_1 = 0$$

(13)

Since $\hat{M}_1 = GJ\hat{\kappa}_1 = GJ\hat{\theta}_1'$ and both $\hat{\theta}_1(\ell)$ and $\hat{M}_2(\ell)$ vanish, the second equation can be integrated once to yield

$$\hat{M}_2 = Q\left(1 - \frac{GJ}{EI_3}\right)\hat{\theta}_1 \tag{14}$$

which, when substituted into the first equation, yields

$$-\frac{E\Gamma}{GJ}\hat{\theta}_1'''' + \hat{\theta}_1'' + \beta^2\hat{\theta}_1 = 0 \tag{15}$$

Letting $\hat{\theta}_1 = \hat{\theta}\exp(\phi x_1)$, we get a characteristic equation of the form

$$-\frac{E\Gamma}{GJ}\phi^4 + \phi^2 + \beta^2 = 0 \tag{16}$$

so that

$$\hat{\theta}_1 = a\sin(\alpha_1 x_1) + b\cos(\alpha_1 x_1) + c\sinh(\alpha_2 x_1) + d\cosh(\alpha_2 x_1) \tag{17}$$

where

$$\alpha_1^2 = \frac{\sqrt{1 + 4\beta^2\frac{E\Gamma}{GJ}} - 1}{2\frac{E\Gamma}{GJ}} = -\phi_1^2$$

$$\alpha_2^2 = \frac{\sqrt{1 + 4\beta^2\frac{E\Gamma}{GJ}} + 1}{2\frac{E\Gamma}{GJ}} = \phi_2^2 \tag{18}$$

The boundary conditions are $\hat{\theta}_1(0) = \hat{\theta}_1(\ell) = 0$ and either specifying $\hat{\theta}'$ or $\hat{\theta}''$ at the ends; the former sets the warping displacement equal to zero, and the latter represents a condition of zero stress. Let us consider here the case of zero stress, so that $\hat{\theta}''(0) = \hat{\theta}''(\ell) = 0$. The boundary conditions at $x_1 = 0$ require that $b = d = 0$. The boundary conditions at $x_1 = \ell$ give rise to the characteristic equation $\left(\alpha_1^2 + \alpha_2^2\right)\sin(\alpha_1\ell)\sinh(\alpha_2\ell)$. Thus, $\alpha_1\ell = \pi$, so that

$$Q_{\mathrm{cr}} = \pm\frac{\pi}{\ell}\sqrt{\frac{GJ\,EI_2}{\left(1 - \frac{GJ}{EI_3}\right)\left(1 - \frac{EI_2}{EI_3}\right)}}\sqrt{1 + \frac{\pi^2 E\Gamma}{\ell^2 GJ}} \tag{19}$$

As expected, this correction raises the critical torque at which buckling occurs. The size of the correction is strongly dependent on the cross-sectional configuration.

9.1.3 CORRECTION FOR INITIAL CURVATURE

Let's add a correction for initial curvature to the formulation. According to Hodges (1999) the constitutive law is

$$\begin{Bmatrix} F_1 \\ M_1 \\ M_2 \\ M_3 \end{Bmatrix} = \begin{bmatrix} EA & 0 & 0 & -\frac{EI_3(1+\nu)}{R} \\ 0 & GJ & 0 & 0 \\ 0 & 0 & EI_2 & 0 \\ -\frac{EI_3(1+\nu)}{R} & 0 & 0 & EI_3 \end{bmatrix} \begin{Bmatrix} \varepsilon \\ \kappa_1 \\ \kappa_2 \\ \kappa_3 \end{Bmatrix} \tag{20}$$

with $I_3 \ll AR^2$.

For the pre-buckled state, $\overline{M}_3 = EI_3\overline{\kappa}_3 = Q$. As before, $\overline{F} = 0$ and $\overline{M} = Qe_3$. However, here

$$\overline{K} = \left(\frac{1}{R} + \overline{\kappa}_3\right)e_3 \tag{21}$$

The perturbation equations then become

$$\hat{F}' + \left(\frac{1}{R} + \overline{\kappa}_3\right)\tilde{e}_3\,\hat{F} = 0$$

$$\hat{M}' + \left(\frac{1}{R} + \overline{\kappa}_3\right)\tilde{e}_3\hat{M} - Q\tilde{e}_3\hat{\kappa} + \tilde{e}_1\,\hat{F} = 0 \tag{22}$$

As before the perturbation force equation contributes mainly the result that $\hat{F}_1 \equiv 0$, and the first two of the moment equations become

$$\hat{M}_1' + \left(\frac{Q}{EI_2} - \frac{Q}{EI_3} - \frac{1}{R}\right)\hat{M}_2 = 0$$

$$\hat{M}_2' - \left(\frac{Q}{GJ} - \frac{Q}{EI_3} - \frac{1}{R}\right)\hat{M}_1 - \hat{F}_3^{\,0} = 0 \tag{23}$$

These equations can be combined into a single equation of the form

$$\hat{M}_2'' + \beta^2\hat{M}_2 = 0 \tag{24}$$

where

$$\beta^2 = \left(\frac{Q}{EI_2} - \frac{Q}{EI_3} - \frac{1}{R}\right)\left(\frac{Q}{GJ} - \frac{Q}{EI_3} - \frac{1}{R}\right) \tag{25}$$

With the vanishing of $\hat{M}_2$ at each end, it follows that

$$\beta^2\ell^2 = \pi^2 \tag{26}$$

so that

$$\pi^2 = \frac{Q^2\ell^2}{GJ\,EI_2}\left(1 - \frac{EI_2}{EI_3} - \frac{EI_2}{QR}\right)\left(1 - \frac{GJ}{EI_3} - \frac{GJ}{QR}\right) \tag{27}$$

The critical torque can be found by solving this expression for Q, yielding two roots. For zero initial curvature (i.e. infinitely large R), these roots are the same as those of Eqs. (10). When the initial curvature is small so that $R\alpha = \ell$ and $\alpha \ll 1$, the roots can be written as

$$Q_{\mathrm{cr}} = Q_0\left[\pm 1 + \frac{(A + B - 2AB)\alpha}{2\sqrt{AB}(1 - A)(1 - B)\pi}\right] \tag{28}$$

where

$$Q_0 = \frac{\pi}{\ell}\sqrt{\frac{GJ\,EI_2}{\left(1 - \frac{GJ}{EI_3}\right)\left(1 - \frac{EI_2}{EI_3}\right)}} \tag{29}$$

$$A \equiv \frac{GJ}{EI_3} \qquad B \equiv \frac{EI_2}{EI_3}$$

In the limit for an extremely deep beam, we have $A \to 0$ and $B \to 0$, so that

$$Q_{cr} = \pm \frac{\pi}{\ell} \sqrt{GJ\, EI_2} + \frac{GJ + EI_2}{2R} \tag{30}$$

The magnitude of the positive root increases because of the presence of initial curvature. However, the magnitude of the negative root decreases because of the presence of initial curvature. To put it more in practical language, a deep beam with initial inplane curvature buckles with a load of smaller magnitude when it is being bent in such a way as to straighten it, whereas it requires a larger load to buckle such a beam when it is being bent so as to bend it more.

9.2   CANTILEVERED BEAM UNDER BENDING MOMENT

As seen in the previous chapter, a torque can be applied in a variety of ways. We can conceive of a mechanism that applies a moment, the direction of which follows the end cross section of the deformed beam, and which induces bending. We expect a static analysis of this nonconservative load to be inconclusive. That is, we know that a kinetic analysis is necessary to ascertain the nature of the instability. Despite this, for completeness we consider a cantilevered beam under a torque created by follower forces applied at the free end. Such a torque could be realized by equal and opposite jets in the longitudinal direction applied at the ends of rigid rods attached to the end of the beam, oriented normal to the axis of the deformed beam; see Fig. 9.2.

For such a beam, the pre-buckled state can be found as

$$\overline{F} = 0$$
$$\overline{M} = Qe_3 \tag{31}$$

where Q is the magnitude of the follower moment. The perturbation equations become

$$\hat{F}' + \widetilde{\overline{K}}\hat{F} = 0$$
$$\hat{M}' + \widetilde{\overline{K}}\hat{M} - \widetilde{\overline{M}}\hat{K} + \tilde{e}_1\hat{F} = 0 \tag{32}$$

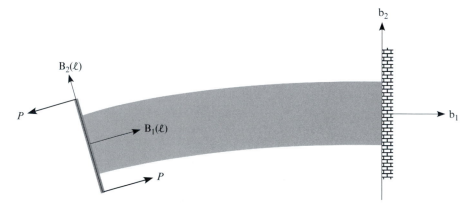

FIGURE 9.2   Cantilevered beam under bending induced by equal and opposite follower forces

The first perturbation equation shows that $\hat{F}'_3 = 0$. Thus, because $\hat{F}_3 = 0$ at the boundary where the load is applied, we have $\hat{F}_3 \equiv 0$. The second equation gives way to two moment equations for out-of-plane bending and torsion, viz.,

$$\hat{M}'_1 + \left(\frac{Q}{EI_2} - \frac{Q}{EI_3} \right) \hat{M}_2 = 0$$

$$\hat{M}'_2 - \left(\frac{Q}{GJ} - \frac{Q}{EI_3} \right) \hat{M}_1 - \hat{F}'^0_3 = 0 \tag{33}$$

which leads to a single equation for $\hat{M}_1$ as

$$\hat{M}''_1 + \beta^2 \hat{M}_1 = 0 \tag{34}$$

where

$$\beta^2 = \left(\frac{Q}{EI_2} - \frac{Q}{EI_3} \right) \left(\frac{Q}{GJ} - \frac{Q}{EI_3} \right) \tag{35}$$

Because $\hat{M}_1 = \hat{M}_2 = 0$ at the end where the load is applied, one finds that $\beta \equiv 0$, so that only the trivial solution exists. Therefore, buckling is not possible. Exactly the same conclusion follows from a moment whose components are constant in the space-fixed frame.

Thus, we focus here on conservatively applied torques at the free end. To do so, we can make use of the equations above, specifically, Eqs. (34), together with equations derived in Chapter 8 for quasi- and semi-tangential torques and for torques applied through a Hooke joint. Because of the direction of the applied moment here, we permute the indices associated with the moments of Chapter 8 by the following replacement rules: $1 \to 3$, $2 \to 1$, and $3 \to 2$. The simplest case is the semi-tangential torque, which in the deformed-beam cross-sectional basis can be written as

$$\begin{aligned}
\mathbf{M} &= Q\left[-\frac{1}{2}(C_{31}C_{11} + C_{32}C_{12})\mathbf{B}_1 - \frac{1}{2}(C_{31}C_{21} + C_{32}C_{22})\mathbf{B}_2 + \mathbf{B}_3 \right] \\
&= Q\left(-\frac{1}{2}\hat{\theta}_2\mathbf{B}_1 + \frac{1}{2}\hat{\theta}_1\mathbf{B}_2 + \mathbf{B}_3 \right) \\
&= Q\left(\frac{1}{2}\hat{u}'_3\mathbf{B}_1 + \frac{1}{2}\hat{\theta}_1\mathbf{B}_2 + \mathbf{B}_3 \right)
\end{aligned} \tag{36}$$

which are independent of the pre-buckling deformation. Using the second of these relations, one finds that the perturbed moment components in the deformed beam cross-sectional frame can simply be written as

$$\hat{M}_1(\ell) = -\frac{Q}{2}\hat{\theta}_2(\ell)$$

$$\hat{M}_2(\ell) = \frac{Q}{2}\hat{\theta}_1(\ell) \tag{37}$$

or, in column matrix form, as

$$\hat{M} = \frac{Q}{2}\tilde{e}_3\hat{\theta} \tag{38}$$

In addition we'll need at least the rotational kinematics since the boundary conditions involve θ. Recalling Eqs. (3.137), one can write the curvature to first order as

$$\hat{\kappa} = \hat{\theta}' + \widetilde{\widehat{\kappa}}\hat{\theta}$$

$$= \hat{\theta}' + \frac{Q}{EI_3}\tilde{e}_3\hat{\theta} \tag{39}$$

To keep the analysis tractable, we need to ignore the pre-buckling curvature and rotation, both here and in Eq. (34), so that

$$\hat{M}_1' + \frac{Q}{EI_2}\hat{M}_2 = 0$$

$$\hat{M}_2' - \frac{Q}{GJ}\hat{M}_1 = 0 \tag{40}$$

with

$$\hat{\kappa} = \hat{\theta}' \tag{41}$$

Letting

$$\beta^2 = \frac{Q^2}{GJEI_2} \tag{42}$$

the solution can be found as $\beta\ell = \pi$ or

$$Q_{cr} = \frac{\pi\sqrt{GJEI_2}}{\ell} \tag{43}$$

The quasi-tangential torque and a torque applied through a Hooke joint have identical expressions for the case when pre-buckling curvature and rotation are ignored, given by

$$\mathbf{M} = Q\left[-(C_{31}\cos\alpha + C_{32}\sin\alpha)(\cos\alpha\mathbf{B}_1 + \sin\alpha\mathbf{B}_2) - \mathbf{B}_3\right]$$

$$= Q\left[-(\hat{\theta}_2\cos\alpha - \hat{\theta}_1\sin\alpha)(\cos\alpha\mathbf{B}_1 + \sin\alpha\mathbf{B}_2) - \mathbf{B}_3\right] \tag{44}$$

so that

$$\hat{M}_1(\ell) = Q\cos\alpha(\hat{\theta}_1\sin\alpha - \hat{\theta}_2\cos\alpha)$$

$$\hat{M}_2(\ell) = Q\sin\alpha(\hat{\theta}_1\sin\alpha - \hat{\theta}_2\cos\alpha) \tag{45}$$

One can express the solution as

$$\beta\ell = \pm\cot^{-1}\left[\sin\alpha\cos\alpha\left(\sqrt{\frac{GJ}{EI_2}} - \sqrt{\frac{EI_2}{GJ}}\right)\right] \tag{46}$$

To get an idea of how this varies with α for a particular case, consider an isotropic beam with deep rectangular cross section, so that

$$\frac{GJ}{EI_2} = \frac{1}{2(1+\nu)}\frac{\frac{1}{3}bh^3}{\frac{1}{12}bh^3} = \frac{2}{1+\nu} \tag{47}$$

For this case

$$Q_{cr} = \pm\frac{\sqrt{GJ\,EI_2}}{\ell}\cot^{-1}\left[\frac{(1-\nu)\cos\alpha\sin\alpha}{\sqrt{2}\sqrt{1+\nu}}\right] \tag{48}$$

A plot of $\beta\ell/\pi$ versus α for $\nu = 1/3$ is shown in Fig. 9.3 where it is noted that the solution is valid for positive or negative values of Q and where for $\alpha = 0$ or $90°$ its

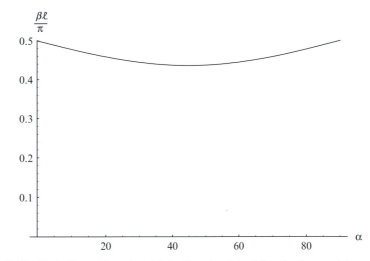

FIGURE 9.3 Plot of nondimensional lateral-torsional buckling load caused by applied quasi-tangential torque versus α

value is $\pi/2$. This means that the lateral-torsional buckling load for a semi-tangential torque is at least twice that for a quasi-tangential torque. Remember that these torques differ from each other by an infinitesimal amount! Thus, the *lateral-torsional buckling moment for a cantilever* under an applied bending moment *is very sensitive to small differences in the way the moment is applied*: tangentially, no buckling is possible; semi-tangentially, $\beta\ell = \pi$; and quasi-tangentially, $\beta\ell \leq \pi/2$.

9.3 CANTILEVERED BEAM UNDER TRANSVERSE FORCE

The original work on the lateral-torsional stability of cantilevered beams goes back to the pioneering work of Michell (1899), Prandtl (1899), and H. Reissner (1904). Michell formulated the linear model for deep beams, and Prandtl and Reissner independently developed closed-form approximations for the buckling load that included some effects of bending prior to buckling. Hodges and Peters (1975) corrected some small errors in the earlier work. The corrected equations turned out to be simpler than the original ones, and Hodges and Peters used an asymptotic expansion to develop a closed-form buckling load formula that included the effect of bending prior to buckling in an asymptotically correct way. The method of asymptotic expansions was also used by E. Reissner (1979) to include the effect of shear deformations on the buckling of cantilever beams.

Since 1979, little further attention was given to the lateral buckling of cantilever beams until the advent of practical applications with composite materials. Work which attempts to address the potential of composites to improve the lateral-torsional stability of I-beams was presented by Pandey et al. (1995) based on the composite I-beam theory of Bauld and Tzeng (1984). Unfortunately, the theory of Bauld and Tzeng has been shown to fail in certain situations by comparison with numerical solutions and with asymptotically exact treatments of composite I-beams, such as the one developed by Volovoi et al. (1999). Here we will present the theory in terms of the stiffnesses of the beam model of Section 3.7, Eq. (3.117) in particular, augmented with the Vlasov correction (see Section 3.8.1). In this Section we will present a derivation of the governing equations for the lateral-torsional buckling of a cantilevered composite beam. In the governing equations the effects of pre-buckling deflections, offset of the applied load from the centroid, and elastic coupling will be included. For the strip-beam case an approximate, closed-form solution is presented for the buckling load taking into account all of these phenomena. The equations including the Vlasov effect are much more complicated. Results obtained from numerical solutions without some of the secondary effects will be included. For formulae that govern the more general case, the reader is referred to Hodges and Peters (2001).[1]

9.3.1 DERIVATION OF GOVERNING EQUATIONS

Consider a thin-walled composite I-beam. When the flange length is zero, the beam has a thin rectangular cross section (herein termed a "strip-beam"), so the treatment here is designed to treat both cases. Let the beam be cantilevered and loaded at its free end with a transverse dead load. The point of application for the load is in the plane of greatest flexural rigidity and at a distance $\bar{e}$ above the centroidal axis; see Fig. 9.4. The load is directed vertically downward, perpendicular to the undeformed beam axis and parallel to the plane of the beam's greatest flexural rigidity. In the figure $\mathbf{B}_1$, $\mathbf{B}_2$, and $\mathbf{B}_3$ are unit vectors fixed in the cross-sectional frame. The beam is spanwise uniform and made of generally anisotropic materials, the only restriction being that the plane of greatest flexural rigidity is assumed to be a plane of symmetry for the beam geometry, material, and loading. This implies that pre-buckling deflections of the beam axis occur in this plane and that in the pre-buckling state there is neither torsion nor out-of-plane bending. Thus, the cross-sectional frame rotates about $\mathbf{b}_3 = \mathbf{B}_3$ by the angle $\theta_3(x_1)$, where x_1 is the beam axial coordinate.

Ignoring the stretching of the beam, this means that there is elastic coupling only between torsion and bending about the weak flexural axis. Asymptotically-exact formulae for the cross-sectional stiffness constants of thin-walled anisotropic strip- and I-beams can be found in Volovoi et al. (1999). The only difference in the two cases is that the Vlasov term (see Fig. 9.4) is absent in the strip-beam case but very important for the I-beam. For the present analysis, the constitutive law is only needed for the moment stress resultants. Letting x_2 and x_3 be its cross-sectional principal axes for bending, it is of the form

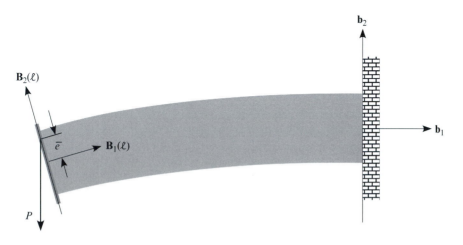

FIGURE 9.4   Schematic of end-loaded cantilever

[1] Portions of this material are based on Hodges and Peters (2001), used by permission.

$$M_1 = D_{11}\kappa_1 + D_{12}\kappa_2$$
$$M_2 = D_{12}\kappa_1 + D_{22}\kappa_2$$
$$M_3 = D_{33}\kappa_3 \tag{49}$$
$$Q_1 = D_{44}\kappa_1'$$

where $(\ )'$ denotes the derivative with respect to x_1. Here Latin indices vary from 1 to 3, and Greek ones from 2 to 3. It is noted that $\mathbf{B}_i' = \boldsymbol{\kappa} \times \mathbf{B}_i$ with $\kappa_i = \boldsymbol{\kappa} \cdot \mathbf{B}_i$. Thus, κ_1 is the elastic twist per unit length, and κ_α are the elastic bending curvatures. The cross-sectional moment stress resultant is $\mathbf{M}$, with $M_i = \mathbf{M} \cdot \mathbf{B}_i$; Q_1 is the bi-moment. For isotropic beams $D_{12} = 0$; D_{11}, D_{22}, D_{33}, and D_{44} are usually written as GJ, EI_2, EI_3, and $E\Gamma$ with G as the shear modulus, E as the Young's modulus, J as the Saint-Venant torsion constant, I_α as area moments of inertia, and Γ as the Vlasov constant (sometimes called warping rigidity). The only difference between the models for the two types of beams considered is the presence (absence) of the stiffness constant D_{44} in the case of the I-beam (strip-beam).

The present analysis will be based on the geometrically-exact equilibrium equations of classical[1] beam theory as augmented with the Vlasov effect. One can write these equations in a compact form as demonstrated in Section 3.7 and easily augment them with the Vlasov effect to yield

$$F_1' - F_2\kappa_3 + F_3\kappa_2 = 0$$
$$F_2' - F_3\kappa_1 + F_1\kappa_3 = 0$$
$$F_3' - F_1\kappa_2 + F_2\kappa_1 = 0$$
$$M_1' - M_2\kappa_3 + M_3\kappa_2 - Q_1'' = 0 \tag{50}$$
$$M_2' - M_3\kappa_1 + M_1\kappa_3 - (1+\varepsilon)F_3 - \kappa_3 Q_1' = 0$$
$$M_3' - M_1\kappa_2 + M_2\kappa_1 + (1+\varepsilon)F_2 + \kappa_2 Q_1' = 0$$

where the cross-sectional stress resultant is $\mathbf{F}$, and $F_i = \mathbf{F} \cdot \mathbf{B}_i$. For small-strain analysis, it is permissible to drop the stretching strain ε compared to unity in the last two equations.

For lateral-torsional buckling analysis, one can let

$$\kappa_i = \overline{\kappa}_i + \hat{\kappa}_i$$
$$F_i = \overline{F}_i + \hat{F}_i$$
$$M_i = \overline{M}_i + \hat{M}_i \tag{51}$$
$$Q_1 = \overline{Q}_1 + \hat{Q}_1$$

where the ˆ quantities are regarded as infinitesimal, and write two sets of equations from the above. The first set, which contains no ˆ quantities, can be described as the pre-buckling equations of equilibrium. The second set which is linear in the ˆ quantities is the set of equations which govern the stability.

The pre-buckling equilibrium equations only involve F_1, F_2, and M_3 and are given by

$$\overline{F}_1' - \overline{F}_2\overline{\kappa}_3 = 0$$
$$\overline{F}_2' + \overline{F}_1\overline{\kappa}_3 = 0 \tag{52}$$
$$\overline{M}_3' + \overline{F}_2 = 0$$

---

[1] The term "classical" is used to refer to the fact that transverse shear deformation is neglected.

where $\overline{\kappa}_3' = \overline{\theta}_3'$. Letting $\overline{F}_1 = P\sin\overline{\theta}_3$ and $\overline{F}_2 = P\cos\overline{\theta}_3$, one can reduce the pre-buckling equations to one equation

$$D_{33}\overline{\theta}_3'' + P\cos\overline{\theta}_3 = 0 \tag{53}$$

subject to boundary conditions $\overline{\theta}_3(\ell) = 0$ and $\overline{M}_3(0) = -\overline{e}P\sin\overline{\theta}_3(0)$.

The required perturbation equations for buckling analysis can be written as

$$\hat{F}_3' - P\sin\overline{\theta}_3\hat{\kappa}_2 + P\cos\overline{\theta}_3\hat{\kappa}_1 = 0$$
$$\hat{M}_1' - \overline{\kappa}_3\hat{M}_2 + \overline{M}M_3\hat{\kappa}_2 - \hat{Q}_1'' = 0 \tag{54}$$
$$\hat{M}_2' - \overline{M}_3\hat{\kappa}_1 + \overline{\kappa}_3\hat{M}_1 - \hat{F}_3 - \overline{\kappa}_3\hat{Q}_1' = 0$$

subject to boundary conditions $\hat{F}_3(\ell) = 0$, $\hat{\kappa}_1(\ell) = 0$, $\hat{\kappa}_1'(0) = 0$, $\hat{M}_1(0) - \hat{Q}_1'(0) - e\hat{F}_3(0) = 0$, and $\hat{M}_2(0) = 0$.

The exact solution to these equations is unknown. The approach of Hodges and Peters (1975) provides an approximate formula for the buckling load in the presence of a variety of secondary phenomena. To undertake this approach in the present context, one must introduce a set of small parameters. First, the square of the pre-buckling rotation is assumed to be small compared to unity, so that $\overline{\theta}_3^2 \ll 1$. Thus, the pre-buckling rotation equation, Eq. (53), can be simplified to

$$D_{33}\overline{\theta}_3'' + P = 0 \tag{55}$$

Using arguments similar to those of Hodges and Peters (1975) one can easily show that the following small parameters are all of the same order

$$\max(\overline{\theta}_3) = \frac{P\ell^2}{D_{33}}$$
$$e = \frac{\overline{e}}{\ell}$$
$$A = \frac{D_{11}}{D_{33}} \tag{56}$$
$$B = \frac{D_{22}}{D_{33}}$$

Thus, ignoring terms that are second order in these small parameters, one can solve Eq. (55) to obtain

$$\overline{\theta}_3 = \frac{P(\ell^2 - x_1^2)}{D_{33}} \tag{57}$$

It should be noted that Hodges and Peters (1975) did not consider the parameter e.

Introducing ζ and γ such that

$$D_{12} = \zeta\sqrt{D_{11}D_{22}}$$
$$D_{44} = \gamma D_{11}\ell^2 = \gamma AD_{33}\ell^2 \tag{58}$$

one can write the entire constitutive law in terms of D_{33} and the non-dimensional parameters A, B, ζ, and γ. Following Hodges and Peters (1975), we note that the combination of the second of Eqs. (54) plus $\overline{\theta}_3$ times the third yields an equation which, when small terms are consistently neglected, is a perfect differential of the following equation:

$$k + B\hat{\kappa}_2 + A\hat{\kappa}_1\bar{\theta}_3 + \sqrt{A}\sqrt{B}\zeta(\hat{\kappa}_1 + \hat{\kappa}_2\bar{\theta}_3) + \frac{\hat{F}_3\bar{\theta}_3'}{P} - A\ell^2\gamma\bar{\theta}_3\hat{\kappa}_1'' = 0 \qquad (59)$$

where k is an arbitrary constant to be determined later. This equation can be solved along with the first of Eqs. (54) for $\hat{\kappa}_1$ and $\hat{\kappa}_2$. When small terms are discarded consistently, one obtains

$$\hat{\kappa}_1 = -\phi' + \frac{\bar{\theta}_3\left(\sqrt{A}\sqrt{B}\zeta\phi' - \phi\bar{\theta}_3'\right)}{B}$$

$$\hat{\kappa}_2 = \frac{-B\kappa + \sqrt{A}B^{\frac{3}{2}}\zeta\phi' - B\phi\bar{\theta}_3' + \sqrt{A}\,\bar{\theta}_3\left[\sqrt{A}B(1 - 2\zeta^2)\phi' + 2\sqrt{B}\zeta\phi\bar{\theta}_3' - \sqrt{A}B\ell^2\gamma\phi'''\right]}{B^2}$$

$$(60)$$

where the nondimensional perturbation out-of-plane shear force $\phi = \hat{F}_3/P$. Thus, the $\hat{M}_1$ equation, the second of Eqs. (54), and all the boundary conditions can now be expressed completely in terms of ϕ and its derivatives with respect to x_1.

To facilitate the writing of this equation, we introduce a nondimensional buckling load β such that $P = \beta\sqrt{D_{11}D_{22}}/\ell^2 = \beta\sqrt{AB}D_{33}/\ell^2$ and a nondimensional axial coordinate x such that $\ell x = x_1$. By using the boundary conditions the constant k can be found to be

$$k = \frac{A^{\frac{3}{2}}\sqrt{B}\beta\left[(1 - \zeta^2)\phi'(0) - \gamma\phi'''(0)\right]}{2} \qquad (61)$$

where the prime now indicates a derivative with respect to x. However, use of the boundary conditions leads to homogeneous equations with boundary values of the unknown in them. It turns out that the boundary conditions must be multiplied by appropriate constants in order to make them variationally consistent. The details of this operation are straightforward and not given here. The resulting equation is

$$\gamma\left[1 - A(1 - x^2)\beta\zeta\right]\phi'''' + 4Ax\beta\gamma\zeta\phi'''$$

$$-\left\langle 1 - \zeta^2 - \frac{A\beta}{2}\left\{3(3x^2 - 1)\beta\gamma + 2\zeta\left[3\gamma + (1 - x^2)(1 - \zeta^2)\right]\right\}\right\rangle\phi''$$

$$+Ax\beta(9\beta\gamma - 2\zeta + 2\zeta^3)\phi' \qquad (62)$$

$$-\beta\left\{(1 - B)x^2\beta - \zeta - A\beta\left[3\gamma + x^2(1 - x^2)\beta\zeta + (1 - 3x^2)\left(\frac{1}{2} - \zeta^2\right)\right]\right\}\phi = 0$$

and the boundary conditions are

$$\phi(1) = \phi'(1) = 0$$

$$\gamma\left[(1 - A\beta\zeta)\phi''(0) - \frac{A\beta^2\phi(0)}{2}\right] = 0 \qquad (63)$$

$$\gamma(1 - A\beta\zeta)\phi'''(0) - \left[(1 - \zeta^2)(1 - A\beta\zeta) - A\beta\gamma(\zeta - \beta)\right]\phi'(0) - \frac{\sqrt{B}e\beta\phi(0)}{\sqrt{A}} = 0$$

For stability analysis it is convenient to work in terms of an energy functional in terms of ϕ and its derivatives and boundary values. This functional can be shown to be of the form

$$L = \frac{1}{2}\int_0^1 \left(L_0\phi^2 + L_1\phi'^2 + L_2\phi''^2\right)dx + \frac{A\beta^2\gamma\phi(0)\phi'(0)}{2} - \frac{\sqrt{B}e\beta\phi^2(0)}{2\sqrt{A}} \qquad (64)$$

where

$$L_0 = -\beta\left[(1-B)x^2\beta - \zeta\right] + \frac{A\beta^2}{2}\left[1 + 6\gamma - 2x^4\beta\zeta - 2\zeta^2 - x^2\left(3 - 2\beta\zeta - 6\zeta^2\right)\right]$$
$$L_1 = 1 - \zeta^2 - \frac{A\beta}{2}\left\{3\left(3x^2 - 1\right)\beta\gamma + 2\zeta\left[\gamma + \left(1 - x^2\right)\left(1 - \zeta^2\right)\right]\right\} \tag{65}$$
$$L_2 = \gamma\left[1 - A\left(1 - x^2\right)\beta\zeta\right]$$

The above differential equation, boundary conditions, and functional reduce to the corresponding expressions given by Hodges and Peters (1975) when $\gamma = e = \zeta = 0$.

For both the strip- and I-beam cases, the load offset parameter e is multiplied by $\sqrt{B/A}$ in its only appearance in the formulation. Somewhat complicating the procedure from here on, is the fact that this quantity may be large under certain circumstances, thus magnifying the influence of the e parameter. In such cases, the assumption that e is small has to be relaxed.

9.3.2   APPROXIMATE SOLUTION

Case of Zero Warping Rigidity

For thin beams having rectangular cross sections—e.g., strip-beams—the Vlasov effect may be ignored even for anisotropic beams according to Volovoi et al. (1999). For the case of zero warping rigidity, i.e., $\gamma = 0$, the above equations and energy functional are greatly simplified. The main reason for the simplification mathematically is that the order of the governing equation and number of boundary conditions are reduced. The differential equation becomes

$$\left[1 - A\beta\zeta\left(1 - x^2\right)\right]\left(1 - \zeta^2\right)\phi'' + 2A\beta\zeta x\left(1 - \zeta^2\right)\phi'$$
$$+ \beta\left\{(1-B)x^2\beta - \zeta - A\beta\left[x^2\left(1 - x^2\right)\beta\zeta + \left(1 - 3x^2\right)\left(\frac{1}{2} - \zeta^2\right)\right]\right\}\phi = 0 \tag{66}$$

and the boundary conditions are

$$\phi(1) = \left(1 - \zeta^2\right)\left(1 - A\beta\zeta\right)\phi'(0) + \frac{\sqrt{B}e\beta\phi(0)}{\sqrt{A}} = 0 \tag{67}$$

The corresponding energy functional is

$$L = \frac{1}{2}\int_0^1 \left(L_0\phi^2 + L_1\phi'^2\right)dx - \frac{\sqrt{B}e\beta\phi^2(0)}{2\sqrt{A}} \tag{68}$$

where

$$L_0 = -\beta\left[(1-B)x^2\beta - \zeta\right] + \frac{A\beta^2}{2}\left[1 - 2x^4\beta\zeta - 2\zeta^2 - x^2\left(3 - 2\beta\zeta - 6\zeta^2\right)\right]$$
$$L_1 = 1 - \zeta^2 - \frac{A\beta}{2}\left\{2\zeta\left[\gamma + \left(1 - x^2\right)\left(1 - \zeta^2\right)\right]\right\} \tag{69}$$

An analytical solution of this boundary-value problem is not known to the authors. However, using the solution to the case $A = B = e = \zeta = 0$ to obtain an accurate comparison function, as done by Hodges and Peters (1975), one obtains a useful result. To do so, one notes that the simplified equation is

$$\phi'' + \beta^2 x^2 \phi = 0 \tag{70}$$

with boundary conditions $\phi(1) = \phi'(0) = 0$. The solution is given by

$$\phi = \sqrt{x} J_{-\frac{1}{4}}\left(\frac{4.01260x^2}{2}\right) \tag{71}$$

where the factor 4.01260 is the smallest positive zero of the Bessel function $J_{-1/4}(\beta/2)$. Using this function in the energy functional allows one to obtain an approximate solution for β, given by

$$\beta = 4.01260\left(1 + 0.642365A + \frac{B}{2} - 1.02543\sqrt{\frac{B}{A}}e\right) + 2.78473\zeta - 1.04000\zeta^2 \tag{72}$$

where ζ^2 is taken to be of the same order as A, B, and e. Subject to a few limitations, one can use this expansion to find the buckling load of uniform strip-beams. The pre-buckling deflection effects are associated with A and B and, while accurately captured by the formula, tend to be small corrections to the standard result of $\beta = 4.01260$. Because they tend to raise the buckling load, they are typically ignored. The effects of ζ and e do not fall into this category, as both can lower the buckling load.

Fortunately, the expansion in ζ is valid for $|\zeta| \leq 0.3$ or so, which is sufficiently large for most practical purposes. To illustrate the utility of the buckling formula of Eq. (72), in Fig. 9.5 we compare results from it with those of the exact solution of Eq. (66) with all secondary effects set equal to zero except ζ. Note that the exact solution for this special case can be expressed in terms of the Kummer confluent hypergeometric function; see Abramowitz and Stegun (1970). As long as ζ is small, the formula and the exact solution agree quite well. The accuracy of the formula for A and $B < 0.1$ is comparable to its accuracy for $|\zeta| < 0.3$.

However, when $e\sqrt{B/A}$ is not small of the same order as A, B, and e, the expansion in Eq. (72) breaks down. As an example, this can happen in anisotropic strip-beams when the effective ratio of extension modulus to shear modulus (E/G)

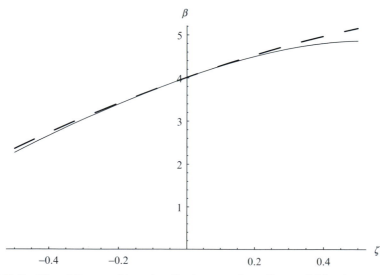

FIGURE 9.5   Plot of β versus ζ ignoring all other secondary effects; solid line is exact solution of Eq. (66) with A, B and e set to zero, and dashed line is from Eq. (72)

becomes large, which tends to magnify the influence of the load offset. The solution for the case in which $e\sqrt{B/A} \neq 0$ but where otherwise $A = B = \zeta = 0$ is

$$\phi = \sqrt{x}\left[c_1 J_{-\frac{1}{4}}\left(\frac{x^2\beta}{2}\right) + c_2 J_{\frac{1}{4}}\left(\frac{x^2\beta}{2}\right)\right] \tag{73}$$

subject to the boundary conditions

$$\phi(1) = 0$$
$$\phi'(0) + \frac{\sqrt{B}e\beta\phi(0)}{\sqrt{A}} = 0 \tag{74}$$

which leads to the characteristic equation

$$\frac{4e\sqrt{\frac{B}{A}}\sqrt{\beta}}{\Gamma\left(\frac{3}{4}\right)}J_{\frac{1}{4}}\left(\frac{\beta}{2}\right) - \left[\frac{4}{\Gamma\left(\frac{1}{4}\right)} + \frac{1}{\Gamma\left(\frac{5}{4}\right)}\right]J_{-\frac{1}{4}}\left(\frac{\beta}{2}\right) = 0 \tag{75}$$

which can be solved numerically for β given any value of $e\sqrt{B/A}$. A plot of the solution is shown in Fig. 9.6. Indeed, in the limit of large and positive $e\sqrt{B/A}$, the result approaches the case of applying a downward compressive load at the end of a long, rigid rod extending upward from the beam axis (instead of from within the beam cross section). This buckling load β clearly approaches zero as the length of the rigid rod increases. On the other hand, when $e\sqrt{B/A}$ is large and negative, the buckling load approaches that for a downward tensile load applied at the end of a long, rigid rod extending downward (i.e., in the direction of the load). In this case the buckling load approaches that for the case of $\phi(0) = 0$, for which $\beta = 5.56178$, the smallest positive zero of $J_{1/4}(\beta/2)$. Obviously no linear function of e such as Eq. (9.72) can capture this behavior outside of a small range near $e\sqrt{B/A} = 0$. Eq. (72) is then not valid in the cases when $e\sqrt{B/A}$ is not small, but it can serve to indicate the trend of β versus e. One can develop a comparison function which contains e as a parameter, but the resulting buckling load formula turns out to be very complex.

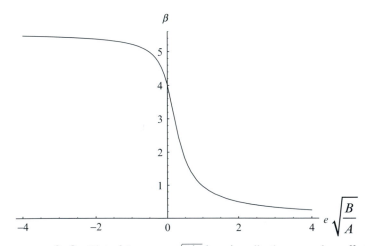

FIGURE 9.6   Plot of β versus $e\sqrt{B/A}$ ignoring all other secondary effects

Case of Nonzero Warping Rigidity

The case of nonzero warping rigidity is far more difficult. The governing equations and boundary conditions with all secondary effects set to zero are

$$\gamma \phi'''' - \phi'' - x^2 \beta^2 \phi = 0$$
$$\phi(1) = \phi'(1) = \phi''(0) = \gamma \phi'''(0) - \phi'(0) = 0 \tag{76}$$

Other than an infinite series solution, results of which are presented by Timoshenko and Gere (1961) (and also following), the solution to this deceptively simple-looking boundary-value problem is not known to the authors. Without a one-term solution which has an explicit behavior in γ, one has no exact "zeroth-order" solution to use as the assumed mode in the energy to find the buckling load in the presence of the small parameters A, B, e, and ζ. The only alternative seems to be an approximate comparison function.

In Eq. (76), the parameter, γ, can vary from being small compared to unity to being considerably larger than unity. In order to obtain a better understanding of the behavior of the equation, it is interesting to first look at the limiting behaviors for cases of γ small and γ large. A natural means of doing this is the method of matched asymptotic expansions. In that method, the buckling load parameter, β, must be expanded in some selected powers of γ as must the solution, ϕ, itself. Furthermore, there must be an expansion of the independent variable, x, in terms of powers of γ at the fixed and free ends (the boundary layers) in order to capture the correct limiting behavior. The solutions in the boundary layers (inner solutions) must be matched (as the inner variables become large) to the general solution away from the ends (outer expansion) as the outer variable approaches either end. This yields a unique solution for the buckling load and the buckling mode shape. Since the term with γ multiplies ϕ'''', and the next highest derivative is ϕ'', all expansions for small γ must be done in powers of $\gamma^{1/2}$ in order to allow the inner and outer solutions to be matched. As it turns out, only the fixed end ($x = 1$) has a boundary layer that affects the outer solution and buckling load to order $\gamma^{1/2}$ or γ^1. The resultant expansion shows that the boundary layer dies out into the outer solution as $\exp[(1 - x)/\gamma^{1/2}]$. The next term in the outer expansion (and the next term in the expression for the approximate buckling load) can then be found by solving a nonhomogeneous, second-order equation for the outer solution. This involves integrals of the Bessel functions that are rather involved. For our purposes here, it is enough to know how the fixed-end boundary layer behaves and, particularly, how it decays into the outer region.

The large γ solution can be written in terms of confluent, hypergeometric functions, described by Abramowitz and Stegun (1970). Since no derivatives are lost as γ becomes large, there are no boundary layers or matching required in that case. One can do a straightforward expansion in terms of the small parameter, $1/\gamma$. The first expansion term in that solution involves fairly complicated integrals of the hypergeometric functions and is not tractable in closed form. Nevertheless, the fact that there are no boundary layers in that solution indicates that an approximation for the entire mode, fairly accurate at all values of γ, might be constructed from simple polynomials plus the crucial boundary layer terms. In contrast to the asymptotic expansions, in which boundary conditions are satisfied only to the order of the expansion that is taken, here, we use the exponential decay term for the fixed end and add the simplest possible polynomial that will match the other boundary conditions exactly. We expect that this comparison function will provide accurate results.

Whether or not this expectation is realized can only be determined by comparison with numerical solutions to the exact equations.

Taking a cubic polynomial along with the exponential, one finds the simplest comparison function of that form (i.e., one which satisfies all the boundary conditions in Eqs. 9.76) to be

$$\theta = (1 - x)^2(1 + 2x - 6\gamma)e^{-\frac{1}{\sqrt{\gamma}}} + 6e^{-\frac{1-x}{\sqrt{\gamma}}}\gamma(1 + 2\gamma) \\ + 2\sqrt{\gamma}\left[1 - x^3 - 3\sqrt{\gamma} + 6(1 - x)\gamma - 6\gamma^{\frac{3}{2}}\right] \tag{77}$$

Simplifying the energy functional of Eq. (64) by setting A, B, e, and ζ equal to zero, one can obtain an approximate closed-form expression for the nondimensional buckling load β, given by

$$\beta_0^2 = \frac{\int_0^1 \left(\phi'^2 + \gamma\phi''^2\right)dx}{\int_0^1 x^2\phi^2 dx} \tag{78}$$

The accuracy of this predicted buckling load as a function of γ can be regarded as a measure of how well θ performs as a one-term approximation of the actual buckling mode ϕ. The resulting nondimensional buckling load can be written as

$$\beta_0 = \sqrt{\frac{\eta_0 + \eta_1 e^{-\frac{1}{\sqrt{\gamma}}} + \eta_2 e^{-\frac{2}{\sqrt{\gamma}}}}{\delta_0 + \delta_1 e^{-\frac{1}{\sqrt{\gamma}}} + \delta_2 e^{-\frac{2}{\sqrt{\gamma}}}}} \tag{79}$$

where

$$\eta_0 = 1512\gamma\left(3 - 15\sqrt{\gamma} + 40\gamma - 60\gamma^{\frac{3}{2}} + 60\gamma^2 - 60\gamma^{\frac{5}{2}}\right)$$

$$\eta_1 = 756\sqrt{\gamma}\{3 - 10\gamma[1 - 12\gamma(1 + 2\gamma)]\}$$

$$\eta_2 = 756\left\{1 - 10\gamma^{\frac{3}{2}}\left[3 + 4\left(2\sqrt{\gamma} + 3\gamma + 3\gamma^{\frac{3}{2}} + 3\gamma^2\right)\right]\right\}$$

$$\delta_0 = 2\gamma\left[140 - 9\sqrt{\gamma}\left(140 - 520\sqrt{\gamma} + 70\gamma + 8862\gamma^{\frac{3}{2}} - 47355\gamma^2 \\ + 131880\gamma^{\frac{5}{2}} - 215460\gamma^3 + 224280\gamma^{\frac{7}{2}} - 243180\gamma^4\right)\right] \tag{80}$$

$$\delta_1 = \sqrt{\gamma}\left[133 - 18\left(28\sqrt{\gamma} - 3\gamma - 28\gamma^{\frac{3}{2}} - 2436\gamma^2 + 21672\gamma^{\frac{5}{2}} \\ - 75600\gamma^3 + 124320\gamma^{\frac{7}{2}} + 40320\gamma^4 + 161280\gamma^{\frac{9}{2}} + 362880\gamma^5\right)\right]$$

$$\delta_2 = 19 - 18\gamma\langle 7 - 3\gamma\{4 - 35\sqrt{\gamma}(1 + 2\gamma)[8 - 3\gamma(111 + 416\sqrt{\gamma} + 190\gamma)]\}\rangle$$

Indeed, one finds an error of less than 1% over a wide range of γ, certainly within an acceptable range of error for design purposes. In Fig. 9.7, the value of β_0 is plotted versus γ over a wide range. The symbols are the exact solutions at discrete values of γ taken from Table 6.3 of Timoshenko and Gere (1961), page 259, corrected values of which are presented in Table 9.1. Notice that the symbols are on top of the curve, to within plotting accuracy, over the whole range of γ shown. Moreover, for $\gamma \to 0$, the relative error from Eq. (79) is only 0.3% off the exact solution of 4.01260. For $\gamma \to \infty$, the limiting value of β_0 is 13.9473γ, only about 1% off the exact value of 13.8058γ.

When $\phi = \theta$ is substituted into the energy, Eq. (64), one can make use of symbolic computational tools to solve for β in terms of γ and the small parameters A, B, e, and ζ. The resulting formula is quite lengthy (see Hodges and Peters, 2001) but can be programmed in a spreadsheet to give rapid estimation of the lateral-torsional buckling load over a wide range of γ and small values of A, B, e, and ζ.

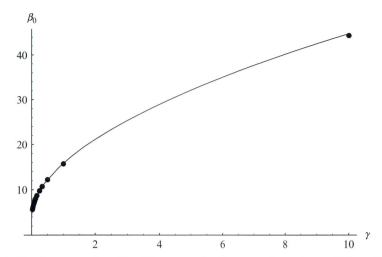

FIGURE 9.7　β_0 versus γ from Eq. (79) compared with exact solution (symbols) from Timoshenko and Gere (1961)

TABLE 9.1　Corrected results from Timoshenko and Gere (1961).

γ	β
10	44.3391
1	15.7078
1/2	12.1650
1/3	10.6487
1/4	9.75474
1/6	8.69273
1/8	8.05211
1/10	7.60915
1/12	7.27860
1/14	7.01961
1/16	6.80964
1/24	6.24835
1/32	5.91400
1/40	5.68755

To give some indications of the behavior of the solution, numerical results from the exact perturbation equations are presented in Figs. 9.8–9.18. In Figs. 9.8–9.10, the effect of pre-buckling deflections on the buckling load is shown. This effect was shown by Hodges and Peters (1975) to be reflected in the parameters A and B. It should be noted that since the buckling load increases significantly for larger A and B, these cases tend to become of less practical significance. In Figs. 9.11 and 9.12 the variation of the buckling load with the offset of the load, e, is shown for small and moderately large values of A and B. Clearly this is a very significant parameter and shifts the buckling load radically. In Fig. 9.13 we look at the effect of elastic coupling. It is shown that bending-torsion coupling can *strongly* affect the buckling load, either positively or negatively. The behavior with ζ appears to be nearly linear for this case. It is clear that elastic coupling could be used to tailor the structure to have a larger buckling load without changing the bending or torsional stiffnesses. In Figs. 9.14 and

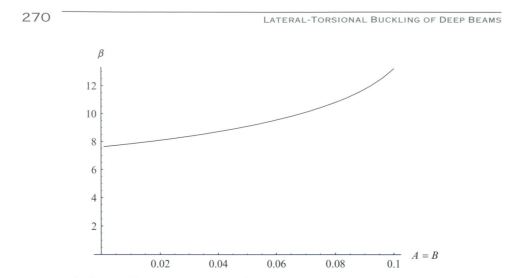

FIGURE 9.8   Buckling load versus $A = B$ for $e = 0$, $\zeta = 0$, $\gamma = 0.1$ from a numerical solution of exact equations

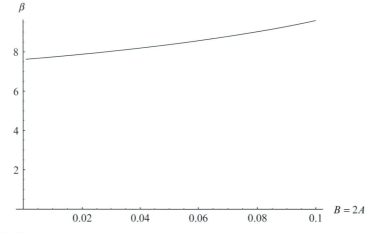

FIGURE 9.9   Buckling load versus $B = 2A$ for $e = 0$, $\zeta = 0$, $\gamma = 0.1$ from a numerical solution of exact equations

9.15 the effects of the Vlasov term are depicted, with zero and nonzero values of all other parameters, respectively.

For I-beams as well as for the strip-beam, when $e\sqrt{B/A}$ is not small of the same order as A, the above approximate solution breaks down. For I-beams, however, the value of $\sqrt{B/A}$ can be large for two different reasons. First, just as for strip-beams, the effective G/E can be small for anisotropic I-beams causing A to be smaller than B. Second, even for isotropic I-beams, B/A can be large because the flange width is typically larger than the wall thickness. This means that the effect of load offset for composite I-beams can be *much* more significant than for isotropic strip-beams. To see this more clearly, we consider the composite beam example presented by Pandey et al. (1995). Results obtained therein will also be compared with the present results.

The I-beam has a uniform wall thickness of 0.00953 m, a depth of 0.2032 m, and a flange width of 0.1016 m. The length considered here is $\ell = 1.2192$ m. In the undertaking of this comparison, the following difficulties were encountered. First, the 3-D

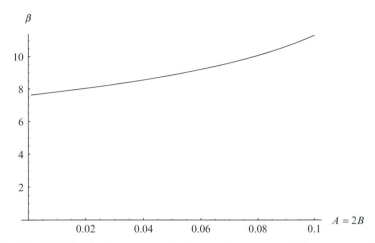

FIGURE 9.10   Buckling load versus $A = 2B$ for $e = 0$, $\zeta = 0$, $\gamma = 0.1$ from a numerical solution of exact equations

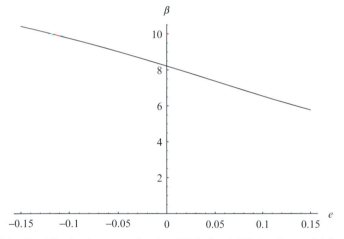

FIGURE 9.11   Buckling load versus e for $A = 0.025$, $B = 0.025$, $\zeta = 0$, $\gamma = 0.1$ from a numerical solution of exact equations

elastic constants for the material system considered by Pandey et al. (1995) were not given directly. Only the fiber and matrix properties were given, along with the statement that the fiber volume fraction was taken to be 60%. No statement was given as to which model was used to obtain the 3-D elastic constants, nor was their actual value given. By comparison with their results, however, it was ascertained that the simple strength of materials model was used. This gives elastic constants of $E_{11} = 42.7 \times 10^9 \, \text{N/m}^2$, $E_{22} = 8.02 \times 10^9 \, \text{N/m}^2$, $G_{12} = 3.10 \times 10^9 \, \text{N/m}^2$, and $\nu_{12} = 0.248$.

The second difficulty was that the plots of torsional rigidity D_{11}, bending stiffness D_{22}, and warping rigidity D_{44} in Pandey et al. (1995) are at variance with those obtained from the asymptotically correct formulation of Volovoi et al. (1999) as well as with the basic physics of the problem. The I-beam considered was such that the upper and lower flanges were unidirectional with $0°$ ply angle. Only the web contained off-axis fibers, being basically a single layer of material with ply angle θ. Since

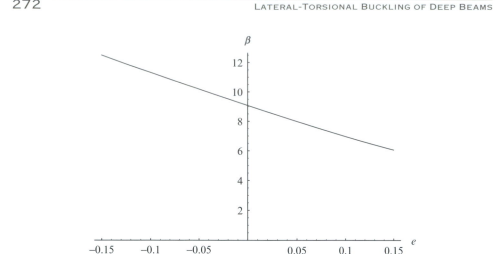

FIGURE 9.12 Buckling load versus e for $A = 0.05$, $B = 0.05$, $\zeta = 0$, $\gamma = 0.1$ from a numerical solution of exact equations

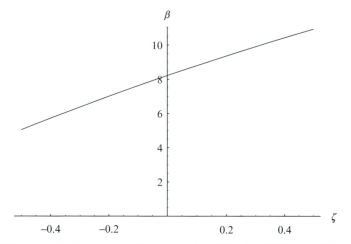

FIGURE 9.13 Buckling load versus ζ for $e = 0$, $A = 0.025$, $B = 0.025$, $\gamma = 0.1$ from a numerical solution of exact equations

the web does not contribute to the smallest bending rigidity, it should not be a function of ply angle. This is not at all reflected in the plots of D_{22} and D_{44} in Pandey et al. (1995). For both these elastic constants, the asymptotic analysis shows them to be constant with ply angle, as expected, with values of $D_{22} = 71200$ N-m^2 and $D_{44} = 735$ N-m^4. Although our values for these constants at $0°$ ply angle are very close to theirs (see their Fig. 9), they show both D_{22} and D_{44} to be sharply decreasing from their maximum values at $0°$ ply angle to very small values at a ply angle of $90°$. It is also noted that the units given by Pandey et al. (1995) are incorrect for D_{44}, the correct units being N-m^4 rather than kN-m.

Finally, Pandey et al. (1995) show the torsional rigidity, D_{11}, on the same plot as D_{22}, reaching a maximum value at ply angle of $45°$. However, results from the asymptotic analysis for D_{11} are three orders of magnitude smaller than theirs; moreover, from the asymptotic analysis the correct peak value is not nearly as

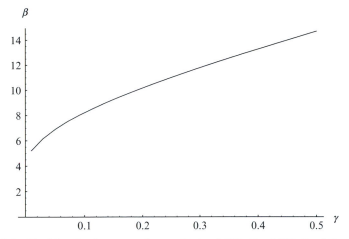

FIGURE 9.14   Buckling load versus γ for $e = 0$, $A = 0.025$, $B = 0.025$, $\zeta = 0$ from a numerical solution of exact equations

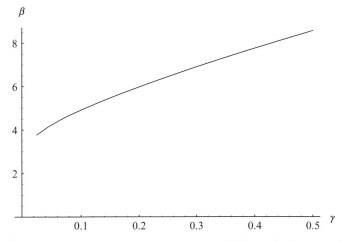

FIGURE 9.15   Buckling load versus γ for $e = 0.1$, $A = 0.0125$, $B = 0.025$, $\zeta = -.2$ from a numerical solution of exact equations

large relative to the $0°$ ply angle value as the one they obtained. For ply angle of $0°$, we obtain $D_{11} = 364$ N-m^2, and for ply angle of $45°$ we obtain $D_{11} = 610$ N-m^2. For completeness, we note that for $0°$ ply angle $D_{33} = 1.14 \times 10^6$ N-m^2 and for ply angle of $45°$ we get $D_{33} = 912000$ N-m^2.

The cases with ply angles of $0°$ and $45°$ are chosen for presentation here. For both cases $\zeta = 0$; indeed, we have found it difficult to conceive of an I-beam configuration which will yield a large value of ζ. For the case with ply angle equal to $0°$, the nondimensional constants are $A = 0.000319$, $B = 0.0625$, and $\gamma = 1.36$. For the case with ply angle of $45°$, $A = 0.000669$, $B = 0.0780$, and $\gamma = 0.810$. The sensitivity with respect to e is large for both cases because of the large values of $\sqrt{B/A}$, resulting in highly nonlinear behavior of β relative to e as shown in Figs. 9.16 and 9.17. As with the strip-beam, when $\sqrt{B/A}$ is large, one must *not* expect a linear trend with respect to e to be valid. Moreover, for the case in which $e < 0$, the buckling load increases, so the use of the value for $e = 0$ is conservative.

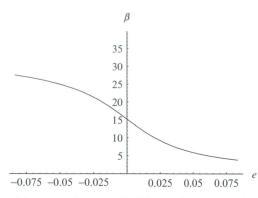

FIGURE 9.16   Plot of β versus e for composite I-beam of Pandey *et al.* (1995) with ply angle of $0°$ from a numerical solution of the exact equations using elastic constants from Volovoi *et al.* (1999)

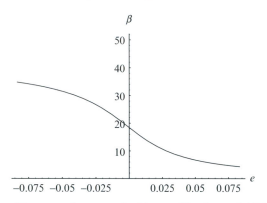

FIGURE 9.17   Plot of β versus e for composite I-beam of Pandey *et al.* (1995) with ply angle of $45°$ from a numerical solution of the exact equations using elastic constants from Volovoi *et al.* (1999)

Given the significant discrepancies in the cross-sectional constants used in Pandey et al. (1995), one should not expect our results to be in agreement with theirs. At $0°$ ply angle, their results give a buckling load of approximately 72,000 N, versus 63,500 N from our analysis. At $45°$ they obtained 54,000 N, whereas our results give 67,500 N. Indeed, not only are the numerical values very different; more significantly, the trend versus ply angle is the opposite. Their results are maximum at $0°$, while ours peak at $45°$; their results are minimum at $45°$ but ours at $0°$. From the above discussion of the section properties, there are significant reasons to believe that the results presented by Pandey et al. (1995) are incorrect. On the other hand, although we do not have any results from a truly independent approach to validate ours (such as 3-D finite elements), there are good reasons to suggest the present results are correct: the well-validated asymptotic formulae used for the section constants and the trends in these constants following the expected behavior.

Accurate treatment of the cases where $e\sqrt{B/A}$ is not small compared to unity is far more problematic. Somehow one must remove the restriction that e is a small parameter, because it always appears multiplied by $\sqrt{B/A}$ in the boundary conditions. Alternatively, it could mean treating A as a much smaller parameter than all others. Proceeding as above with these kinds of changes would make the already long formulae above far more complicated. For the case in which all pre-buckling deflections are ignored, an approach based on the treatment of Rayleigh's quotient with a free parameter can be developed as follows; see Hodges (1997).

The governing equation is the same as Eq. (76), but with a different boundary condition at the free end, viz.,

$$\gamma\phi'''' - \phi'' - x^2\beta^2\phi = 0$$

$$\phi(1) = \phi'(1) = \phi''(0) = \gamma\phi'''(0) - \phi'(0) - \frac{\sqrt{B}e\beta\phi(0)}{\sqrt{A}} = 0 \tag{81}$$

The simplest possible admissible function that satisfies the first three boundary conditions and contains both the exponential term and a free parameter α is used, given by

$$\theta = -(1-x)^2 e^{-\frac{1}{\sqrt{\gamma}}} + 2(1-x)\sqrt{\gamma} + 2\gamma e^{-\frac{1-x}{\sqrt{\gamma}}} - 2\gamma\left[1 - \alpha\left(2 - 3x + x^3\right)\right] \tag{82}$$

The energy functional depends on ϕ and its derivatives and is of the form

$$L = \frac{1}{2}\int_0^1 \left(\gamma\phi''^2 + \phi'^2 - \beta^2 x^2\phi^2\right)dx - \frac{\sqrt{B}e\beta\phi^2(0)}{2\sqrt{A}} \tag{83}$$

One first substitutes $\phi = \theta$ into L, then minimizes L with respect to α, and finally solves the resulting quartic equation for β. This yields a closed-form approximate formula for the buckling load. Although it is a very complicated expression (too long to print here), it is not too difficult for computerized symbolic manipulation to handle. Results from this approximate closed-form expression for the buckling load are compared with a numerical solution of Eqs. (81) in Fig. 9.18. The agreement is excellent. To add the pre-buckling deflections may present difficulties for some symbolic computational tools. However, as shown above, these effects generally make the predicted buckling load larger and frequently may be ignored for design purposes.

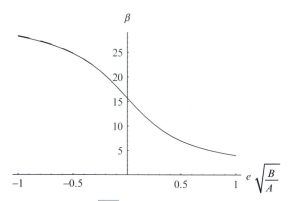

FIGURE 9.18 Plot of β versus $e\sqrt{B/A}$ for an I-beam with $\gamma = 1$, ignoring pre-buckling deflections; solid line is from the numerical solution of Eqs. (81), and dashed line (nearly coincident) is from minimum over α of L (Eq. 83) with $\phi = \theta$ from Eq. (82)

PROBLEMS

1. Determine the Vlasov correction for lateral-torsional buckling of a pinned-pinned I-beam with zero warping displacement at both ends.

2. For an initially curved isotropic strip-beam, determine the relative contribution of the initial curvature case as a function of the geometric parameters (b, h, ℓ, and R). Choose $\nu = 1/3$ and consider effects of the nondimensional cross-sectional aspect ratio (thickness to depth) and slenderness ratio (length to depth).

3. For an isotropic strip-beam, consider the formula developed for the lateral-torsional buckling of a deep beam subjected to a moment applied at the right end (with the sense of an inplane bending moment) with zero inplane rotation at the left end, zero torsional rotation at both ends, and free out-of-plane rotation at both ends. Determine for a beam with rectangular cross section the relative increase ($\beta/4.01260$) in the critical load that arises due to our taking into account the pre-buckling *deformation* as a function of the thickness ratio of the cross section (say b/a where $b > a$). Discuss the implications of your results.

4. For a thin-walled, isotropic I-beam, determine the relative contribution of pre-buckling deformations for the cantilever case as a function of the geometric parameters. Choose $\nu = 1/3$ and consider the cross-sectional aspect ratio (width to depth), the thinness parameter (constant wall thickness to depth) and slenderness ratio (length to depth).

5. Carry out the analysis described below Eq. (83) to obtain the plot shown in Fig. 9.18.

6. Consider a deep beam with rectangular cross section, which is attached to spherical bearings at each end that are only free to move closer to each other and loaded by a semi-tangential bending moment of magnitude Q at $x_1 = \ell$. Let the undeformed beam lie in a plane parallel to the $\mathbf{a}_1$-$\mathbf{a}_2$ plane. Ignoring pre-buckling curvature and rotation (but not bending moment!), determine the buckling load. Discuss the implications of your results.

7. Consider a deep beam with rectangular cross section, which is attached to spherical bearings at each end that are only free to move closer to each other and loaded by a quasi-tangential bending moment of magnitude Q at $x_1 = \ell$. Let the undeformed beam lie in a plane parallel to the $\mathbf{a}_1$-$\mathbf{a}_2$ plane. Ignoring pre-buckling curvature and rotation (but not bending moment!), determine how the buckling load depends on the angle α which defines the direction of the forces comprising the quasi-tangential moment. Discuss the implications of your results.

REFERENCES

Abramowitz, M. and Stegun, I., eds. (1970). *Handbook of Mathematical Functions*. National Bureau of Standards, Washington, D.C.

Bauld, N. R., Jr. and Tzeng, L.-S. (1984). A Vlasov theory for fiber-reinforced beams with thin-walled open cross sections. *International Journal of Solids and Structures* 20, 277–297.

Hodges, D. H. (1997). Improved approximations via Rayleigh's quotient. *Journal of Sound and Vibration* 199, 155–164.

Hodges, D. H. (1999). Non-linear inplane deformation and buckling of rings and high arches. *International Journal of Non-linear Mechanics* 34, 723–737.

Hodges, D. H. and Peters, D. A. (1975). On the lateral buckling of uniform slender cantilever beams. *International Journal of Solids and Structures* 11, 1269–1280.

Hodges, D. H. and Peters, D. A. (2001). Lateral-torsional buckling of cantilevered elastically coupled strip- and I-beams. *International Journal of Solids and Structures* 38, 1585–1603.

Michell, A. G. M. (1899). Elastic stability of long beams under transverse forces. *Philosophical Magazine* 48, 298–309.

Pandey, M., Kabir, M. and Sherbourne, A. (1995). Flexural-torsional stability of thin-walled composite I-section beams. *Composites Engineering* 5, 321–342.

Prandtl, L. (1899). Kipperscheinungen. Doctoral dissertation, Universität München.

Reissner, E. (1979). On lateral buckling of end-loaded cantilever beams. *Journal of Applied Mathematics and Physics (ZAMP)* 30, 31–40.

Reissner, H. (1904). Über die Stabilität der Biegung. *Sitz.-Ber. der Berliner Mathem. Gesellschaft* III. Jhg., 53–56. Beilage zum Archiv der Mathem. u. Physik.

Timoshenko, S. P. and Gere, J. M. (1961). *Theory of Elastic Stability*. McGraw-Hill Book Company, New York, 2nd edition.

Volovoi, V. V., Hodges, D. H., Berdichevsky, V. L., and Sutyrin, V. (1999). Asymptotic theory for static behavior of elastic anisotropic I-beams. *International Journal of Solids and Structures* 36, 1017–1043.

10

INSTABILITIES ASSOCIATED WITH ROTATING BEAMS

When slender structural members such as beams are forced to rotate in specific ways, at least portions of the structure may be put under compression. As expected, this situation may induce buckling. However, there are also situations involving rotation that induce static instability in which only tensile forces are involved. We will examine both situations in this chapter. The first section deals with axial instability of rotating rods that are oriented perpendicular to the axis of rotation and hence referred to as "radial" rods; we define a rod as a special case of a beam that only undergoes stretching caused by an axial force. The second section looks at the buckling instability of radial beams.[1]

10.1   AXIAL INSTABILITY OF RADIAL RODS

For strain sufficiently small such that Hooke's Law is valid, we will show that only a linear model for axial deformation of rotating rods can be derived. This linear model exhibits an instability when the angular speed reaches a certain critical value. However, unless this linear model is valid for large strain, it is impossible to determine whether this instability occurs in reality. This is because the strain ceases to be small well short of the critical speed as the angular speed increases. In order to understand this situation in more detail, we will undertake the analysis of axial deformation of rotating rods using two strain energy functions to model *nonlinear elastic* behavior. The first of these functions is the usual quadratic strain energy function augmented with a cubic term. With this model it is shown that no instability exists if the nonlinearity is stiffening (i.e., if the coefficient of the cubic term is positive), although the strain can become large. However, if the nonlinearity is of the softening variety, then the critical angular speed drops as the degree of softening increases. Still, the strains are large enough that, except for rubber-like materials, a nonlinear elastic model is not likely to be appropriate. The second strain energy function is based on the square of the logarithmic strain and yields a softening model. It quite accurately models the behavior of certain rubber rods which exhibit the instability within the validated range of elongation.

---

[1] Portions of this material are based on Hodges and Bless (1994), used by permission.

Since the system is conservative, the static approach is adequate to study the nature of the instability. The instability was noted first by Bhuta and Jones (1963) and independently by Brunelle (1971). Although Brunelle makes it clear that non-linearities need to be considered, it is not clearly indicated how nonlinear elastic models would behave. In particular, when pure extensional motion is considered, one must include material nonlinearities in order to obtain a physically meaningful non-linear rod model.

10.1.1 LINEAR ELASTIC MODEL: QUADRATIC ENERGY

We consider a prismatic rod of cross-sectional area A undergoing purely axial deformation in the range of validity of Hooke's law. If the rod is made of homogeneous, isotropic material with Young's modulus E, Saint-Venant's interior solution for the stress field is uniaxial so that the strain energy per unit length, normalized by EA, is given by

$$\Psi = \frac{\varepsilon^2}{2} \tag{1}$$

where ε is the average longitudinal strain of the cross section. Were the rod subject to bending and torsional deformation, one would expect Ψ to contain three additional quadratic terms in appropriate bending and torsional generalized strain measures. This looks like a linear theory except that the extensional, bending, and torsional generalized strain measures would, in general, be nonlinear functions of displacement and rotation variables. When Ψ contains only terms of second degree in nonlinear generalized strain measures, we call this a "geometrically" nonlinear theory. This sort of theory is reasonable for modeling slender beams, which can undergo large bending and torsional deflections without large strain. However, in the special case of small-strain, purely axial deformation, geometrically nonlinear expressions for ε in terms of the axial displacement are inappropriate, since all such measures reduce to the elongation when it is small compared to unity. The only consistent way to obtain a nonlinear structural model for pure axial displacement is to retain terms in Ψ which are of cubic and higher degree in the elongation. This type of model, often called a nonlinear elastic or a "physically" nonlinear model, is normally intended for treating strains which are larger than those typically encountered in the linear range. Theories based on such models we can call "physically" nonlinear.

Consider a rotating radial rod that undergoes stretching caused by the axial centrifugal force. The rod remains normal to and rotates about an axis that is both fixed in space and passes through the rod root end at point O. The undeformed rod length is denoted by ℓ, and the longitudinal displacement of the cross section at $x_1 = \ell x$ by $u_1 = \ell u(x)$. Thus, one can write the elongation as

$$\varepsilon = u' \tag{2}$$

where $(\ )'$ represents the derivative with respect to x.

Let the rod rotate with angular speed Ω as depicted in Fig. 10.1. As the rod stretches, its cross-sectional area would decrease (for rods made of materials with Poisson's ratio $\nu > 0$) as depicted in the Figure. Letting $x = 0$ be at the center of rotation (point O) and ρ be the mass per unit undeformed volume of the rod material, one can represent the action of centrifugal forces as a distributed tensile force $EA\omega^2(x + u)$ where $\omega^2 = \rho\Omega^2\ell^2/E$. Thus, it is clear that the deformation is governed by the minimization of the total energy functional

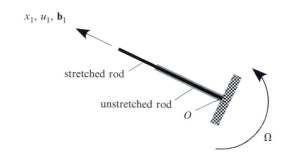

$x_1, u_1, \mathbf{b}_1$

stretched rod

unstretched rod

O

Ω

FIGURE 10.1   Flexible radial rod rotating about fixed point O

$$\mathcal{F}_1 = \int_0^1 \left[\frac{u'^2}{2} - \frac{\omega^2}{2}(x+u)^2 \right] dx \tag{3}$$

subject to the geometric boundary condition that $u(0) = 0$, as shown in Fig. 10.1. Note that the first term represents Ψ and that the axial force in the rod is EAT where

$$T = \frac{\partial \Psi}{\partial u'} = u' \tag{4}$$

indicating a linear structural model. The minimum is obtained when the Euler-Lagrange equation

$$u'' + \omega^2(x+u) = 0 \tag{5}$$

and the geometric boundary condition $u(0) = 0$ and the natural boundary condition $u'(1) = 0$ are satisfied. The solution is

$$u = \frac{\sin(\omega x)}{\omega \cos \omega} - x \tag{6}$$

The mathematical model exhibits an instability reflected in the blowing up of this expression when ω reaches the lowest critical value $\pi/2$. However, Eq. (6) was derived based on Hooke's law, in which the strain is small compared to unity. Thus, its validity is limited. For small strain $u' \ll 1$, and a simple perturbation analysis reveals that

$$u' = \frac{\cos(\omega x)}{\cos \omega} - 1 = \frac{\omega^2}{2}(1 - x^2) + \ldots \tag{7}$$

Obviously, the strain is $O(\omega^2/2) \ll 1$, and as ω is increased from zero, Hooke's law ceases to be valid *long* before the instability at $\omega = \pi/2$ ever comes into play. In other words, the instability predicted by the mathematical model is far outside the domain of its applicability.

If the solution is approximated as

$$u \approx \omega^2 \left(\frac{x}{2} - \frac{x^3}{6} \right) \tag{8}$$

then it should be noted that this is the solution of

$$u'' + \omega^2 x = 0 \tag{9}$$

which is a consistent approximation of Eq. (5) taking into account the observation that $u' \ll 1$ implies $u \ll x$ for all $x > 0$.

It should be noted that if the strain energy has the behavior reflected in the first term of Eq. (3) even when the strain is large, then the instability would exist. This would mean that the axial force in the rod would be linear with u' even for large strain—a situation which is rare at best.

10.1.2 NONLINEAR ELASTIC MODELS

The linear model is based on the simple quadratic energy per unit length of Eq. (1). There are two ways to extend this energy per unit length to a physically nonlinear problem: (a) add a term of cubic or higher degree in ε, or (b) write the strain energy per unit length in terms of a nonlinear function which, when expanded, will agree with the appropriate physically linear model for small values of ε. In this section we will look at one of each, because the physical behaviors of these two approaches are quite different from each other.

Case I: Quadratic and Cubic Energy

It is more reasonable to assume a normalized strain energy per unit length of the form

$$\Psi = \frac{u'^2}{2} + \frac{\beta u'^3}{3} \qquad u' \geq 0 \tag{10}$$

where β is a nondimensional constant which depends on the material and section geometry. Since the axial force $T = u' + \beta u'^2$, it is straightforward to determine β experimentally for an actual rod, but a theoretical determination would be problematic. With this model the existence of an instability of the type described above as ω increases depends on the value of β, and thus cannot be determined on the basis of linear theory.

The behavior of this model is governed by minimization of the functional

$$\mathcal{F}_2 = \int_0^1 \left[\frac{u'^2}{2} + \frac{\beta u'^3}{3} - \frac{\omega^2}{2}(x+u)^2 \right] dx \tag{11}$$

which gives the Euler-Lagrange equation

$$(1 + 2\beta u')u'' + \omega^2(x+u) = 0 \tag{12}$$

with boundary conditions $u(0) = u'(1) = 0$, unchanged. For small ω^2 the perturbation solution is appropriate

$$u = \omega^2 u^{(1)} + \omega^4 u^{(2)} + \dots \tag{13}$$

where

$$u^{(1)} = \frac{x}{2} - \frac{x^3}{6}$$
$$u^{(2)} = \frac{(5 - 6\beta)x}{24} - \frac{(1 - 2\beta)x^3}{12} + \frac{(1 - 6\beta)x^5}{120} \tag{14}$$
$$\vdots$$

Since exact solutions of Eq. (12) have not been published to date, we will seek to minimize $\mathcal{F}_2$ with one- and two-term approximations to gain a qualitative understanding of the behavior. An essentially exact numerical solution is also obtained to ensure that the one-term approximations are accurate.

In accordance with extensive studies by Geer and co-workers, e.g. Geer and Andersen (1989), the solutions to perturbation equations make excellent trial functions. Indeed, as long as $\omega^2 < \pi^2/4$, a one-term approximation based on the form of $u^{(1)}$ above, such that

$$u = 3u_1 \left(\frac{x}{2} - \frac{x^3}{6} \right) \tag{15}$$

where $u_1 = u(1)$ is the tip displacement, gives excellent agreement with a two-term approximation based on functions of the form of $u^{(1)}$ and $u^{(2)}$ above and with the "exact" solution. The one-term solution is simply

$$u_1 = \frac{\sqrt{1764 - 1428\omega^2 + 3024\beta\omega^2 + 289\omega^4} - 42 + 17\omega^2}{108\beta} \tag{16}$$

for the relevant one of the two roots (the irrelevant root has a minus instead of a plus before the square root but will in one case be plotted below for completeness). From examining the behavior of this equation, we must differentiate between a nonstiffening model ($\beta \leq 0$) and a stiffening model ($\beta > 0$).

Nonstiffening Model: When $\beta \leq 0$ there is an instability. If $\beta = 0$, the structural model is linear and the instability is the same as that reported for the linear model. It is encountered at $\omega = \sqrt{42/17}$, which differs from the exact solution $\omega = \pi/2$ by only 0.065%. This small error is indicative of the accuracy of the one-term approximation for $\omega^2 < \pi^2/4$. As ω^2 tends toward this value the displacement u_1 blows up.

The behavior of the one-term solution is shown in Fig. 10.2 for $\beta = -0.1$, $\beta = 0$, and $\beta = 0.1$. The instability for $\beta < 0$ is exhibited by the turning of the solution back to the left, a so-called limit-point instability so named because the value of u_1 is finite at the nose of the curve. The long-dashed part of the $\beta < 0$ curve is the "irrelevant" root, shown for completeness. As ω^2 is increased to the point where the slope is vertical, the effective stiffness goes to zero. That equilibrium point is unstable, and no

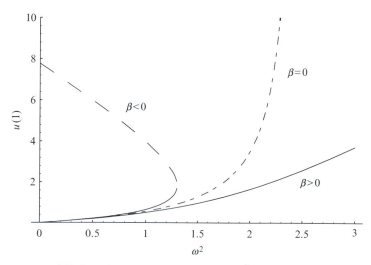

FIGURE 10.2   Tip displacement $u(1)$ versus ω^2 for $\beta = -0.1$, 0, and 0.1

equilibrium exists for values of ω^2 greater than that critical value. This is somewhat different from the situation when $\beta = 0$ for which the vertical slope is only asymptotically reached. Thus, if the angular speed is increased toward the critical value, no equilibrium will be reached at or beyond the critical value of ω^2 for $\beta = 0$. For $\beta \neq 0$, to plotting accuracy, the relevant roots are virtually indistinguishable from the "exact" numerical solution. See below for discussion of the stiffening model where $\beta > 0$.

The stability boundary for $\beta \leq 0$ can be found by solving for the value of ω^2 at which the quantity under the radical in Eq. (16) vanishes. Denoting this by ω_c^2, one obtains

$$\omega_c^2 = \frac{42}{289}\left[17 - 36\beta + 6\sqrt{2\beta(18\beta - 17)}\right] \quad \beta \leq 0 \tag{17}$$

This stability boundary is shown in Fig. 10.3 versus β. Clearly, for larger negative values of β the instability is encountered at lower values of angular speed. This limit-point instability is analogous to tensile instabilities as encountered in load-controlled experiments. The one-term approximation is quite accurate throughout the range of $\omega^2 \leq \omega_c^2$ for $-\beta = O(1)$, and the instability still occurs at large strains in this range.

Stiffening Model: It is seen from Fig. 10.2 that if $\beta > 0$, there is no instability, but the strain can become large when ω^2 becomes large. The tip displacement is shown in Figs. 10.4 and 10.5 for two different values of β. The dashed lines represent asymptotes for small and large ω^2. When β is small, the two asymptotes cross near the value of ω^2 where the tip displacement begins to grow more rapidly and depart significantly from the small ω^2 value of

$$u_1 = \frac{\omega^2}{3} \quad \text{small } \omega^2 \text{ asymptote} \tag{18}$$

and approach the large ω^2 asymptote, given by

$$u_1 = \frac{14}{17} - \frac{7}{9\beta} + \frac{17\omega^2}{54\beta} \quad \text{large } \omega^2 \text{ asymptote} \tag{19}$$

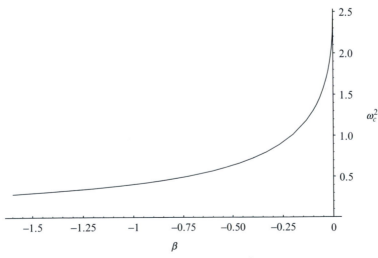

FIGURE 10.3   Critical value of ω^2 versus $\beta < 0$

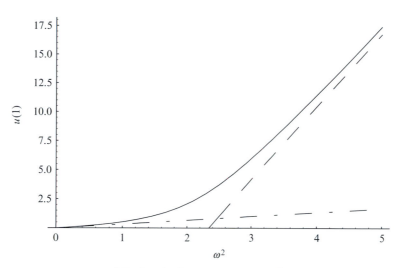

FIGURE 10.4 Tip displacement versus ω^2 for $\beta = .05$

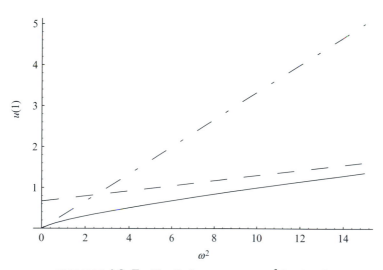

FIGURE 10.5 Tip displacement versus ω^2 for $\beta = 5$

Both asymptotes can be easily extracted from the one-term solution. Note that for small β the tip displacement (and strain) become large as ω^2 is increased. For large β, on the other hand, ω^2 must become much larger to reach a given value of tip displacement than with small β. The suddenness with which the strain increases is exacerbated as β becomes small compared to unity due to the appearance of β in the denominator. In the limit as $\beta \to 0$, the unstable behavior is reached as the large ω^2 asymptote becomes vertical.

Note that the one-term solution exhibits linear behavior when $\beta = 17/18$, where the asymptotes match exactly. Linearity is a condition that is unlikely to occur in the exact solution since the equilibrium equation is nonlinear when $\beta \neq 0$. Thus, for positive β one is led to expect, and indeed finds, growing differences between the one-term solution based on Eq. (15) and the exact solution when ω^2 becomes large.

This is reasonable since a trial function which provides an accurate approximate solution for small ω^2 cannot be expected to work well when ω^2 is large. Thus, an improved trial function is needed.

A trial function which gives excellent results when ω^2 is large compared to unity can be determined by the following change of variable. Let $u = \phi\omega^2/\beta$ and multiply the equilibrium equation through by β/ω^4. This leads to a new governing equation

$$2\phi'\phi'' + \phi + \frac{(\phi'' + \beta x)}{\omega^2} = 0 \tag{20}$$

To find the appropriate trial function we note that first approximation for the large ω^2 solution is then governed by the homogeneous equation

$$2\phi'\phi'' + \phi = 0 \tag{21}$$

subject to $\phi(0) = \phi'(1) = 0$. This equation can be rewritten as

$$\left(\phi'^2\right)' + \phi = 0 \tag{22}$$

which, upon use of the boundary condition $\phi'(1) = 0$ can be written in first-order form as

$$\phi'^2 = \int_x^1 \phi(\xi)d\xi \tag{23}$$

This nonlinear equation can be used in an iterative sense to improve a trial function. The simplest possible admissible function, satisfying only the essential boundary condition, $\phi(0) = 0$, is given by $\phi = x$. Substituting this into the right hand side of Eq. (23), solving for ϕ, and normalizing so that $u(1) = u_1$, one obtains a comparison function

$$u = \frac{2u_1}{\pi}\left[x\sqrt{1 - x^2} + \arcsin(x)\right] \tag{24}$$

which, as a trial function, yields the lowest value of the energy functional from which Eq. (22) can be derived among all those tried. The result is

$$u_1 = \frac{16384 - 2880\pi^2 + 1215\beta\pi^2}{72\beta(45\pi^2 - 256)}$$

$$+ \frac{\omega^2\left(45\pi^2 - 256\right)}{540\beta} \quad \text{improved large } \omega^2 \text{ asymptote} \tag{25}$$

for the tip displacement at large ω^2.

An exact solution for Eq. (22) is derived as follows. Consider a functional of the form

$$\mathcal{F}_3 = \int_0^1 \left(\frac{\phi'^3}{3} - \frac{\phi^2}{2}\right)dx \tag{26}$$

Note that the minimization of $\mathcal{F}_3$ subject to $\phi(0) = 0$ yields Eq. (22). Introducing the relation $\tau = \phi'$ and a Lagrange multiplier λ to enforce it, one can write

$$\mathcal{F}_3^* = \int_0^1 \left[\frac{\tau^3}{3} - \frac{\phi^2}{2} + \lambda(\tau - \phi')\right]dx \tag{27}$$

The minimization of $\mathcal{F}_3$ now takes on the form of an optimal control problem; see Bryson and Ho (1975). Variation with respect to τ shows that

$$\lambda = -\tau^2 \tag{28}$$

which leads to an energy integral (analogous to the Hamiltonian) for Eq. (22)

$$H = \frac{2\tau^3}{3} + \frac{\phi^2}{2} = \text{constant} \tag{29}$$

Since $\tau = 0$ at $x = 1$, it is clear that

$$\frac{4\tau^3}{3} = \phi_1^2 - \phi^2 \tag{30}$$

This can be written as a first-order equation

$$\phi' = \left[\frac{3}{4}\left(\phi_1^2 - \phi^2\right)\right]^{\frac{1}{3}} \tag{31}$$

Using $\phi(0) = 0$ and $\phi(1) = \phi_1$, we can write this as a simple quadrature relation

$$\left(\frac{3}{4\phi_1}\right)^{\frac{1}{3}} = \int_0^1 \frac{d\xi}{\left(1 - \xi^2\right)^{\frac{1}{3}}} \tag{32}$$

yielding

$$\phi_1 = \frac{3}{4\left[\frac{\sqrt{\pi}\Gamma\left(\frac{2}{3}\right)}{2\Gamma\left(\frac{7}{6}\right)}\right]^3} \approx 0.3465 \tag{33}$$

The one-term approximation based on $\phi = \frac{2\phi_1}{\pi}\left[x\sqrt{1 - x^2} + \arcsin(x)\right]$ gives

$$\phi_1 = \frac{45\pi^2 - 256}{540} \approx 0.3484 \quad \text{one-term approximation} \tag{34}$$

for an error of 0.55%.

Two features of the improved asymptote in Eq. (25) are worth noting. First, the slope of this asymptote is quantitatively only about 10% different from the earlier one in Eq. (19) (based on the small ω^2 trial function), but it differs from the exact, closed-form solution of Eq. (22) by only 0.55%. Second, an almost unnoticeable qualitative difference is exhibited in the possibility of finding a value of β which yields a linear solution. The value $\beta = 1.004$ which makes the intercept vanish yields a slope of 0.3470, which is slightly different from the slope for small ω^2 of 1/3. Thus, there seems to be no value of β for which the solution behaves exactly linearly, but the nonlinear behavior is almost negligible when $\beta = 1.004$.

In summary, the instability occurs only for $\beta \leq 0$, but only at large strains, even for $\beta = -1.5$. Strains of such magnitudes are not encountered within the range of elastic deformation except in rubber-like materials. So it may be useful to look at a "model which performs well for rubber rods.

Nonlinear Elastic Model II: Hencky Strain Energy

One such model is the Hencky strain energy model, shown by Degener *et al.* (1988) to give excellent correlation with experimental data for stretching of rubber rods (tubes) up to values of elongation of the order of 0.5. For axial deformation alone, this model reduces to finding the minimum of the functional

$$\mathcal{F}_4 = \int_0^1 \left[\frac{\log^2 (1 + u')}{2} - \frac{\omega^2}{2}(x + u)^2 \right] dx \tag{35}$$

Letting $z = x + u$ so that $z' = 1 + u' = v$, one can rewrite $\mathcal{F}_4$ as

$$\mathcal{F}_4^* = \int_0^1 \left[\frac{\log^2 v}{2} - \frac{\omega^2 z^2}{2} + \lambda(v - z') \right] dx \tag{36}$$

Again, λ can be found as

$$\lambda = -\frac{\log v}{v} \tag{37}$$

The constancy of the Hamiltonian and the natural boundary condition $v(1) = 1$ show that

$$\frac{\log^2 v}{2} - \log v + \frac{\omega^2}{2} \left(z_1^2 - z^2 \right) = 0 \tag{38}$$

where $z_1 = z(1)$. Thus,

$$\log v = \log z' = 1 - \sqrt{1 - \omega^2 \left(z_1^2 - z^2 \right)} \tag{39}$$

The minus is taken in front of the square root because $v(1) = 1$. Using the above procedure and $z(0) = 0$, one can show that

$$z_1 = \frac{e}{\int_0^1 e^{\sqrt{1 - \omega^2 z_1^2 \left(1 - \xi^2 \right)}} d\xi} \tag{40}$$

(We note here that a similar procedure can be used with the quadratic/cubic model. Since the resulting procedure is computationally intensive, it was used only to verify the accuracy of the one-term approximation.) Thus, u_1 is governed by the exact relation

$$1 + u_1 = \frac{e}{\int_0^1 e^{\sqrt{1 - \omega^2 (1 + u_1)^2 \left(1 - \xi^2 \right)}} d\xi} \tag{41}$$

Numerical evaluation is done by first picking values of $a^2 = \omega^2 (1 + u_1)^2$, finding the integral by numerical quadrature (which can be done to any accuracy desired by using *Mathematica*), finding $1 + u_1$ from Eq. (41), and finding ω^2 as $a^2/(1 + u_1)^2$. The largest value of a^2 chosen for plotting was 0.8592, where the slope of u_1 versus ω^2 is infinite.

Results are shown in Fig. 10.6, and clearly this model has similar behavior to that of our simple nonstiffening model above. The instability is at $\omega^2 = 0.4210$ at which $u(1) = 0.4285$. However, the maximum elongation at the root is $u'(0) = 0.8667$. Although this is somewhat larger than the maximum elongations observed in the experiments of Degener *et al.* (1988), it is possible that this instability could occur for some rubber rods whose force-deformation relation is of the form

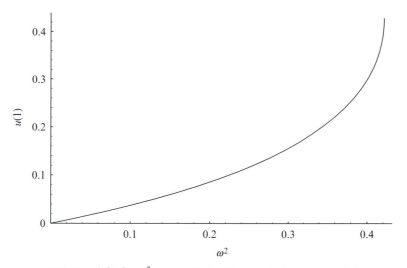

FIGURE 10.6 ω^2 versus $u(1)$ for Hencky strain energy model

$$T = \frac{\log(1 + u')}{1 + u'} \tag{42}$$

if the allowable strain were this large.

10.2   BUCKLING OF ROTATING RADIAL BEAMS

In Fig. 10.7 we have a flexible beam cantilevered to the rim of a rotating wheel and directed toward the center of rotation. (Note that in this example the beam has distributed mass and thus differs from the case depicted in Fig. 3.9 in which the beam is massless and has a tip mass.) Thus, as the wheel spins, the beam is compressed axially. The transverse displacement along the beam is $u_2(x, t)$ and the longitudinal displacement is $u_1(x, t)$. Only longitudinal and inplane bending displacements are shown in the figure, but the out-of-plane displacement u_3 is also possible. Thus, the position vector to any point along the beam is

$$\mathbf{r} = (x_1 + u_1)\mathbf{b}_1 + u_2\mathbf{b}_2 + u_3\mathbf{b}_3 \tag{43}$$

where the unit vectors $\mathbf{b}_1$ and $\mathbf{b}_2$ are both parallel to the plane of rotation, along and perpendicular to the undeformed beam, respectively, and $\mathbf{b}_3$ is parallel to the axis of rotation (normal to the plane of the figure).

Thus, in the case of static deformation in the rotating frame, the additional kinetic energy caused by rotation is

$$T = \frac{\Omega^2}{2} \int_0^\ell \mu \left[(-R + x_1 + u_1)^2 + u_2^2 \right] dx_1 \tag{44}$$

where μ is the mass per unit length. The beam is to be treated as inextensible, and for the pre-buckled state the beam properties are assumed to be such that it remains in the

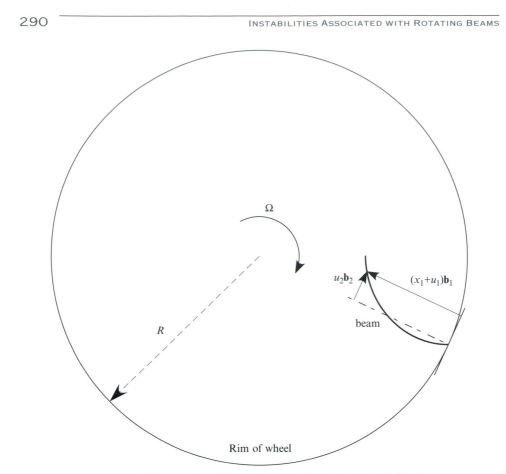

FIGURE 10.7   Flexible beam cantilevered to the rim of a rigid wheel

radial orientation. Thus, $\bar{u}_1 = \bar{u}_2 = \bar{u}_3 = 0$, and the strain energy for the pre-buckled state zero. Thus, the weak form of the equation governing the pre-buckled state is

$$\delta \int_0^\ell \left[\overline{F}_1 \bar{u}_1' - \mu \Omega^2 (-R + x_1 + \bar{u}_1)^2 \right] dx_1 = 0 \tag{45}$$

where $(\ )'$ is the derivative with respect to the axial coordinate and $\overline{F}_1$ is a Lagrange multiplier that enforces zero axial strain in the pre-buckled state. Physically $\overline{F}_1$ is the steady-state axial force, which is not constant with x_1. The resulting Euler-Lagrange equation is

$$\overline{F}_1' + \mu \Omega^2 (-R + x_1 + \bar{u}_1) = 0 \tag{46}$$

and boundary condition that $\overline{F}_1(\ell) = 0$. Therefore, for constant mass per unit length μ and zero strain in the pre-buckled state such that $\bar{u}_1 = 0$, the axial force is

$$\overline{F}_1 = \mu \Omega^2 \ell^2 \left[\frac{1}{2} \left(1 - x^2 \right) - \alpha (1 - x) \right] \tag{47}$$

with $x = x_1/\ell$ and $\alpha = R/\ell$. Notice that if $\alpha < 0$, then the beam is facing radially outward and is entirely in tension. If $0 < \alpha < 1$, then the beam passes through the center of rotation and sticks out on the other side of it; part of the beam is in tension and part in compression. If $\alpha > 1$ the beam is entirely in compression. The problem of interest is to determine the stability of small perturbations about the steady-state solution.

Assuming small perturbations about the pre-buckled state, the unit vector tangent to the beam can be written as

$$\mathbf{B}_1 = \mathbf{r}' = \left(1 + \hat{u}_1'\right)\mathbf{b}_1 + \hat{u}_2'\mathbf{b}_2 + \hat{u}_3'\mathbf{b}_3 \tag{48}$$

Since the beam is assumed to be inextensible, we can identify this as a unit vector, so that

$$\left(1 + \hat{u}_1'\right)^2 + \hat{u}_2'^2 + \hat{u}_3'^2 = 1 \tag{49}$$

which leads to a constraint on $\hat{u}_1$ of the form

$$\hat{u}_1' = \sqrt{1 - \hat{u}_2'^2 - \hat{u}_3'^2} - 1 \tag{50}$$

For small deflections, this means that the axial displacement is a second-order quantity (but one which we need) given by

$$\hat{u}_1' \approx -\frac{1}{2}\left(\hat{u}_2'^2 + \hat{u}_3'^2\right) \tag{51}$$

Dropping all terms of degree three and higher in the unknowns, one finds that the kinetic energy associated with the perturbations is

$$
\begin{aligned}
T &= \frac{\mu\Omega^2}{2} \int_0^\ell \left[(-R + x_1 + \hat{u}_1)^2 + \hat{u}_2^2\right] dx_1 \\
&= \int_0^\ell \left(\frac{\mu\Omega^2 \hat{u}_2^2}{2} - \overline{F}_1' \hat{u}_1\right) dx_1 \\
&= \int_0^\ell \left(\frac{\mu\Omega^2 \hat{u}_2^2}{2} + \overline{F}_1 \hat{u}_1'\right) dx_1 \\
&= \frac{1}{2} \int_0^\ell \left[\mu\Omega^2 \hat{u}_2^2 - \overline{F}_1\left(\hat{u}_2'^2 + \hat{u}_3'^2\right)\right] dx_1
\end{aligned}
\tag{52}
$$

where the boundary condition $\hat{u}_1(0) = 0$ was used. The strain energy for the perturbed state is

$$U = \frac{1}{2} \int_0^\ell \left(EI_2 \hat{u}_3''^2 + EI_3 \hat{u}_2''^2\right) dx_1 \tag{53}$$

Interestingly enough, some of the terms that originated as kinetic energy terms have how become potential energy-like terms associated with the load $\overline{F}_1$, so that the negative of the Lagrangean can be written as

$$U - T = \frac{1}{2} \int_0^\ell \left[EI_2 \hat{u}_2''^2 + EI_3 \hat{u}_3''^2 + \overline{F}_1\left(\hat{u}_2'^2 + \hat{u}_3'^2\right) - \mu\Omega^2 \hat{u}_2^2\right] dx_1 \tag{54}$$

and the resulting weak forms are decoupled:

$$\delta \int_0^\ell \left(EI_3 \hat{u}_2''^2 + \overline{F}_1 \hat{u}_2'^2 - \mu\Omega^2 \hat{u}_2^2 \right) dx_1 = 0$$

$$\delta \int_0^\ell \left(EI_2 \hat{u}_3''^2 + \overline{F}_1 \hat{u}_3'^2 \right) dx_1 = 0 \tag{55}$$

We first look at the out-of-plane case in terms of nondimensional parameters. Using the $(\ )'$ to now represent the derivative with respect to x, and introducing $w = \hat{u}_3/\ell$ and

$$\omega^2 = \frac{\mu\Omega^2 \ell^4}{EI_2} \tag{56}$$

one can write the weak form as

$$\delta \int_0^1 \left\{ \frac{w''^2}{2} - \omega^2 \left[\alpha(1-x) - \frac{1}{2}(1-x^2) \right] \frac{w'^2}{2} \right\} dx = 0 \tag{57}$$

the Euler-Lagrange equation of which is

$$w'''' + \omega^2 \left\{ \left[\alpha(1-x) - \frac{1}{2}(1-x^2) \right] w' \right\}' = 0 \tag{58}$$

with corresponding boundary conditions $w(0) = w'(0) = w''(1) = w'''(1) = 0$. The equation can be integrated once and written in terms of $\beta = w'$ to yield

$$\beta'' + \omega^2 \left[\alpha(1-x) - \frac{1}{2}(1-x^2) \right] \beta = 0 \tag{59}$$

with $\beta(0) = \beta'(1) = 0$. This equation has a solution in terms of Kummer confluent hypergeometric functions; see Abramowitz and Stegun (1970). The value of ω_{cr} depends on α: Buckling is not possible for $\alpha \leq \sqrt{\pi}/3$, and when $\alpha = \sqrt{\pi}/3$, $\omega_{cr} \to \infty$. The behavior of ω_{cr} versus α is depicted in Fig. 10.8.

Although we have an exact solution, it is a very complicated one, and a good numerical approximation may be of more value for use by designers. We can write down a weak form for β directly from Eq. (57) by substituting β for w', yielding

$$\delta \int_0^1 \left\{ \frac{\beta'^2}{2} - \omega^2 \left[\alpha(1-x) - \frac{1}{2}(1-x^2) \right] \frac{\beta^2}{2} \right\} dx = 0 \tag{60}$$

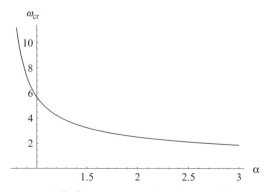

FIGURE 10.8 ω_{cr} versus α for the out-of-plane case

Thus, a one-term Rayleigh quotient approximation for the critical angular speed can be written as

$$\omega_{\text{cr}}^2 \le R(\phi) = \frac{\int_0^1 \phi'^2 dx}{\int_0^1 \left[\alpha(1-x) - \frac{1}{2}(1-x^2)\right]\phi^2 dx} \tag{61}$$

where ϕ is an admissible function, at least satisfying the essential boundary condition $\phi(0) = 0$, and the inequality is an indication that the quantity on the right hand side is an upper bound. We can also see that buckling is not possible when

$$\alpha \le \alpha^* = \frac{\int_0^1 (1-x^2)\phi^2 dx}{2\int_0^1 (1-x)\phi^2 dx} \tag{62}$$

where the same function must be used in the Eq. (62) as in Eq. (61); it has not been proven that the latter is an upper bound, but it generally does turn out to be larger than the exact value.

The simplest admissible function is $\phi = x$, which is also a comparison function (i.e. it satisfies all the boundary conditions). The result is

$$\omega_{\text{cr}} = \sqrt{\frac{60}{5\alpha - 4}} \tag{63}$$

At $\alpha = 1$, for example, this yields $\omega_{\text{cr}} = 7.74597$ compared to the exact value of $\omega_{\text{cr}} = 5.67467$, which is the lowest root of the Bessel function $J\left(-\frac{1}{4}, \frac{\omega}{2\sqrt{2}}\right) = 0$. Unfortunately, the value of $\alpha^* = 0.8$ is well above the exact value of $\sqrt{\pi}/3 \approx 0.590818$.

Let us employ the technique of Stodola and Vianello (see Hodges, 1997), which allows us to construct improved trial functions by iteration. We solve the governing differential equation for the highest derivative term, yielding

$$\phi_{i+1}'' = -\omega^2 \left[\alpha(1-x) - \frac{1}{2}(1-x^2)\right]\phi_i \quad i = 0, 1, \ldots \tag{64}$$

where $\phi_0 = x$. Using the boundary condition $\phi_i(0) = 0$, one finds the results for ϕ_i for $i > 0$ to be more involved polynomial comparison functions. A sampling of the numerical results may be found in Table 10.1. The results for the critical angular speed are excellent, converging to the exact solution to four places in only two iterations. The size of the resulting polynomial functions is easily manageable with Mathematica. However, as well as this technique may work for a wide class of problems, it falls short of achieving a good approximation for α^*. The reason for this is important to note: As $\alpha \to \alpha^*$ from above, ω_{cr} tends to infinity, which creates a

TABLE 10.1   Results for out-of-plane buckling.

Iteration number	ω_{cr} for $\alpha = 1$	α^*
0	7.74597	0.800000
1	5.69011	0.778794
2	5.67501	0.772134
3	5.67468	0.768790
4	5.67467	0.766788
exact	5.67467	0.590818

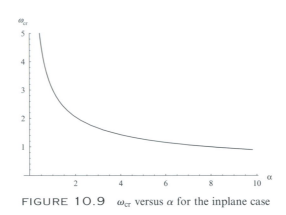

FIGURE 10.9 ω_{cr} versus α for the inplane case

boundary layer for the solution of the differential equation. The curvature becomes infinitely large near the point $x = 0$, making it more and more difficult to capture the behavior with a polynomial function. Admissible functions that have an exponential term may be useful to combat this problem.

The inplane problem is governed by a more complex equation. Introducing $v = \hat{u}_2/\ell$ and

$$\omega^2 = \frac{\mu \Omega^2 \ell^4}{EI_3} \tag{65}$$

one can write the weak form as

$$\delta \int_0^1 \left\{ \frac{v''^2}{2} - \omega^2 \left[\alpha(1-x) - \frac{1}{2}(1-x^2) \right] \frac{v'^2}{2} - \frac{\omega^2 v^2}{2} \right\} dx = 0 \tag{66}$$

the Euler-Lagrange equation of which is

$$v'''' + \omega^2 \left\{ \left[\alpha(1-x) - \frac{1}{2}(1-x^2) \right] v' \right\}' - \omega^2 v = 0 \tag{67}$$

with corresponding boundary conditions $v(0) = v'(0) = v''(1) = v'''(1) = 0$. This is a much more difficult equation to solve, and an exact solution is unknown to the authors. The converged result after five iterations for ω_{cr} by the Stodola and Vianello method is $\omega_{\mathrm{cr}} = 2.99391$ for $\alpha = 1$, which is close to the published numerical solutions of Lakin and Nachman (1979). However, the best result for $\alpha^* = 0.184937$ is nowhere near the exact value of $\alpha^* = 0$, known from asymptotic considerations. The behavior of the inplane critical angular speed for the inplane case is shown in Fig. 10.9.

PROBLEMS

1. Apply the method of Rayleigh's quotient with a free parameter (as described in Chapter 9) to find the critical angular speed for out-of-plane buckling of an inwardly-directed, rotating, uniform beam, the tip of which is at the center of rotation. Try to get the best one-term approximation.

2. Apply the method of Rayleigh's quotient with a free parameter (as described in Chapter 9) to find the critical angular speed for inplane buckling of an inwardly-directed, rotating, uniform beam, the tip of which is at the center of rotation. Try to get the best one-term approximation.

3. Apply the method of Rayleigh's quotient with the Stodola-Vianello method (as described in this chapter) to find the critical angular speed for inplane buckling of an inwardly-directed, rotating, uniform beam, the tip of which is at the center of rotation. Try to get the best one-term approximation. Compare the results and the effort to that of Problem.

4. Apply the method of Ritz or the Galerkin method with multiple modes to find the critical angular speed for inplane buckling of an inwardly-directed, rotating, uniform beam, the tip of which is at the center of rotation. Compare the results and the effort to those of Problems 2 and 3.

REFERENCES

Abramowitz, M. and Stegun, I., eds., (1970). *Handbook of Mathematical Functions.* National Bureau of Standards, Washington, D.C.

Bhuta, P. G. and Jones, J. P. (1963). On axial vibrations of a whirling bar. *The Journal of the Acoustical Society of America* 35.

Brunelle, E. J. (1971). Stress redistribution and instability of rotating beams and disks. *AIAA Journal* 9, 758–759.

Bryson, A. E., Jr. and Ho, Y.-C. (1975). *Applied Optimal Control.* Blaisdell Publishing Company, Waltham, Massachusetts.

Degener, M., Hodges, D. H., and Petersen, D. (1988). Analytical and experimental study of beam torsional stiffness with large axial elongation. *Journal of Applied Mechanics* 110, 171–178.

Geer, J. F. and Andersen, C. M. (1989). A hybrid perturbation galerkin technique with applications to slender body theory. *SIAM Journal of Applied Mathematics* 49, 344–361.

Hodges, D. H. (1997). Improved approximations via Rayleigh's quotient. *Journal of Sound and Vibration* 199, 155–164.

Hodges, D. H. and Bless, R. R. (1994). Axial instability of rotating rods revisited. *International Journal of Non-Linear Mechanics* 29, 879–887.

Lakin, W. D. and Nachman, A. (1979). Vibration and buckling of rotating flexible rods at transitional parameter values. *Journal of Engineering Mathematics* 13, 339–346.

Peters, D. A. and Hodges, D. H. (1980). In-plane vibration and buckling of a rotating beam clamped off the axis of rotation in-plane vibration and buckling of a rotating beam clamped off the axis of rotation. *Journal of Applied Mechanics* 47, 398–402.

11

Nonconservative Systems

11.1 PRELIMINARY REMARKS

All of the previous chapters have dealt with the stability of conservative elastic structural systems under static loads. A classification of loads and reactions, when dealing with all mechanical systems, is given by Ziegler (1968). A system is conservative when subjected only to conservative forces (see Chapter 1). One example of nonconservative forces is the follower force. A follower force follows the deformations of the body in some manner such that the work done by the force is path-dependent. Consider the system in Fig. 11.1. It is easily seen that the applied force P, which follows the orientation of the upper rod, is nonconservative. Let us consider two different sequences of deflection away from the starting point when $q_1 = q_2 = 0$. First, as the load is applied in that state, zero work is done. Then, let q_2 move from zero to $q_2 = \hat{q}_2$. In this motion zero work is done. Then let q_1 move from zero to $q_1 = \hat{q}_1$. During this motion, the work done by P is also zero. So, the total work done to get in this first way from $q_1 = q_2 = 0$ to $q_1 = \hat{q}_1$ and $q_2 = \hat{q}_2$ is zero. Now the second scenario is very similar but simply reversed in order. The load is again applied when $q_1 = q_2 = 0$, where zero work is done. Then the system moves so that $q_1 = \hat{q}_1$ so that the work done is $P\ell(1 - \cos\hat{q}_1) \approx P\ell\hat{q}_1^2/2$. Finally, the system moves again so that $q_2 = \hat{q}_2$, for which the work done is zero. So, the total work done to get in this second way from $q_1 = q_2 = 0$ to $q_1 = \hat{q}_1$ and $q_2 = \hat{q}_2$ is approximately $P\ell\hat{q}_1^2/2$. Additional scenarios with still different answers for the work done are not hard to conceive. Thus, it is quite clear that the follower force in Fig. 11.1 is nonconservative. Another aspect of the properties of such a force is that it does not possess a potential energy function which, when varied, will give the negative of the force's virtual work. To put it another way, the virtual work of the forces cannot be "integrated" to provide the negative of the force's potential energy. Follower forces are typically nonconservative in this sense, but the distributed follower forces in Section 7.7 provide an interesting exception.

As pointed out by Bolotin (1963, 1964), the study of the stability of structures under follower force systems apparently started with work by Nikolai in the late 1920s. In addition to the books by Bolotin and others, there are also many papers devoted to this subject; see, for example, the work of Leipholz (1978), Celep (1979),

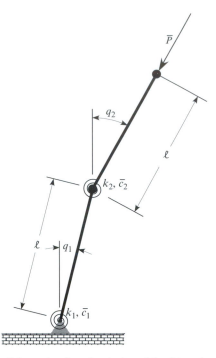

FIGURE 11.1   Schematic of mechanical model subjected to a follower force.

Park (1987), Chen and Ku (1992), and Higuchi (1994). Much of the analytical research to date has focused on the stability of beams subjected to various types of follower forces and examination of the effects of various physical phenomena, such as damping and transverse shear deformation.

Analytical examples of solved follower force problems help to clarify the nature of these systems and their analysis. For example, it is now commonly understood that static analysis of elastic systems subjected to follower forces may erroneously show that the system is free of instability. In order to ascertain whether a system subjected to follower forces is stable requires a kinetic analysis. For problems that do in fact lose their stability by buckling, the kinetic method will predict that one of the system natural frequencies will tend to zero as the critical load is approached (see Problem 1). However, for nonconservative systems one may also find flutter instabilities in addition to possible buckling instabilities. By this we mean that small perturbations about the static equilibrium state oscillate with increasing amplitude.

In this chapter we will present several examples and then present an alternative solution method based on the fully intrinsic equations of beam vibration. The first of the examples is a mechanical analog to the so-called Beck column, which consists of a cantilevered beam undergoing a compressive concentrated follower force at its free end; see Beck (1952). Next, both exact and approximate analyses of the Beck column itself are presented. Then, a column undergoing a compressive and uniformly distributed follower force, as analyzed by Leipholz (1975), is treated. The next example is a shaft subject to a tangential follower torque, previously considered by the static method in Chapter 8. The final example is a deep cantilevered beam with a lateral follower force applied at the tip and in the plane of greatest flexural rigidity. These and other follower force problems can be also solved by use of the fully intrinsic equations, derived in the final section of this chapter and used to set up two of the follower force example problems.

11.2   MECHANICAL FOLLOWER FORCE MODEL

Let us recall the simple mechanical models discussed in Chapter 2, in particular Model A. For this the kinetic method yields a differential equation of the form (see Eq. 2 in Section 2.2)

$$I\ddot{\theta} + \left(ka^2 - P\ell\right)\theta = 0 \tag{1}$$

Assuming a solution of the form $\theta = \breve{\theta}\exp(\bar{s}t)$, we find the characteristic equation to be

$$s^2 + 1 - p = 0 \tag{2}$$

where $s^2 = I\bar{s}^2/\left(ka^2\right)$ and $p = P/P_{cr} = P\ell/\left(ka^2\right)$. In Figs. 11.2 and 11.3 one finds the real and imaginary parts of s, respectively, versus p, showing that the real parts of both roots become nonzero when the applied force exceeds the critical load. Since the real part for one of the two roots is positive when $p > 1$, the perturbations about the static equilibrium state grow in amplitude. However, it is also interesting to note that for $p \geq 1$ the imaginary part is identically zero. This is characteristic of all systems that lose their stability by buckling: one of the natural frequencies of oscillations

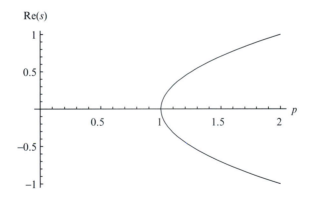

FIGURE 11.2   Real part of s for mechanical Model A (see Chapter 2) subjected to a nondimensional dead force p.

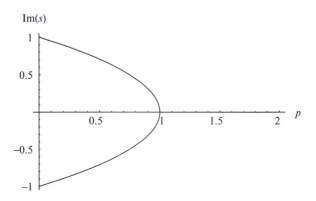

FIGURE 11.3   Imaginary part of s for mechanical Model A (see Chapter 2) subjected to a nondimensional dead force p.

about the static equilibrium state becomes zero as the critical load is approached. This is one of the differences that will be seen for nonconservatively loaded systems.

In Fig. 11.1 a mechanical model of a simple system loaded by a follower force is depicted. The system is comprised of two particles of mass m joined together with massless rigid rods of length ℓ. The rods are joined to each other with a rotational hinge, and one of the rods is also joined to the ground with a rotational hinge. The motion of the system takes place in a plane, and the hinges are spring- and damper-restrained with elastic and damping constants equal to k_α and $\bar{c}_\alpha$, respectively, with $\alpha = 1$ and 2. This system is a mechanical model that behaves in a manner similar to Beck's column, treated in Section 11.3.

Here we will use Lagrange's equations to derive equations of motion for this system. For small angles q_1 and q_2, the kinetic and potential energies are

$$T = \frac{m\ell^2}{2}\dot{q}_1^2 + \frac{m\ell^2}{4}(\dot{q}_2^2 + 2\dot{q}_1\dot{q}_2)$$
$$V = \frac{k}{2}q_1^2 + \frac{k}{2}(q_2 - q_1)^2 \tag{3}$$

The virtual work of the nonconservative applied and damping forces is

$$\overline{\delta W} = -\overline{P}\ell(q_2 - q_1)\delta q_1 - [(\bar{c}_1 + \bar{c}_2)\dot{q}_1 - \bar{c}_2\dot{q}_2]\delta q_1 - \bar{c}_2(\dot{q}_2 - \dot{q}_1)\delta q_2 \tag{4}$$

Thus, the equations of motion are

$$\begin{bmatrix} m\ell^2 & \dfrac{m\ell^2}{2} \\ \dfrac{m\ell^2}{2} & \dfrac{m\ell^2}{2} \end{bmatrix}\begin{Bmatrix} \ddot{q}_1 \\ \ddot{q}_2 \end{Bmatrix} + \begin{bmatrix} \bar{c}_1 + \bar{c}_2 & -\bar{c}_2 \\ -\bar{c}_2 & \bar{c}_2 \end{bmatrix}\begin{Bmatrix} \dot{q}_1 \\ \dot{q}_2 \end{Bmatrix} + \begin{bmatrix} 2k - \overline{P}\ell & \overline{P}\ell - k \\ -k & k \end{bmatrix}\begin{Bmatrix} q_1 \\ q_2 \end{Bmatrix} = \begin{Bmatrix} 0 \\ 0 \end{Bmatrix} \tag{5}$$

First, we consider only the static terms in the equation, viz.,

$$\begin{bmatrix} 2k - \overline{P}\ell & \overline{P}\ell - k \\ -k & k \end{bmatrix}\begin{Bmatrix} q_1 \\ q_2 \end{Bmatrix} = \begin{Bmatrix} 0 \\ 0 \end{Bmatrix} \tag{6}$$

From this one sees that a nontrivial solution can only exist when $2k^2 - \overline{P}k\ell + \overline{P}k\ell - k^2 = 0$, which cannot happen for nonzero k. Thus, no matter how large a force $\overline{P}$ is applied, the mechanism does not exhibit a static buckling instability.

To better treat the dynamic case, we introduce nondimensional variables for time, $\tau = t\sqrt{k/(m\ell^2)}$; force, $P = \overline{P}\ell/k$; and damping parameters, $c_\alpha = \bar{c}_\alpha/\sqrt{km}$. Then one can write the equations of motion more simply as

$$\begin{bmatrix} 1 & \dfrac{1}{2} \\ \dfrac{1}{2} & \dfrac{1}{2} \end{bmatrix}\begin{Bmatrix} q_1'' \\ q_2'' \end{Bmatrix} + \begin{bmatrix} c_1 + c_2 & -c_2 \\ -c_2 & c_2 \end{bmatrix}\begin{Bmatrix} q_1' \\ q_2' \end{Bmatrix} + \begin{bmatrix} 2 - P & P - 1 \\ -1 & 1 \end{bmatrix}\begin{Bmatrix} q_1 \\ q_2 \end{Bmatrix} = \begin{Bmatrix} 0 \\ 0 \end{Bmatrix} \tag{7}$$

where $(\ )'$ represents the derivative with respect to τ. Letting $q_\alpha = \check{q}_\alpha \exp(s\tau)$, we find that a nontrivial solution only exists when

$$\begin{vmatrix} s^2 + (c_1 + c_2)s + 2 - P & \dfrac{s^2}{2} - c_2 s + P - 1 \\ \dfrac{s^2}{2} - c_2 s - 1 & \dfrac{s^2}{2} + c_2 s + 1 \end{vmatrix} \tag{8}$$

If the damping is ignored for now, the characteristic equation becomes

$$s^4 + 4(3 - P)s^2 + 4 = 0 \tag{9}$$

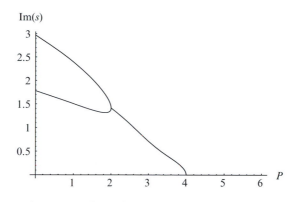

FIGURE 11.4 Imaginary part of nondimensional eigenvalue s versus P for double mechanical pendulum model without damping.

Notice that $s = 0$ is not a root, so a loss of stability by buckling (i.e., passing from a stable system directly to a buckled one) is not possible. The quartic equation has four roots such that

$$s^2 = -2(3-P) \pm 2\sqrt{8-6P+P^2} \tag{10}$$

The first sign change of the radicand is at $P = 2$. If $P \le 2$ the real parts of all roots are zero, as shown in Fig. 11.5; the real part of one root becomes positive when $P > 2$, which means that there is a loss of stability. Since $\mathrm{Im}(s) \ne 0$ when $P > 2$, the unstable motion is oscillatory with increasing amplitude. This type of instability is usually referred to as flutter in the mechanics literature and is closely related mathematically to the flutter instability of aeroelasticity.[1] When $P > 4$ all roots are real and there is a strong buckling instability; but the system always first loses stability by flutter.

The addition of damping forces to the model of a nongyroscopic conservative system will generally stabilize the system. Such is not the case with either gyroscopic conservative systems or with nonconservative systems. The potential destabilizing effect of damping is often exacerbated when there is a strong disparity in the amount of damping in the various degrees of freedom. For example, in Fig. 11.6 the real part of s is plotted versus P and a loss of stability is observed for $P > 0.401928$. Such a dramatic change in the stability boundary can lead to catastrophic failure if not

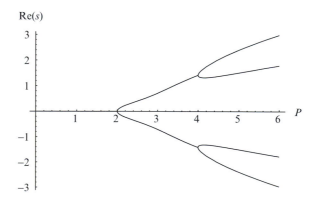

FIGURE 11.5 Real part of nondimensional eigenvalue s versus P for double mechanical pendulum model without damping.

[1] The physical connection is weak, however, in that unsteady aerodynamics are involved in the aeroelasticity problem.

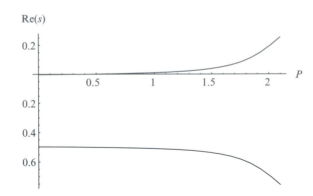

FIGURE 11.6   Real part of nondimensional eigenvalue s versus P for double mechanical pendulum model with damping parameters $c_1 = 0.0001$ and $c_2 = 0.1$.

properly accounted for in the design of a system undergoing nonconservative forces. See Herrmann (1967) for further discussion of this point.

11.3   BECK'S COLUMN

Consider a cantilevered beam of length ℓ undergoing planar deformation and having bending stiffness EI and mass per unit length μ that are constant along the beam. The beam is subjected to a force of constant magnitude $\overline{P}$, the line of action of which passes through the elastic axis of the beam at the end cross section and remains tangent to the beam in its deformed state. The addition of damping, which should be included in order to adequately account for the physics of follower forces, is left as an exercise for the reader; see Problem 3 at the end of this chapter.

The beam is depicted in Fig. 11.7 and the transverse displacement along the beam is $u_2(x_1,t)$ and the longitudinal displacement is $u_1(x_1,t)$. Thus, the position vector to any point along the beam reference line is

$$\mathbf{r} = (x_1 + u_1)\mathbf{b}_1 + u_2\mathbf{b}_2 \tag{11}$$

where the unit vectors $\mathbf{b}_1$ and $\mathbf{b}_2$ are both in the plane of deformation, along and perpendicular to the undeformed beam, respectively. Considering the beam as inextensible, one finds the unit vector tangent to the beam to be

$$\mathbf{B}_1 = \frac{\partial \mathbf{r}}{\partial x_1} = \left(1 + \frac{\partial u_1}{\partial x_1}\right)\mathbf{b}_1 + \frac{\partial u_2}{\partial x_1}\mathbf{b}_2 \tag{12}$$

Since this is a unit vector, the length of the vector must be equal to unity, so that

$$\left(1 + \frac{\partial u_1}{\partial x_1}\right)^2 + \left(\frac{\partial u_2}{\partial x_1}\right)^2 = 1 \tag{13}$$

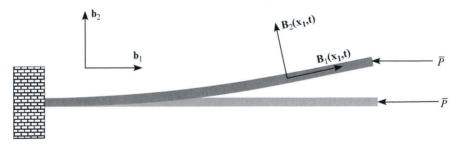

FIGURE 11.7   Schematic of beam undergoing compressive axial follower force.

which leads to a constraint on u_1 of the form

$$\frac{\partial u_1}{\partial x_1} = \sqrt{1 - \left(\frac{\partial u_2}{\partial x_1}\right)^2} - 1 \tag{14}$$

For small deflections, this means that

$$\frac{\partial u_1}{\partial x_1} \approx -\frac{1}{2}\left(\frac{\partial u_2}{\partial x_1}\right)^2 \tag{15}$$

or

$$u_1(x_1,t) = -\frac{1}{2}\int_0^{x_1}\left[\frac{\partial u_2(\xi, t)}{\partial \xi}\right]^2 d\xi \tag{16}$$

Now the relationship between the unit tangent vector $\mathbf{B}_1$ and the unit vectors of the inertial frame $\mathbf{b}_i$ may be written as

$$\mathbf{B}_1 = \frac{\partial \mathbf{r}}{\partial x_1} \approx \left[1 - \frac{1}{2}\left(\frac{\partial u_2}{\partial x_1}\right)^2\right]\mathbf{b}_1 + \frac{\partial u_2}{\partial x_1}\mathbf{b}_2 \tag{17}$$

The strain energy of the beam is

$$U = \frac{1}{2}\int_0^{\ell} EI\left(\frac{\partial^2 u_2}{\partial x_1^2}\right)^2 dx_1 \tag{18}$$

and we find the kinetic energy, ignoring the higher-order longitudinal motion, to be

$$T = \frac{1}{2}\int_0^{\ell} \mu\left(\frac{\partial u_2}{\partial t}\right)^2 dx_1 \tag{19}$$

Ignoring higher-order terms in the section rotation $\partial u_2/\partial x_1$, the virtual work of the applied force is

$$\begin{aligned}
\overline{\delta} W &= -\overline{P}\mathbf{B}_1(\ell, t) \cdot [\delta u_1(\ell, t)\mathbf{b}_1 + \delta u_2(\ell, t)\mathbf{b}_2] \\
&= -\overline{P}\mathbf{B}_1(\ell, t) \cdot \left[-\mathbf{b}_1\int_0^{\ell}\frac{\partial u_2}{\partial x_1}\frac{\partial \delta u_2}{\partial x_1}dx_1 + \delta u_2(\ell, t)\mathbf{b}_2\right] \\
&= \overline{P}\left[\int_0^{\ell}\frac{\partial u_2}{\partial x_1}\frac{\partial \delta u_2}{\partial x_1}dx_1 - \frac{\partial u_2}{\partial x_1}(\ell, t)\delta u_2(\ell, t)\right]
\end{aligned} \tag{20}$$

Clearly, the first term of the last line can be expressed as the variation of a potential energy functional and is the standard term one finds in energy treatments of strings and beams undergoing axial forces. However, the second term cannot be derived from a potential. Using integration by parts and the root boundary condition that $u_2(0, t) = 0$, one may write the virtual work in the simplest form as

$$\overline{\delta W} = -\overline{P}\int_0^{\ell}\frac{\partial^2 u_2}{\partial x_1^2}\delta u_2 dx_1 \tag{21}$$

Applying Hamilton's principle to obtain the equation of motion and boundary conditions, one first obtains

$$\int_{t_1}^{t_2}\left(\delta U - \delta T - \overline{\delta W}\right)dt = 0 \tag{22}$$

or

$$\int_{t_1}^{t_2} \left[\int_0^\ell \left(EI \frac{\partial^2 u_2}{\partial x_1^2} \frac{\partial^2 \delta u_2}{\partial x_1^2} - \mu \frac{\partial u_2}{\partial t} \frac{\partial \delta u_2}{\partial t} + \overline{P} \frac{\partial^2 u_2}{\partial^2 x_1} \delta u_2 \right) dx_1 \right] dt = 0 \qquad (23)$$

Integrating by parts in time and setting the virtual displacement $\delta u_2(x_1, t)$ equal to zero at $t = t_1$ and t_2, one finds that the time integral is no longer necessary. The result is a weak form of the equation of motion

$$\int_0^\ell \left[EI \frac{\partial^2 u_2}{\partial x_1^2} \frac{\partial^2 \delta u_2}{\partial x_1^2} + \left(\mu \frac{\partial^2 u_2}{\partial t^2} + \overline{P} \frac{\partial^2 u_2}{\partial x_1^2} \right) \delta u_2 \right] dx_1 = 0 \qquad (24)$$

Integrating by parts in x_1, one now finds

$$\int_0^\ell \left[\frac{\partial^2}{\partial x_1^2} \left(EI \frac{\partial^2 u_2}{\partial x_1^2} \right) + \mu \frac{\partial^2 u_2}{\partial t^2} + \overline{P} \frac{\partial^2 u_2}{\partial x_1^2} \right] \delta u_2 dx_1$$

$$+ \left[EI \frac{\partial^2 u_2}{\partial x_1^2} \frac{\partial \delta u_2}{\partial x_1} - \frac{\partial}{\partial x_1} \left(EI \frac{\partial^2 u_2}{\partial x_1^2} \right) \delta u_2 \right] \Bigg|_0^\ell = 0 \qquad (25)$$

The virtual displacement and rotation are arbitrary everywhere except at the beam root where they both vanish. In order for this expression to vanish, the integrand must vanish. The result is the Euler-Lagrange partial differential equation of motion

$$\frac{\partial^2}{\partial x_1^2} \left(EI \frac{\partial^2 u_2}{\partial x_1^2} \right) + \mu \frac{\partial^2 u_2}{\partial t^2} + \overline{P} \frac{\partial^2 u_2}{\partial x_1^2} = 0 \qquad (26)$$

and the boundary conditions are

$$u_2(0, t) = \frac{\partial u_2}{\partial x_1}(0, t) = EI(\ell) \frac{\partial^2 u_2}{\partial x_1^2}(\ell, t) = \frac{\partial}{\partial x_1} \left(EI \frac{\partial^2 u_2}{\partial x_1^2} \right)(\ell, t) = 0 \qquad (27)$$

For constant EI, one may simplify these to

$$EI \frac{\partial^4 u_2}{\partial x_1^4} + \mu \frac{\partial^2 u_2}{\partial t^2} + \overline{P} \frac{\partial^2 u_2}{\partial x_1^2} = 0 \qquad (28)$$

with

$$u_2(0, t) = \frac{\partial u_2}{\partial x_1}(0, t) = \frac{\partial^2 u_2}{\partial x_1^2}(\ell, t) = \frac{\partial^3 u_2}{\partial x_1^3}(\ell, t) = 0 \qquad (29)$$

This problem can be solved exactly. Below we will also present an approximate solution by the Ritz method, but here we consider exact solutions to both the static and dynamic problems to illustrate an important point. As with the mechanical system in the last section, we consider the static terms first to explore the possibility of buckling, which is governed by

$$EI \frac{\partial^4 u_2}{\partial x_1^4} + \overline{P} \frac{\partial^2 u_2}{\partial x_1^2} = 0 \qquad (30)$$

the solution of which is

$$u_2 = a_1 + a_2 x_1 + a_3 \sin(kx_1) + a_4 \cos(kx_1) \qquad (31)$$

Making use of the boundary conditions, we obtain

$$
\left\{
\begin{array}{c}
u_2(0) \\
u_2'(0) \\
u_2''(\ell) \\
u_2'''(\ell)
\end{array}
\right\}
=
\begin{bmatrix}
1 & 0 & 0 & 1 \\
0 & 1 & k & 0 \\
0 & 0 & -k^2\sin(k\ell) & -k^2\cos(k\ell) \\
0 & 0 & -k^3\cos(k\ell) & k^3\sin(k\ell)
\end{bmatrix}
\left\{
\begin{array}{c}
a_1 \\
a_2 \\
a_3 \\
a_4
\end{array}
\right\}
=
\left\{
\begin{array}{c}
0 \\
0 \\
0 \\
0
\end{array}
\right\}
\tag{32}
$$

with a_i being arbitrary constants. A nontrivial solution can only exist if the determinant of the coefficient matrix

$$
\begin{vmatrix}
1 & 0 & 0 & 1 \\
0 & 1 & k & 0 \\
0 & 0 & k^2\sin(k\ell) & k^2\cos(k\ell) \\
0 & 0 & -k^3\cos(k\ell) & k^3\sin(k\ell)
\end{vmatrix}
= k^5
\tag{33}
$$

vanishes. Clearly this value cannot vanish except when $k = 0$ (i.e., for a trivial solution). Thus, no matter how large the applied force is, buckling will not take place.

So, now let us consider the dynamic case. Introducing $x = x_1/\ell$, $P = \overline{P}\ell^2/EI$, and nondimensional time $\tau = t\sqrt{EI/(m\ell^4)}$, and substituting $u_2 = v(x)\exp(s\tau)$ into the equation of motion, Eq. (28), one finds the governing equation reduces to

$$
v'''' + Pv'' + s^2 v = 0
\tag{34}
$$

where $(\ )' = d(\ )/dx$. The general solution is

$$
v = a_1\sin(\alpha x) + a_2\cos(\alpha x) + a_3\sinh(\beta x) + a_4\cosh(\beta x)
\tag{35}
$$

where a_i are arbitrary constants and

$$
\alpha = \frac{\sqrt{\sqrt{P^2 - 4s^2} + P}}{\sqrt{2}}
$$
$$
\beta = \frac{\sqrt{\sqrt{P^2 - 4s^2} - P}}{\sqrt{2}}
\tag{36}
$$

Using the boundary conditions as above, one finds that

$$
\left\{
\begin{array}{c}
u_2(0) \\
u_2'(0) \\
u_2''(\ell) \\
u_2'''(\ell)
\end{array}
\right\}
=
\begin{bmatrix}
0 & 1 & 0 & 1 \\
\alpha & 0 & \beta & 0 \\
-\alpha^2\sin\alpha & -\alpha^2\cos\alpha & \beta^2\sinh\beta & \beta^2\cosh\beta \\
-\alpha^3\cos\alpha & \alpha^2\sin\alpha & \beta^3\cosh\beta & \beta^3\sinh\beta
\end{bmatrix}
\left\{
\begin{array}{c}
a_1 \\
a_2 \\
a_3 \\
a_4
\end{array}
\right\}
=
\left\{
\begin{array}{c}
0 \\
0 \\
0 \\
0
\end{array}
\right\}
\tag{37}
$$

or

$$
\begin{vmatrix}
0 & 1 & 0 & 1 \\
\alpha & 0 & \beta & 0 \\
-\alpha^2\sin\alpha & -\alpha^2\cos\alpha & \beta^2\sinh\beta & \beta^2\cosh\beta \\
-\alpha^3\cos\alpha & \alpha^2\sin\alpha & \beta^3\cosh\beta & \beta^3\sinh\beta
\end{vmatrix}
= 0
\tag{38}
$$

which reduces to

$$
\alpha^4 + \beta^4 + \alpha\beta[2\,\alpha\,\beta\cos\alpha\cosh\beta + (\alpha^2 - \beta^2)\sin\alpha\sinh\beta] = 0
\tag{39}
$$

This can be simplified by noting that

$$
\alpha^4 + \beta^4 = P^2 - 2s^2
$$
$$
\alpha^2 - \beta^2 = P
\tag{40}
$$
$$
\alpha^2\beta^2 = -s^2
$$

so that

$$P^2 - 2s^2(1 + \cos\alpha\cosh\beta) \pm iPs\sin\alpha\sinh\beta = 0 \qquad (41)$$

Since the solution is represented in the form of $\exp(s\tau)$, the characteristic equation must be independent of the sign of the imaginary part of s. Thus, we can further simplify the last term yielding

$$P^2 - 2s^2(1 + \cos\alpha\cosh\beta) - i\,P\,s\,\mathrm{sgn}[\mathrm{Im}(s)]\sin\alpha\sinh\beta = 0 \qquad (42)$$

The roots of Eq. (42) must be found by numerical methods. When $P = 0$, the roots are all pure imaginary and are equal to $s = \pm i1.87510^2,\ \pm i4.69409^2,\ldots$. The curves in Figs. 11.8 and 11.9 show the variation of the lowest two roots with P. As P tends toward $P = P_{cr} = 20.0510$, the imaginary parts of the first two roots coalesce near $\mathrm{Im}(s) = 11.0156$ and the real parts remain zero. When P exceeds P_{cr}, the real parts of the two roots being tracked suddenly become nonzero, while the imaginary parts lock onto one another becoming only one curve. When $P > P_{cr}$, one root will always have a positive real part, which means that we have vibrations with growing amplitude. This is, then, a flutter instability just as in the case of the mechanical analog above.

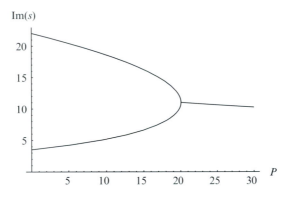

FIGURE 11.8   Exact solution for imaginary part of nondimensional eigenvalue s versus nondimensional force P for Beck's column.

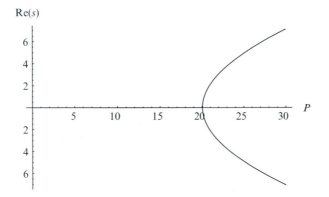

FIGURE 11.9   Exact solution for real part of nondimensional eigenvalue s versus nondimensional force P for Beck's column.

It is very tedious to work with such equations as Eq. (42). Thus, it will prove to be convenient to solve the problem approximately using the method of Ritz in conjunction with the principle of virtual displacements. To do so, we assume that the displacement can be written as a truncated series involving a complete set of basis functions. For example, here we may write

$$u_2(x_1,t) = \sum_{i=1}^{N} \xi_i(t)\phi_i(x_1) \tag{43}$$

where ϕ_i are the uniform cantilever beam free-vibration mode shapes given by

$$\phi_i = \cosh(\alpha_i x_1) - \cos(\alpha_i x_1) - \beta_i[\sinh(\alpha_i x_1) - \sin(\alpha_i x_1)] \tag{44}$$

with

$$\beta_i = \frac{\cosh(\alpha_i \ell) + \cos(\alpha_i \ell)}{\sinh(\alpha_i \ell) + \sin(\alpha_i \ell)} \tag{45}$$

Note that $\cos(\alpha_i \ell)\cosh(\alpha_i \ell) + 1 = 0$, so that $\alpha_1 \ell = 1.87510$, $\beta_1 = 0.734096$, $\alpha_2 \ell = 4.69409$, $\beta_2 = 1.01847$, etc.; the natural frequencies of the unforced system are equal to $(\alpha_i \ell)^2 \sqrt{EI/(\mu \ell^4)}$. The mode shapes are normalized such that, for all i,

$$\int_0^\ell \phi_i^2 dx_1 = \ell \tag{46}$$

For application of the Ritz method, we substitute this series into the weak form, yielding

$$\sum_{i=1}^{N} \delta\xi_i \left[\sum_{j=1}^{N} \xi_j \int_0^\ell \left(EI\phi_i''\phi_j'' + \overline{P}\phi_i\phi_j'' \right) dx_1 + \sum_{j=1}^{N} \ddot{\xi}_j \int_0^\ell \mu\phi_i\phi_j dx_1 \right] = 0 \tag{47}$$

which, for arbitrary $\delta\xi_i$ yields a system of linear, second-order, ordinary differential equations of the form

$$\sum_{j=1}^{N} \left(M_{ij}\ddot{\xi}_j + K_{ij}\xi_j \right) = 0 \quad i = 1, 2, \ldots, N \tag{48}$$

or, in matrix form,

$$[M]\{\ddot{\xi}\} + [K]\{\xi\} = 0 \tag{49}$$

where

$$K_{ij} = \int_0^\ell \left(EI\phi_i''\phi_j'' + \overline{P}\phi_i\phi_j'' \right) dx_1$$

$$M_{ij} = \int_0^\ell \mu\phi_i\phi_j dx_1 \tag{50}$$

We note that $[K]$ is not symmetric, thus allowing for the possibility of complex eigenvalues. The matrix $[K]$ carries the elastic forces, proportional to EI, and the applied forces, proportional to $\overline{P}$. The matrix $[M]$ carries the inertial forces, proportional to μ. The system of governing equations is linear with constant coefficients.

Letting $\{\xi\} = \{\check{\xi}\} \exp(\bar{s}t)$, $s^2 = \mu\ell^4\bar{s}^2/EI$, $P = \overline{P}\ell^2/EI$, and $x = x_1/\ell$, one finds that the equation can be expressed as

$$\left[s^2\begin{bmatrix}\ddots&\\&1\end{bmatrix}+\begin{bmatrix}\ddots&\\&(\alpha_i\ell)^4\end{bmatrix}+P[A]\right]\{\check{\xi}\}=0 \tag{51}$$

where

$$A_{ij}=\int_0^1\phi_i(x)\phi_j''(x)dx \tag{52}$$

and where for $N=2$, $A_{11}=0.858244$, $A_{12}=-11.7432$, $A_{21}=1.87385$, $A_{22}=-13.2943$.

This system has only a trivial solution for arbitrary values of P and s. For $P=0$, the roots for s are purely imaginary and equal to the values of $\pm i(\alpha_1\ell)^2=\pm1.87510^2i$ and $\pm i(\alpha_2\ell)^2=\pm4.69409^2i$. As P increases, the roots remain imaginary at first as they come together, but there is a point at which the roots coalesce into a pair of double roots, both purely imaginary (see Figs. 11.10 and 11.11). Since the governing eigenvalue problem reduces to a bi-quadratic in s, i.e.,

$$as^4+bs^2+c=0 \tag{53}$$

with $a=1$, $b=(\alpha_1\ell)^2+(\alpha_2\ell)^2+P(A_{11}+A_{22})$, and

$$c=P^2(A_{11}A_{22}-A_{12}A_{21})+PA_{22}(\alpha_1\ell)^4+PA_{11}(\alpha_2\ell)^4+(\alpha_1\ell)^4(\alpha_2\ell)^4 \tag{54}$$

the point of the coalescence can be determined to be when the discriminant, b^2-4ac, vanishes. The value of P at this point is then the value at flutter, denoted by P_{cr} and given by

$$P_{cr}=\frac{(\alpha_2\ell)^4-(\alpha_1\ell)^4}{A_{11}-A_{22}-2\sqrt{-A_{12}}\sqrt{A_{21}}}=20.1048 \tag{55}$$

at which point the nondimensional flutter frequency $\mathrm{Im}(s_{cr})=11.1323$. The plots in Figs. 11.10 and 11.11 not only look very similar to those in Figs. 11.8 and 11.9, but these values are very close to the exact solution, where $P_{cr}=20.0510$ and $\mathrm{Im}(s_{cr})=11.0156$. When $P>P_{cr}$, the imaginary parts of the two roots remain equal to each other, while the real parts of the roots have the same magnitude but opposite signs. This means that when $P>P_{cr}$ there is always a root with a positive real part, which means that the system will oscillate sinusoidally with an exponentially increasing amplitude.

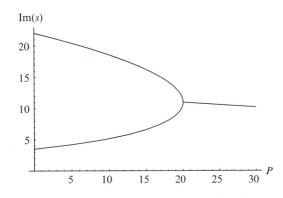

FIGURE 11.10 Approximate solution for imaginary part of nondimensional eigenvalue s versus nondimensional force P for Beck's column.

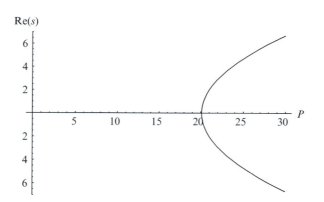

FIGURE 11.11 Approximate solution for real part of nondimensional eigenvalue s versus non-dimensional force P for Beck's column.

Actual follower forces typically have damping associated with them, which will mean that subcritical values of s will not be purely imaginary. In this case we would look for the value of P for which the real part of s crosses the zero axis and becomes positive, as in Fig. 11.6. Herrmann (1967) was evidently the first to point out that the addition of damping for nonconservative systems can destabilize a system. However, it is also known that without some damping mechanism in the model, a follower force may not represent a realistic phenomenon, as pointed out by Langthjem and Sugiyama (2000). Cases with damping are left as exercises for the reader.

11.4   LEIPHOLZ'S COLUMN

Leipholz (1975) has considered the problem of a beam subjected to a uniformly distributed follower force $\bar{p}$ along its length (see Fig. 11.12). This development follows closely that of the last section, the only difference being in the virtual work of the applied load, which is now uniformly distributed along the beam length instead of concentrated at the tip. Thus, assuming the beam to be inextensible, one can express the force on each differential element of length dx_1 as $-\bar{p}\mathbf{B}_1(x_1,t)dx_1$, the total virtual work of which given by

$$\overline{\delta W} = -\bar{p}\int_0^\ell \mathbf{B}_1 \cdot [\delta u_1 \mathbf{b}_1 + \delta u_2 \mathbf{b}_2]dx_1$$

$$= -\bar{p}\int_0^\ell \left\{\mathbf{B}_1 \cdot \left[-\mathbf{b}_1 \int_0^{x_1} \frac{\partial u_2(\xi)}{\partial \xi}\frac{\partial \delta u_2(\xi)}{\partial \xi}dx_1 + \delta u_2 \mathbf{b}_2\right]\right\}dx_1 \quad (56)$$

$$= \bar{p}\int_0^\ell \left[(\ell - x_1)\frac{\partial u_2}{\partial x_1}\frac{\partial \delta u_2}{\partial x_1} - \frac{\partial u_2}{\partial x_1}\delta u_2\right]dx_1$$

where the step from the second line to the last is accomplished by integration by parts.

Taking the contributions from strain and kinetic energy from the previous section, one finds a weak form of the equation of motion to be

$$\int_0^\ell \left[EI\frac{\partial^2 u_2}{\partial x_1^2}\frac{\partial^2 \delta u_2}{\partial x_1^2} + \left(\mu\frac{\partial^2 u_2}{\partial t^2} + \bar{p}\frac{\partial u_2}{\partial x_1}\right)\delta u_2 - \bar{p}(\ell - x_1)\frac{\partial u_2}{\partial x_1}\frac{\partial \delta u_2}{\partial x_1}\right]dx_1 = 0 \quad (57)$$

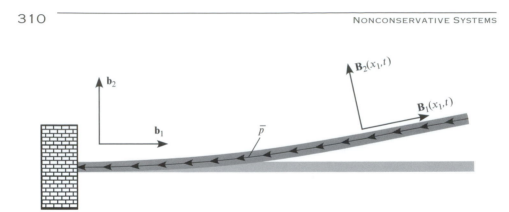

FIGURE 11.12   Uniform beam subjected to uniformly distributed follower force.

or, alternatively,

$$\int_0^\ell \left[EI \frac{\partial^2 u_2}{\partial x_1^2} \frac{\partial^2 \delta u_2}{\partial x_1^2} + \mu \frac{\partial^2 u_2}{\partial t^2} + \bar{p}(\ell - x_1) \frac{\partial^2 u_2}{\partial x_1^2} \delta u_2 \right] dx_1 = 0 \tag{58}$$

To obtain the partial differential equation of motion and boundary conditions, one now integrates by parts in x_1 to obtain

$$\int_0^\ell \left[\frac{\partial^2}{\partial x_1^2} \left(EI \frac{\partial^2 u_2}{\partial x_1^2} \right) + \mu \frac{\partial^2 u_2}{\partial t^2} + \bar{p}(\ell - x_1) \frac{\partial^2 u_2}{\partial x_1^2} \right] \delta u_2 dx_1$$
$$+ \left[EI \frac{\partial^2 u_2}{\partial x_1^2} \frac{\partial \delta u_2}{\partial x_1} - \frac{\partial}{\partial x_1} \left(EI \frac{\partial^2 u_2}{\partial x_1^2} \right) \delta u_2 \right]\Big|_0^\ell = 0 \tag{59}$$

The virtual displacement and rotation are arbitrary everywhere except at the beam root where they both vanish. In order for this expression to vanish, the integrand must vanish. The resulting partial differential equation of motion is then

$$\frac{\partial^2}{\partial x_1^2} \left(EI \frac{\partial^2 u_2}{\partial x_1^2} \right) + \mu \frac{\partial^2 u_2}{\partial t^2} + \bar{p}(\ell - x_1) \frac{\partial^2 u_2}{\partial x_1^2} = 0 \tag{60}$$

and the boundary conditions are

$$u_2(0,\, t) = \frac{\partial u_2}{\partial x_1}(0,\, t) = EI(\ell) \frac{\partial^2 u_2}{\partial x_1^2}(\ell,\, t) = \frac{\partial}{\partial x_1} \left(EI \frac{\partial^2 u_2}{\partial x_1^2} \right)(\ell,\, t) = 0 \tag{61}$$

For constant EI, one may simplify these to

$$EI \frac{\partial^4 u_2}{\partial x_1^4} + \mu \frac{\partial^2 u_2}{\partial t^2} + \bar{p}(\ell - x_1) \frac{\partial^2 u_2}{\partial x_1^2} = 0 \tag{62}$$

with

$$u_2(0,\, t) = \frac{\partial u_2}{\partial x_1}(0,\, t) = \frac{\partial^2 u_2}{\partial x_1^2}(\ell,\, t) = \frac{\partial^3 u_2}{\partial x_1^3}(\ell,\, t) = 0 \tag{63}$$

An exact solution for this problem is not known, so an approximate solution via the Ritz method is in order. As with the Beck column, let

$$u_2(x_1,t) = \sum_{i=1}^{N} \xi_i(t)\phi_i(x_1) \tag{64}$$

where ϕ_i are the uniform cantilever beam free-vibration mode shapes. For application of the Ritz method, we substitute this series into the weak form, yielding

$$\sum_{i=1}^{N} \delta\xi_i \left\{ \sum_{j=1}^{N} \xi_j \int_0^\ell \left[EI\phi_i''\phi_j'' + \bar{p}(\ell - x_1)\phi_i\phi_j'' \right]dx_1 + \sum_{j=1}^{N} \ddot{\xi}_j \int_0^\ell \mu\phi_i\phi_j dx_1 \right\} = 0 \tag{65}$$

which, for arbitrary $\delta\xi_i$ yields a system of linear, second-order, ordinary differential equations of the form

$$\sum_{j=1}^{N} \left(M_{ij}\ddot{\xi}_j + K_{ij}\xi_j \right) = 0 \quad i = 1, 2, \ldots, N \tag{66}$$

or, in matrix form,

$$[M]\{\ddot{\xi}\} + [K]\{\xi\} = 0 \tag{67}$$

where

$$K_{ij} = \int_0^\ell \left[EI\phi_i''\phi_j'' + \bar{p}(\ell - x_1)\phi_i\phi_j'' \right]dx_1$$
$$M_{ij} = \int_0^\ell \mu\phi_i\phi_j dx_1 \tag{68}$$

We note that $[K]$ is not symmetric, thus allowing for the possibility of complex eigenvalues. The matrix $[K]$ carries the elastic forces, proportional to EI, and the applied forces, proportional to $\bar{p}$. The matrix $[M]$ carries the inertial forces, proportional to μ. The system is linear with constant coefficients.

Letting $\{\xi\} = \{\check{\xi}\} \exp(\bar{s}t)$, $s^2 = \mu\ell^4\bar{s}^2/EI$, $p = \bar{p}\ell^3/EI$, and $x = x_1/\ell$, one finds that the equation can be expressed as

$$\left[s^2 \lceil 1 \rfloor + \lceil (\alpha_i\ell)^4 \rfloor + p[A] \right]\{\check{\xi}\} = 0 \tag{69}$$

where

$$A_{ij} = \int_0^1 (1 - x)\phi_i(x)\phi_j''(x)dx \tag{70}$$

with $A_{11} = 0.429122$, $A_{12} = -4.33714$, $A_{21} = 1.18178$, and $A_{22} = -6.64714$. The real and imaginary parts of the roots are plotted versus p in Figs. 11.13 and 11.14, respectively. The two-mode solution reveals that flutter takes place at $p = 40.7746$ and with a nondimensional flutter frequency of $\text{Im}(s) = 11.0531$. These results are comparable to values published by Leipholz (1975).

11.5 CANTILEVERED SHAFT SUBJECT TO TANGENTIAL TORQUE

Here we consider a cantilevered shaft subject to a twisting moment, the direction of which follows the deformation of the shaft as it undergoes perturbations in bending. Because of the complexity of the solution, we will only solve the problem approximately by making use of Galerkin's method.

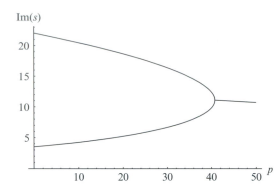

FIGURE 11.13　Imaginary part of nondimensional eigenvalue s versus nondimensional force p for Leipholz's column.

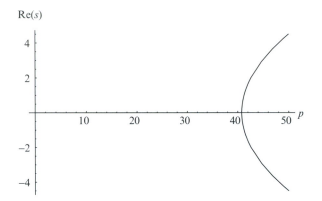

FIGURE 11.14　Real part of nondimensional eigenvalue s versus nondimensional force p for Leipholz's column.

The strain energy for a shaft loaded by a twisting moment has already been developed in Chapter 8, Eq. (8.21). Owing to their being decoupled from the bending perturbations, we ignore the longitudinal and torsional contributions. Here we present the strain energy for the special case of zero axial force from Section 8.2, such that

$$U = \frac{1}{2} \int_0^\ell \left[EI \left(\hat{u}_2''^2 + \hat{u}_3''^2 \right) + Q \left(\hat{u}_2'' \hat{u}_3' - \hat{u}_2' \hat{u}_3'' \right) \right] dx_1 \tag{71}$$

The kinetic energy is

$$T = \frac{1}{2} \int_0^\ell \mu \left(\dot{\hat{u}}_2^2 + \dot{\hat{u}}_3^2 \right) dx_1 \tag{72}$$

The applied load is nonconservative, and hence it does not possess a potential energy. The virtual work of the applied load is easily developed. The applied torque is in the tangential direction and is $Q\mathbf{B}_1(\ell, t)$. The virtual rotation is

$$\overline{\boldsymbol{\delta\psi}} = \overline{\delta\psi}_1 \mathbf{B}_1 + \overline{\delta\psi}_2 \mathbf{B}_2 + \overline{\delta\psi}_3 \mathbf{B}_3 \tag{73}$$

so that the virtual work is

$$\overline{\delta W} = Q\mathbf{B}_1(\ell, t) \cdot \overline{\boldsymbol{\delta\psi}}(\ell, t)$$
$$\overline{\delta W} = Q\overline{\delta\psi}_1(\ell, t)$$
$$= Qe_1^T R\delta\theta|_{x=\ell}$$
$$= \frac{Q}{2}\delta\theta^T \tilde{e}_1\hat{\theta}|_{x=\ell} + \dots$$
$$= \frac{Q}{2}\left[\hat{u}_2'(\ell)\delta u_3'(\ell) - \hat{u}_3'(\ell)\delta u_2'(\ell)\right] + \dots$$

(74)

where the third and fourth lines apply only for small perturbations (indicated by the hats), and the ellipses indicate terms of second and higher orders of the perturbation quantities.

The equations of motion take the form

$$EI\hat{u}_2'''' + Q\hat{u}_3''' + \mu\ddot{\hat{u}}_2 = 0$$
$$EI\hat{u}_3'''' - Q\hat{u}_2''' + \mu\ddot{\hat{u}}_3 = 0$$

(75)

For the clamped-free condition, we have zero displacement and rotation at the root and zero moment and shear at the tip, so that the boundary conditions are

$$\hat{u}_2(0) = \hat{u}_3(0) = \hat{u}_2'(0) = \hat{u}_3'(0) = 0$$
$$\hat{u}_2''(\ell) = \hat{u}_3''(\ell) = \hat{u}_2'''(\ell) = \hat{u}_3'''(\ell) = 0$$

(76)

We now undertake these steps: (a) Let $\breve{u} = \hat{u}_2 + i\hat{u}_3$; (b) introduce the nondimensional axial coordinate $x = x_1/\ell$; (c) let the prime now refer to derivatives with respect to x; (d) let $\breve{u} = u\exp(\bar{s}t)$ with $s = \mu\bar{s}^2\ell^4/EI$. This yields

$$u'''' - iqu''' + s^2 u = 0$$

(77)

with $q = Q\ell/EI$ and

$$u(0) = u'(0) = u''(1) = u'''(1) = 0$$

(78)

The solution to this equation is quite involved, but a simple one-term approximation will suffice to show the result. Let $u = \psi(x)$ with

$$\psi = \cosh(\alpha_1 x) - \cos(\alpha_1 x) - \beta_1[\sinh(\alpha_1 x) - \sin(\alpha_1 x)]$$

(79)

which is the first mode shape of a cantilevered beam undergoing free vibration, where $\alpha_1 = 1.87510$, $\beta_1 = 0.734096$, and

$$\int_0^1 \psi^2 dx = 1$$

(80)

Letting $A = \int_0^1 \psi\psi''' dx = -3.78953$, and using Galerkin's method, the resulting eigenvalue problem becomes a simple quadratic equation in s, given by

$$s^2 + \alpha_1^4 - iAq = 0$$

(81)

the roots of which are

$$s = \pm\sqrt{-\alpha_1^4 + Aiq}$$
$$= \pm i\alpha_1^2 \pm \frac{|q|A}{2\alpha_1^2} + O(q^2)$$
$$= \pm 3.51602i \pm 0.538896|q| + O(q^2)$$

(82)

One of these roots has a positive real part for any nonzero value of q! Moreover, the magnitude of the real part can be shown to be independent of the number of modes taken in the Galerkin approximation to first order in q. The fact that the system is unstable for even infinitesimal values of q is clearly a nonphysical result and has caused some to question the very existence of follower forces. However, the addition of a small amount of viscous damping to the model leads to a region of stability; see Problem 2 at the end of this chapter. Therefore, a more reasonable conclusion is that models of follower forces should include damping in some form; see Langthjem and Sugiyama (2000).

11.6 DEEP CANTILEVER WITH TRANSVERSE FOLLOWER FORCE AT THE TIP

In spite of all the published work on systems with follower forces, there seems to be very little literature concerned with the lateral-torsional stability of deep cantilevered beams loaded by a transverse follower force at the tip. This problem has some practical applications, such as the effect of jet engine thrust on the aeroelastic flutter of a flexible wing. According to Bolotin (1963), this type of system was first considered in Bolotin (1959). Although the analysis presented therein is applicable to the tip-loaded cantilever case, no results specific to that case were presented. Como (1966) analyzed a cantilevered beam subjected to a lateral follower force at the tip. The distributed mass and inertia properties of the beam were neglected, although a concentrated mass and inertia at the tip were included. Without neglecting the distributed mass and inertia properties of the beam, Wohlhart (1971) undertook an extensive study, and results for a wide variation of several parameters were presented. Additional results and observations were presented by Hodges (2001). Barsoum (1971) developed a finite element solution to the problem, which was later revisited by Detinko (2002). Other than these six papers, to the best of the authors' knowledge, this problem appears to have received no further attention in the literature. It is the objective of this section to consider further this nonconservative elastic stability problem and present a few results and observations that go beyond those of Wohlhart (1971).

We first develop a weak form of the partial differential equations of motion for a deep, symmetric beam under the action of a tip follower force acting in the plane of symmetry. Then an approximate solution using cantilever beam bending and torsional modes is obtained. The effects of three parameters are investigated: the ratio of the uncoupled fundamental bending and torsional frequencies, dimensionless parameters reflecting the mass radius of gyration, and the offset from the elastic axis of the mass centroid.

11.6.1 EQUATIONS OF MOTION

Consider a cantilevered beam with torsional stiffness GJ and bending stiffnesses EI_2 and EI_3 with $EI_3 \gg EI_2$. It is noted here that the bending analysis neglects transverse shear deformation and rotary inertia, and the torsional analysis neglects the warping restraint; thus, the beam theory is strictly along classical lines. The Cartesian coordinates x_1, x_2, and x_3 are along the elastic axis and the two transverse directions, respectively, as shown in Fig. 11.15. For the purpose of analysis, we introduce two sets of dextral triads of unit vectors. The unit vectors of the first set,

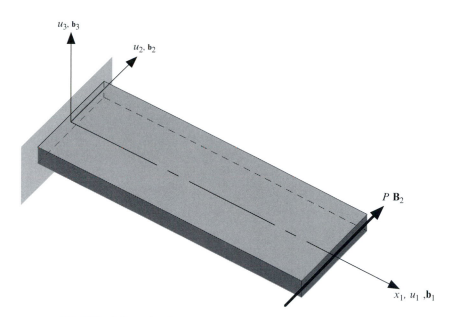

FIGURE 11.15 Schematic of deep beam showing coordinate systems and follower force.

$\mathbf{b}_i$ with $i = 1$, 2, and 3, are parallel to x_i and fixed in an inertial frame. Those of the other set are fixed in the local cross-sectional frame of the deformed beam and denoted by $\mathbf{B}_i(x_1,t)$, with $i = 1$, 2, and 3. Denote the displacements along $\mathbf{b}_i$ as $u_i(x_1,t)$ with $i = 1$, 2, and 3; and denote the section rotation caused by torsion as $\theta_1(x_1,t)$. The load is directed along unit vector $\mathbf{B}_2(\ell,t)$ where $\mathbf{B}_2(\ell,t) = -u_2'(\ell,t)\mathbf{b}_1 + \mathbf{b}_2 + \theta_1(\ell,t)\mathbf{b}_3$; here $(\ )'$ denotes a partial derivative with respect to x_1. Thus, the virtual work done by this force through a virtual displacement is

$$\overline{\delta W} = P\mathbf{B}_2(\ell,t) \cdot [\delta u_1(\ell,t)\mathbf{b}_1 + \delta u_2(\ell,t)\mathbf{b}_2 + \delta u_3(\ell,t)\mathbf{b}_3]$$
$$= P(-u_2'\delta u_1 + \delta u_2 + \theta_1\delta u_3)|_0^\ell \tag{83}$$

In keeping with the nonconservative nature of the follower force, there exists no potential energy which, upon variation, will yield this expression for the virtual work. We will subsequently ignore the longitudinal displacement u_1.

For a beam subject to a bending moment $\overline{M}_3$ that is constant in time but varying in x_1, and in which deflections caused by that moment are ignored (since $EI_3 \gg EI_2$), we need the third component of moment strain to second order. This can be obtained from Eq. (3.137) as

$$\kappa_3 = \bar{\kappa}_3 + u_2'' + \theta_1 u_3'' \tag{84}$$

Thus, the strain energy can written as

$$U = \int_0^\ell \left[\frac{GJ}{2}\theta_1'^2 + \frac{EI_2}{2}u_3''^2 + \overline{M}_3\left(u_2'' + \theta_1 u_3''\right) \right] dx_1 \tag{85}$$

To find the equilibrium state of deformation in the beam, one may consider only the first-order terms in $\delta U - \overline{\delta W}$, such that

$$\int_0^\ell \left(\overline{M}_3\delta u_2'' - P\delta u_2'\right) dx_1 = 0 \tag{86}$$

Thus,

$$\overline{M}_3 = P(\ell - x_1) \tag{87}$$

as expected. To obtain a weak form that governs the behavior of small static perturbations about the equilibrium state, one may set the second-order terms in $\delta U - \overline{\delta W}$ equal to zero, so that

$$\delta U - \overline{\delta W} = \int_0^\ell \left[EI_2 u_3'' \delta u_3'' + GJ\theta_1' \delta\theta_1' + P(\ell - x_1)\left(\theta_1 \delta u_3'' + u_3'' \delta\theta_1\right)\right] dx_1 - P\theta_1 \delta u_3\big|_0^\ell = 0 \tag{88}$$

Integration by parts can eliminate the trailing term, so that

$$\delta U - \overline{\delta W} = \int_0^\ell \left\{ EI_2 u_3'' \delta u_3'' + GJ\theta_1' \delta\theta_1' + P(\ell - x_1) u_3'' \delta\theta_1 + P[(\ell - x_1)\theta_1]'' \delta u_3 \right\} dx_1 = 0 \tag{89}$$

It can be shown that there is no value of P that will result in buckling. To proceed with an investigation of the stability by the kinetic method, one may now add the variation of the kinetic energy and, using Hamilton's principle, consider the stability of small vibrations about the static equilibrium state.

The kinetic energy of the vibrating beam is simply

$$T = \frac{1}{2} \int_0^\ell \left(\mu \dot{u}_2^2 + \mu \dot{u}_3^2 + \mu \overline{\sigma}^2 \dot{\theta}_1^2 + 2\mu \overline{e} \dot{\theta}_1 \dot{u}_3 \right) dx_1 \tag{90}$$

where $(\dot{\ })$ is a partial derivative with respect to time, μ the mass per unit length, $\overline{e}$ the offset in the $\mathbf{b}_2$ direction of the mass centroid from the reference line, and $\overline{\sigma}$ the cross-sectional mass radius of gyration. We now undertake a straightforward application of Hamilton's principle

$$\int_{t_1}^{t_2} \left(\delta U - \overline{\delta W} - \delta T \right) dt = 0 \tag{91}$$

where t_1 and t_2 are fixed times. Integrating by parts in time, setting δu_3 and $\delta\theta_1$ equal to zero at the ends of the time interval, removing the time integration, assuming that the motion variables are proportional to $e^{\overline{s}t}$, and introducing a set of nondimensional variables, such that

$$\begin{aligned}
(\ )' &= \frac{d(\ )}{dx} & x_1 &= x\ell \\
u_3 &= \ell w \exp(\overline{s}t) & \theta_1 &= \theta \exp(\overline{s}t) \\
p &= \frac{P\ell^2}{\sigma EI_2} & s^2 &= \frac{\mu \ell^4 \overline{s}^2}{EI_2} \\
\sigma &= \frac{\overline{\sigma}}{\ell} & e &= \frac{\overline{e}}{\ell} \\
r^2 &= \frac{EI_2 \alpha_1^4 \sigma^2}{GJ\gamma_1^2}
\end{aligned} \tag{92}$$

one obtains a weak form governing the flutter problem

$$\int_0^1 \left\{ w'' \delta w'' + \sigma p[(1-x)\theta]'' \delta w + s^2(w + e\theta)\delta w + \frac{\alpha_1^4}{r^2 \gamma_1^2} \theta' \delta\theta' + \frac{p}{\sigma}(1-x)w'' \delta\theta \right.$$

$$\left. + s^2\left(\frac{e}{\sigma^2}w + \theta\right)\delta\theta \right\} dx = 0 \tag{93}$$

where the dimensionless parameters e and σ govern the offset of the mass centroid and mass radius of gyration, respectively. The dimensionless parameter r is the ratio of the fundamental bending and torsion frequencies of the unloaded beam with $e = 0$.

11.6.2   Approximate Solution and Results

This weak form can be solved approximately by assuming a set of uncoupled cantilever beam free-vibration modes for bending and torsion. To obtain converged results for the range of parameters considered, four of each type were found to be adequate. Specifying values for r, e, and σ, one can solve for the real and imaginary parts of s as functions of p. Depending on the values chosen for these parameters, flutter will occur when the imaginary parts of two modes coalesce. The modes that coalesce can be traced back to modes that for $p = 0$ are either two bending modes, two torsion modes, or one of each.

It can easily be shown that when $e = 0$, the eigenvalues do not depend on σ. This surprising result makes it possible to characterize the critical load in terms of only one parameter, r. It should be noted, however, that the mode shapes are not independent of σ when $e = 0$. In a typical case, the real parts of all eigenvalues are zero for sub-critical values of p. The imaginary parts of the eigenvalues depend only on p and r when $e = 0$, and on p, r, σ and e when $e \neq 0$. At the point when coalescence occurs, the real part of one mode becomes negative, while the real part of another becomes positive.

Example results for $e = 0$ are shown in Figs. 11.16 – 11.20. In Fig. 11.16 the imaginary parts of the four smallest eigenvalues are shown versus p for a large value of $r = 3.8$. In this case, at $p = 0$, two modes that start out as the first two torsional modes coalesce. Notice that the next higher modes, the first bending and third torsion modes at $p = 0$, coalesce for just a bit larger value of p. If r is taken to be a little larger, a complicated pattern emerges with the critical load, because multiple torsional modes occur below the first bending mode, and the critical coalescence may jump up to the second two modes. Fig. 11.17 shows the usual zero real part up to the coalescence and the positive real part thereafter. More realistic values are shown in Figs. 11.18 and 11.19. In the former case, the first bending mode is above the first bending, and in the latter it is below. As r becomes smaller still, the critical load greatly increases. The variation of the critical load versus r for $e = 0$ is shown in Fig. 11.20. Notice that an arbitrarily small force destabilizes the system when $r = 1$; similar observations were made by Bolotin (1964, p. 349) and Wohlhart (1971).

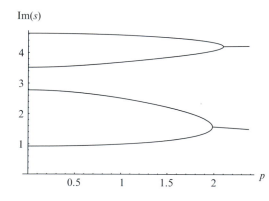

FIGURE 11.16   Imaginary parts of the four smallest eigenvalues versus p for $r = 3.8$ and $e = 0$.

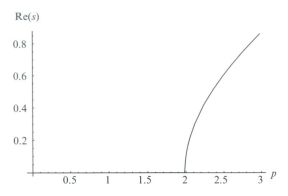

FIGURE 11.17   Real part of the eigenvalue for the unstable mode versus p for $r = 3.8$ and $e = 0$.

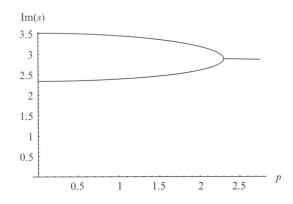

FIGURE 11.18   Imaginary parts of the two smallest eigenvalues versus p for $r = 3/2$ and $e = 0$.

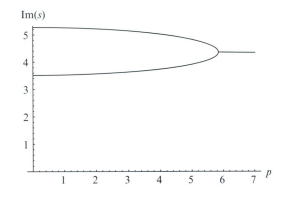

FIGURE 11.19   Imaginary parts of the two smallest eigenvalues versus p for $r = 2/3$ and $e = 0$.

For a typical value of $\sigma = 0.05$, small values of the mass centroid offset parameter e do not change the results qualitatively, except for removing the "cusp" in the plot of p_{cr} versus r at $r = 1$. Rather than a cusp at $r = 1$, when $e \neq 0$ one finds a smooth curve that has a nonzero minimum value. Plots for $e = 0.005$ and $e = -0.005$ are shown in Figs. 11.21 and 11.22, respectively, each also showing results for $e = 0$. One finds a greater qualitative change for positive values of e than for negative values. The variation of the results versus e for typical values of $r = 2/3$ and $r = 3/2$ are shown in Fig. 11.23 for $\sigma = 0.05$ and in Fig. 11.24 for $\sigma = 0.025$. Note the increased sensitivity of the critical load versus e curve for the smaller value of σ and for positive

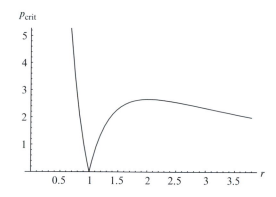

FIGURE 11.20   Critical load for $e = 0$ versus r.

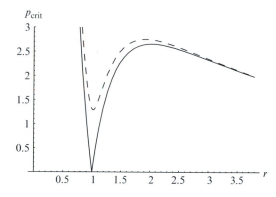

FIGURE 11.21   Critical load versus r for $\sigma = 0.05$ with $e = 0$ (solid line) and $e = 0.005$ (dashed line).

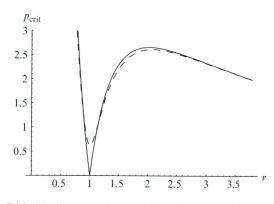

FIGURE 11.22   Critical load versus r for $\sigma = 0.05$ with $e = 0$ (solid line) and $e = -0.005$ (dashed line).

values of e. Finally, we consider the variation of p_{cr} versus σ for typical values of $r = 2/3$ and $r = 3/2$. Fig. 11.25 shows the variation of p_{cr} versus σ for $e = 0.005$ and Fig. 11.26 for $e = -0.005$. One sees p_{cr} decreasing with increasing σ for both positive and negative values of e. For the values of r chosen for these plots, the curve p_{cr} versus σ becomes rather flat as σ becomes large. Depending on the value of r, these flat regions can tend monotonically and asymptotically to the $e = 0$ value of p_{cr} or they may reverse while converging asymptotically. The value of r governs the types of motion that make up the two lowest modes.

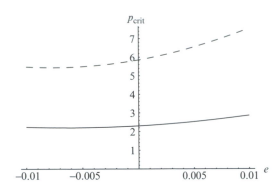

FIGURE 11.23   Critical load versus e for $\sigma = 0.05$ with $r = 3/2$ (solid line) and $r = 2/3$ (dashed line).

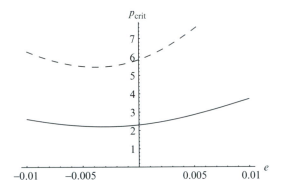

FIGURE 11.24   Critical load versus e for $\sigma = 0.025$ with $r = 3/2$ (solid line) and $r = 2/3$ (dashed line).

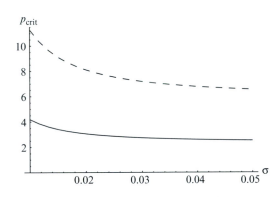

FIGURE 11.25   Critical load versus σ for $e = 0.005$; $r = 3/2$ (solid line) and $r = 2/3$ (dashed line).

11.7   FULLY INTRINSIC FORMULATION FOR BEAMS

The advantages of the intrinsic form of the equations of motion include low-order nonlinearities in the equations of motion and the absence of displacement and finite-rotation variables from the equations of motion. Still, the adjoined kinematical equations must contain displacement and finite-rotation variables in order to be useful in the general case. Although one may find that having such variables appear

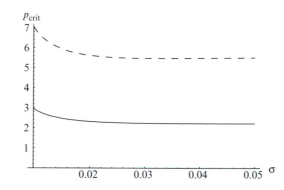

FIGURE 11.26 Critical load versus σ for $e = -0.005$; $r = 3/2$ (solid line) and $r = 2/3$ (dashed line).

only in a subset of the equations an attractive feature, when the finite rotation is represented by orientation angles there are sines and cosines of those angles. When it is represented by Rodrigues parameters, there are rational functions. Both involve infinite-degree nonlinearities. In fact, the kinematical equations must contain infinite-degree nonlinearities unless Euler parameters or the direction cosines themselves are used as finite-rotation variables; see Hodges (1987). These latter approaches add more rotational variables and more Lagrange multipliers, thus creating more unknowns. Finally, the high-order nonlinearities and/or additional unknowns inherent in these approaches make analytical solutions more laborious if not intractable.

This section presents an alternative approach, one in which displacement and finite-rotation variables do not appear. Of course, they or a subset of them can be added to the formulation in order to make possible their recovery; even so, depending on the problem, it may be possible to avoid nonlinearities of order greater than two and still have fewer equations and unknowns than would be necessary for the usual approach applied to the most general case.

In this section, the intrinsic formulation for beams is reviewed, including the equations of motion, the spatial and temporal constitutive equations, and the spatial and temporal kinematical equations needed to close the formulation. From the kinematical equations, the intrinsic kinematical equations are derived. Advantages of this formulation are demonstrated with examples treated by other methods earlier in this chapter.[1]

11.7.1 EQUATIONS OF MOTION

Consider a beam of length ℓ undergoing finite deformation. The displacement vector beam of the reference line is denoted by $\mathbf{u}(x_1,t)$ where x_1 is the running length coordinate along the undeformed beam axis of cross-sectional centroids. The orthogonal set of basis vectors for the cross section of the undeformed beam is denoted by $\mathbf{b}_i(x_1)$, where $\mathbf{b}_1$ is chosen to be tangent to the reference line. The orthogonal set of basis vectors for the cross section of the deformed beam is denoted by $\mathbf{B}_i(x_1,t)$, where $\mathbf{B}_1$ is not in general tangent to the reference line of the deformed beam. The direction cosines are denoted as $C_{ij}(x_1,t) = \mathbf{B}_i \cdot \mathbf{b}_j$. It is important to note that when the beam is in its undeformed state that these two sets of unit vectors coincide so that C reduces to the identity matrix.

The equations of motion in matrix form are given by

---

[1] Portions of this material are based on Hodges (1990, 2003), used by permission.

$$F' + \tilde{K}F + f = \dot{P} + \tilde{\Omega}P$$

$$M' + \tilde{K}M + (\tilde{e}_1 + \tilde{\gamma})F + m = \dot{H} + \tilde{\Omega}H + \tilde{V}P \tag{94}$$

where, as before, $(\ )'$ denotes the partial derivative with respect to the axial coordinate x_1, and $(\dot{\ })$ denotes the partial derivative with respect to the time t; column matrices $F = \lfloor F_1 \ F_2 \ F_3 \rfloor^T, M = \lfloor M_1 \ M_2 \ M_3 \rfloor^T, \gamma = \lfloor \gamma_{11} \ 2\gamma_{12} \ 2\gamma_{13} \rfloor^T, K = \lfloor K_1 \ K_2 \ K_3 \rfloor^T,$ $P = \lfloor P_1 \ P_2 \ P_3 \rfloor^T, H = \lfloor H_1 \ H_2 \ H_3 \rfloor^T, V = \lfloor V_1 \ V_2 \ V_3 \rfloor^T, \Omega = \lfloor \Omega_1 \ \Omega_2 \ \Omega_3 \rfloor^T,$ and $e_1 = \lfloor 1 \ 0 \ 0 \rfloor^T; f = \lfloor f_1 \ f_2 \ f_3 \rfloor^T; m = \lfloor m_1 \ m_2 \ m_3 \rfloor^T; \gamma_{11}$ is the extensional strain of the reference line; $2\gamma_{12}$ and $2\gamma_{13}$ are the transverse shear measures.

All the unknowns are functions of x_1 and t. Recall that the various indexed scalar variables have the following meanings: $F_i = \mathbf{F} \cdot \mathbf{B}_i$ with $\mathbf{F}(x_1,t)$ being the resultant force of all tractions on the cross-sectional face at a particular value of x_1 along the reference line.

$M_i = \mathbf{M} \cdot \mathbf{B}_i$ with $\mathbf{M}(x_1,t)$ being the resultant moment about the reference line at a particular value of x_1 of all tractions on the cross-sectional face, $K_i = \mathbf{K} \cdot \mathbf{B}_i$ with $\mathbf{K}(x_1,t)$ being the curvature of the deformed beam reference line at a particular value of x_1 such that $\mathbf{B}'_i = \mathbf{K} \times \mathbf{B}_i$, $V_i = \mathbf{V} \cdot \mathbf{B}_i$ with $\mathbf{V}(x_1,t)$ being the inertial velocity of a point at a particular value of x_1 on the deformed beam reference line, $\Omega_i = \mathbf{\Omega} \cdot \mathbf{B}_i$ with $\mathbf{\Omega}(x_1,t)$ being the inertial angular velocity of the deformed beam cross-sectional frame such that $\dot{\mathbf{B}}_i = \mathbf{\Omega} \times \mathbf{B}_i$, $P_i = \mathbf{P} \cdot \mathbf{B}_i$ with $\mathbf{P}(x_1,t)$ being the inertial linear momentum of the material points that make up the deformed beam reference cross section at a particular value of x_1, $H_i = \mathbf{H} \cdot \mathbf{B}_i$ with $\mathbf{H}(x_1,t)$ being the inertial angular momentum of all the material points that make up a reference cross section of the deformed beam about the reference line of that cross section at a particular value of x_1, $f_i = \mathbf{f} \cdot \mathbf{B}_i$ with $\mathbf{f}(x_1,t)$ being the applied distributed force per unit length, and $m_i = \mathbf{m} \cdot \mathbf{B}_i$ with $\mathbf{m}(x_1,t)$ being the applied distributed moment per unit length.

11.7.2   CONSTITUTIVE EQUATIONS

It is not necessary to retain all the variables in Eq. (94). For the purposes of the present discussion, the generalized strains and momenta will be eliminated. For beams with solid cross sections (and strip-beams with small axial loading), we have small strain and small local rotation; see Hodges (2006). Thus, the constitutive equations are linear and are written in the form

$$\begin{Bmatrix} \gamma_{11} \\ 2\gamma_{12} \\ 2\gamma_{13} \\ \kappa_1 \\ \kappa_2 \\ \kappa_3 \end{Bmatrix} = \begin{bmatrix} R_{11} & R_{12} & R_{13} & S_{11} & S_{12} & S_{13} \\ R_{12} & R_{22} & R_{23} & S_{21} & S_{22} & S_{23} \\ R_{13} & R_{23} & R_{33} & S_{31} & S_{32} & S_{33} \\ S_{11} & S_{21} & S_{31} & T_{11} & T_{12} & T_{13} \\ S_{12} & S_{22} & S_{32} & T_{12} & T_{22} & T_{23} \\ S_{13} & S_{23} & S_{33} & T_{13} & T_{23} & T_{33} \end{bmatrix} \begin{Bmatrix} F_1 \\ F_2 \\ F_3 \\ M_1 \\ M_2 \\ M_3 \end{Bmatrix} \tag{95}$$

where $\kappa = K - k$ and $k = \lfloor k_1 \ k_2 \ k_3 \rfloor^T$, $k_i = \mathbf{k} \cdot \mathbf{b}_i$, and $\mathbf{k}(x_1)$ is the initial curvature/twist vector at x_1, such that $\mathbf{b}'_i = \mathbf{k} \times \mathbf{b}_i$. Thus, k_1 is the initial twist and k_2 and k_3 are the initial curvature measures. Here the coefficients $R_{11}, R_{12}, \ldots, T_{33}$ are cross-sectional flexibility coefficients. This equation may also be written as

$$\begin{Bmatrix} \gamma \\ \kappa \end{Bmatrix} = \begin{bmatrix} R & S \\ S^T & T \end{bmatrix} \begin{Bmatrix} F \\ M \end{Bmatrix} \tag{96}$$

Similarly, the generalized momentum-velocity relations are

$$\begin{Bmatrix} P_1 \\ P_2 \\ P_3 \\ H_1 \\ H_2 \\ H_3 \end{Bmatrix} = \begin{bmatrix} \mu & 0 & 0 & 0 & \mu\bar{x}_3 & -\mu\bar{x}_2 \\ 0 & \mu & 0 & -\mu\bar{x}_3 & 0 & 0 \\ 0 & 0 & \mu & \mu\bar{x}_2 & 0 & 0 \\ 0 & -\mu\bar{x}_3 & \mu\bar{x}_2 & i_2+i_3 & 0 & 0 \\ \mu\bar{x}_3 & 0 & 0 & 0 & i_2 & i_{23} \\ -\mu\bar{x}_2 & 0 & 0 & 0 & i_{23} & i_3 \end{bmatrix} \begin{Bmatrix} V_1 \\ V_2 \\ V_3 \\ \Omega_1 \\ \Omega_2 \\ \Omega_3 \end{Bmatrix} \tag{97}$$

where μ is the mass per unit length, $\bar{x}_2$ and $\bar{x}_3$ are offsets from the reference line of the cross-sectional mass centroid, and i_2, i_3, and i_{23} are the cross-sectional mass moments and product of inertia, respectively. This equation may also be written as

$$\begin{Bmatrix} P \\ H \end{Bmatrix} = \begin{bmatrix} \mu I & -\mu\widetilde{\bar{x}} \\ \mu\widetilde{\bar{x}} & I \end{bmatrix} \begin{Bmatrix} V \\ \Omega \end{Bmatrix} \tag{98}$$

where $\bar{x} = \lfloor 0 \ \bar{x}_2 \ \bar{x}_3 \rfloor^T$.

11.7.3 CLOSING THE FORMULATION

This formulation can be closed by using a set of kinematical relations. For this we introduce displacement variables $u = \lfloor u_1 \ u_2 \ u_3 \rfloor^T$ and express the change in orientation in terms of the direction cosine matrix C. The kinematical relations are a set of generalized strain-displacement equations that relate γ and κ to u and C and a set of generalized velocity-displacement equations that relate V and Ω to u and C. See Hodges (1990, 2003, 2006) for additional details.

Generalized Strain-Displacement Equations

The generalized strain-displacement relations are of the form

$$\gamma = C\left(e_1 + u' + \widetilde{k}u\right) - e_1 \tag{99}$$

and

$$\widetilde{\kappa} = -C'C^T + C\widetilde{k}C^T - \widetilde{k} \tag{100}$$

Generalized Velocity-Displacement Equations

The generalized velocity-displacement relations are of a similar form, viz.,

$$V = C(v + \dot{u} + \widetilde{\omega}u) \tag{101}$$

and

$$\widetilde{\Omega} = -\dot{C}C^T + C\widetilde{\omega}C^T \tag{102}$$

where v and ω are column matrices which contain measure numbers of the velocity and angular velocity of the frame of reference to which the beam is attached, expressed in the local undeformed beam cross-sectional frame basis. In the present applications these can be set equal to zero, but they are useful for handling rotating beams, for example.

Derivation of Intrinsic Kinematical Equations

Intrinsic kinematical equations can now be derived by careful elimination of all displacement and rotation variables from the generalized strain- and velocity-displacement equations. Starting with differentiation of Ω with respect to x_1, one obtains

$$\widetilde{\Omega}' = -\dot{C}'C^T - \dot{C}C^T + C'\widetilde{\omega}C^T + C\widetilde{\omega}'C^T + C\widetilde{\omega}C'^T \tag{103}$$

where $\omega' = -\widetilde{k}\omega$. Similarly, differentiation of κ with respect to t yields

$$\widetilde{\dot{\kappa}} = -\dot{C}'C^T - C'\dot{C}^T + \dot{C}\widetilde{k}C^T + C\widetilde{k}\dot{C}^T \tag{104}$$

These equations involve C', which can be expressed in terms of K using Eq. (100), and $\dot{C}$, which can be expressed in terms of Ω using Eq. (102). Both equations contain $\dot{C}'$, which can be eliminated with algebraic manipulation to yield a single relation between Ω' and $\dot{\kappa}$ that, when simplified, is given by

$$\Omega' = \dot{\kappa} + \widetilde{\Omega}K \tag{105}$$

In a similar manner, V is differentiated with respect to x_1, yielding

$$V' = C'(v + \dot{u} + \widetilde{\omega}u) + C(v' + \dot{u}' + \widetilde{\omega}'u + \widetilde{\omega}u') \tag{106}$$

and γ is differentiated with respect to t, leading to

$$\dot{\gamma} = \dot{C}\left(e_1 + u' + \widetilde{k}u\right) + C\left(\dot{u}' + \widetilde{k}\dot{u}\right) \tag{107}$$

This time we make use of Eq. (101) to write

$$v + \dot{u} + \widetilde{\omega}u = C^T V \tag{108}$$

and Eq. (99) to write

$$e_1 + u' + \widetilde{k}u = C^T(e_1 + \gamma) \tag{109}$$

and again Eq. (100) and (102) are used to eliminate C' and $\dot{C}$, respectively. We must also use the relation that

$$v' = \widetilde{\omega}e_1 - \widetilde{k}v \tag{110}$$

Now, we can eliminate $\dot{u}'$ and find a single relation between V' and $\dot{\gamma}$, namely,

$$V' = \dot{\gamma} + \widetilde{V}K + \widetilde{\Omega}(e_1 + \gamma) \tag{111}$$

Eqs. (94), (96), (98), (105) and (111) constitute a closed formulation. The beauty of the formulation is striking, especially with regard to the similarity in structure of the left-hand sides of Eq. (94a) and (105) and Eq. (94b) and (111). Moreover, this formulation can be used in the solution of a variety of problems. For example, for situations in which the applied loads f and m and the boundary conditions on F, M, V, and Ω are independent of u and C, these equations allow the solution of nonlinear dynamics problems without finite rotation variables. As these variables are frequently the source of the highest degree nonlinearities and a possible source of singularities or of the need for Lagrange multipliers or trigonometric functions, to be able to avoid finite rotation variables can be quite advantageous. Finally, this formulation leads to explicit expressions for two conservation laws, which may have practical applications in development of computational algorithms; see Hodges (2003, 2006). An example is now presented showing advantages of the formulation for stability problems involving nonconservative forces.

11.7.4 EXAMPLES SHOWING ADVANTAGES OF THE INTRINSIC FORMULATION

In this section the utility of the fully intrinsic formulation will be addressed for problems involving nonconservative forces. Two examples are presented: first, Beck's column, and second, the deep cantilever with a tip transverse follower force.

Considering a prismatic and isotropic beam with the mass centroid coincident with the reference line and the principal axes of the cross section along the $\mathbf{b}_2$ and $\mathbf{b}_3$ directions, the constitutive law becomes

$$
\begin{Bmatrix} \gamma_{11} \\ 2\gamma_{12} \\ 2\gamma_{13} \\ \kappa_1 \\ \kappa_2 \\ \kappa_3 \end{Bmatrix} = \begin{bmatrix} \frac{1}{EA} & 0 & 0 & 0 & 0 & 0 \\ 0 & \frac{1}{GA_2} & 0 & 0 & 0 & 0 \\ 0 & 0 & \frac{1}{GA_3} & 0 & 0 & 0 \\ 0 & 0 & 0 & \frac{1}{GJ} & 0 & 0 \\ 0 & 0 & 0 & 0 & \frac{1}{EI_2} & 0 \\ 0 & 0 & 0 & 0 & 0 & \frac{1}{EI_3} \end{bmatrix} \begin{Bmatrix} F_1 \\ F_2 \\ F_3 \\ M_1 \\ M_2 \\ M_3 \end{Bmatrix} \tag{112}
$$

and the cross-sectional generalized momentum-velocity relations are

$$
\begin{Bmatrix} P_1 \\ P_2 \\ P_3 \\ H_1 \\ H_2 \\ H_3 \end{Bmatrix} = \begin{bmatrix} \mu & 0 & 0 & 0 & 0 & 0 \\ 0 & \mu & 0 & 0 & 0 & 0 \\ 0 & 0 & \mu & 0 & 0 & 0 \\ 0 & 0 & 0 & i_2 + i_3 & 0 & 0 \\ 0 & 0 & 0 & 0 & i_2 & 0 \\ 0 & 0 & 0 & 0 & 0 & i_3 \end{bmatrix} \begin{Bmatrix} V_1 \\ V_2 \\ V_3 \\ \Omega_1 \\ \Omega_2 \\ \Omega_3 \end{Bmatrix} \tag{113}
$$

Beck's Column

For Beck's problem, the constitutive relations, along with a consistent linearization of Eqs. (94), (105), and (111), are used to produce governing equations for the stability of small motions about the static equilibrium state. In the static equilibrium state, the beam is subject to axial compression by a follower force P as depicted in Fig. 11.7, so that the equilibrium state is $\overline{F} = -Pe_1$ and all other variables are equal to zero. Ignoring rotary inertia ($i_2 = i_3 = 0$) except in the torsional equation, considering infinite axial and shearing rigidities ($1/(EA) = 1/(GA_2) = 1/(GA_3) = 0$), and letting $V(x_1,t) = \overline{V}(x_1) + \hat{V}(x_1,t)$ and similarly for all other variables, one obtains the following equations, linearized in the ($\hat{\ }$) quantities:

$$
\begin{aligned}
\hat{V}_2' &= \hat{\Omega}_3 \\
\hat{\Omega}_3' &= \frac{\dot{\hat{M}}_3}{EI_3} \\
\hat{F}_2' &= \mu \dot{\hat{V}}_2 + \frac{P\hat{M}_3}{EI_3} \\
\hat{M}_3' &= -\hat{F}_2
\end{aligned} \tag{114}
$$

where the planar deformation is assumed to take place in the x_1-x_2 plane and EI_3 is the smallest bending stiffness of the beam. These equations can be collapsed into a single equation for any of the variables. When written in terms of $\hat{V}_2$, these equations reduce to a single, fourth-order equation which, together with its boundary conditions, is of identical form to those of the displacement-based analysis in Section 11.3.

Deep Cantilever with Tip Transverse Follower Force

The problem of Section 11.6 provides an interesting illustration of the utility of the subject methodology for follower-force problems. In this problem, a cantilevered beam is loaded with a transverse follower force at its tip, as shown in Fig. 11.15. These relations, along with an exact linearization of Eqs. (94), (105), and (111), are used to produce governing equations for the stability of small motions about the static equilibrium state. Ignoring rotary inertia ($i_2 = i_3 = 0$) except in the torsional equation, considering infinite axial and shearing rigidities ($1/(EA) = 1/(GA_2) = 1/(GA_3) = 0$), and letting $V(x_1,t) = \overline{V}(x_1) + \hat{V}(x_1,t)$ and similarly for all other variables, one obtains the following equations, linearized in the (^) quantities:

$$
\hat{M}_1' + \overline{M}_3 \hat{M}_2 \left(\frac{1}{EI_2} - \frac{1}{EI_3} \right) - (i_2 + i_3)\dot{\hat{\Omega}}_1 = 0
$$

$$
\hat{M}_2' - \overline{M}_3 \hat{M}_1 \left(\frac{1}{GJ} - \frac{1}{EI_3} \right) - \hat{F}_3 = 0
$$

$$
\hat{F}_3' + \frac{\overline{F}_2 \hat{M}_1}{GJ} - \frac{\overline{F}_1 \hat{M}_2}{EI_2} - \mu \dot{\hat{V}}_3 = 0
$$

$$
\hat{\Omega}_1' - \frac{\overline{M}_3 \hat{\Omega}_2}{EI_3} - \frac{\dot{\hat{M}}_1}{GJ} = 0 \tag{115}
$$

$$
\hat{\Omega}_2' + \frac{\overline{M}_3 \hat{\Omega}_1}{EI_3} - \frac{\dot{\hat{M}}_2}{EI_2} = 0
$$

$$
\hat{V}_3' + \hat{\Omega}_2 = 0
$$

where the equilibrium state is governed by three first-order equations given by

$$
\overline{F}_1' - \frac{\overline{M}_3 \overline{F}_2}{EI_3} = 0
$$

$$
\overline{F}_2' + \frac{\overline{M}_3 \overline{F}_1}{EI_3} = 0 \tag{116}
$$

$$
\overline{M}_3' + \overline{F}_2 = 0
$$

The compactness and ease of derivation are noteworthy, as are the beauty and symmetry of the final equations. To appreciate the simplicity of the above formulation, one should compare it with the equations of Detinko (2002). It is clear that the present formulation is considerably simpler. The simplicity of the present formulation for this problem stems from the boundary conditions' independence of displacement and orientation variables, a property typical of follower force problems.

PROBLEMS

1. Show that a pinned-pinned shaft loaded by tangential torques on both ends only loses its stability by buckling as predicted by the static method in Section 8.4.2.
2. Determine the effect of viscous damping on the stability of a shaft undergoing twist from a tangential torque. For a one-term approximation, show that the system is stable for $|q| \le 2\zeta\alpha_1^4/A$ where ζ is the damping ratio and $A = -\int_0^1 \psi\psi''' dx = 3.78953$.

3. Add viscous damping terms to each discrete equation in the two-mode model for Beck's column derived in the text. In particular, show that damping can be destabilizing if one of the damping ratios is sufficiently large (or small) relative to the other.

4. Consider a free-free beam loaded axially in compression by a follower force P applied at one end. Show that the axial force is $P(\ell - x_1)$ and that the boundary conditions are $u_2''(0,t) = u_2'''(0,t) = u_2''(\ell,t) = u_2'''(\ell,t) = 0$. Using a four-term Galerkin approximation (two rigid-body modes and two "elastic" modes), determine the critical load P_{cr} and the nature of the instability.

5. Consider a cantilevered beam lodaded by a transverse follower force at its tip, as shown in Fig. 11.15 Using Eqs. (116) with $\overline{F}_1(\ell) = \overline{M}_1(\ell) = 0$ and $\overline{F}_2(\ell) = P$, determine the steady-state solution. Compare with the approximate steady-state solution found in Eq. (87). Next, using Eqs. (115), determine the critical value of P. Compare your results to those in Section 11.6.2.

6. Using the fully intrinsic method, analyze a cantilevered deep beam undergoing a follower moment at the free end that leads to static, steady-state bending in the plane of greatest flexural rigidity. Determine the value of the moment at which flutter occurs using a Galerkin approximation with one term for each of the unknowns.

REFERENCES

Barsoum, R. S. (1971). Finite element method applied to the problem of stability of a nonconservative system. *International Journal for Numerical Methods in Engineering* 3, 63–87.

Beck, M. (1952). Die Knicklast des einseitig eingespannten, tangential gedrückten Stabes. *ZAMP* 3, 225–288.

Bolotin, V. V. (1959). On vibrations and stability of bars under the action of non-conservative forces. In Kolebaniia v turbomashinakh, 23–42. USSR.

Bolotin, V. V. (1963). *Nonconservative Problems of the Theory of Elastic Stability*. Pergamon Press, New York.

Bolotin, V. V. (1964). *Dynamic Stability of Elastic Systems*. Holden-Day, Inc., San Francisco.

Celep, Z. (1979). On the lateral stability of a cantilever beam subjected to a non-conservative load. *Journal of Sound and Vibration* 64, 173–178.

Chen, L.-W., and Ku, D.-M. (1992). Eigenvalue sensitivity in the stability analysis of Beck's column with a concentrated mass at the free end. *Journal of Sound and Vibration* 153, 403–411.

Como, M. (1966). Lateral buckling of a cantilever subjected to a transverse follower force. *International Journal of Solids and Structures* 2, 515–523.

Detinko, F. M. (2002). Some phenomena for lateral flutter of beams under follower load. *International Journal of Solids and Structures* 39, 341–350.

Herrmann, G. (1967). Stability of equilibrium of elastic systems subjected to nonconservative forces. *Applied Mechanics Reviews* 103–108.

Higuchi, K. (1994). An experimental model of a flexible free-free column in dynamic instability due to an end thrust. In Proceedings of the 35th Structures, Structural Dynamics, and Materials Conference, Hilton Head, South Carolina, 2402–2408. AIAA, Reston, Virginia.

Hodges, D. H. (1987). Finite rotation and nonlinear beam kinematics, *Vertica*, 11, 297–307.

Hodges, D. H. (1990). A mixed variational formulation based on exact intrinsic equations for dynamics of moving beams, *International Journal of Solids and Structures*, 26, 1253–1273.

Hodges, D. H. (2001). Lateral-torsional flutter of a deep cantilever loaded by a lateral follower force at the tip. *Journal of Sound and Vibration* 247, 175–183.

Hodges, D. H. (2003). Geometrically-exact, intrinsic theory for dynamics of curved and twisted anisotropic beams. *AIAA Journal* 41, 1131–1137.

Hodges, D. H. (2006). *Nonlinear Composite Beam Theory for Engineers*. American Institute of Aeronautics and Astronautics, Reston, VA. Chapts. 4 and 5.

Langthjem, M. A., and Sugiyama, Y. (2000). Dynamic stability of columns subjected to follower loads: a survey. *Journal of Sound and Vibration* 238, 809–851.

Leipholz, H. H. E. (1975). On the influence of damping on the stability of elastic beams subjected to uniformly distributed compressive forces. In Proceedings of the Third Brazilian Congress of Mechanical Engineering.

Leipholz, H. H. E. (1978). On variational principles for non-conservative mechanical systems with follower forces. In "Variational Methods in the Mechanics of Solids," Nemat-Nasser, S., ed., 151–155. Pergamon Press.

Park, Y. P. (1987). Dynamic stability of a free Timoshenko beam under a controlled follower force. *Journal of Sound and Vibration* 113, 407–415.

Wohlhart, K. (1971). Dynamische Kippstabilität eines Plattenstreifens unter Folgelast. *Zeitschrift für Flugwissenschaften* 19, 291–298.

Ziegler, H. (1968). *Principles of Structural Stability*. Blaisdell Publishing Company, Waltham, Massachusetts.

12

DYNAMIC STABILITY*

12.1   INTRODUCTION AND FUNDAMENTAL CONCEPTS

Dynamic stability or instability of elastic structures has drawn considerable attention in the past 40 years. The beginning of the subject can be traced to the investigation of Koning and Taub (1933), who considered the response of an imperfect (half-sine wave), simply supported column subjected to a sudden axial load of specified duration. Since then, many studies have been conducted by various investigators on structural systems that are either suddenly loaded or subjected to time-dependent loads (periodic or nonperiodic), and several attempts have been made to find common response features and to define critical conditions for these systems. As a result of this, the term *dynamic stability* encompasses many classes of problems and many different physical phenomena; in some instances the term is used for two distinctly different responses for the same configuration subjected to the same dynamic loads. Therefore, it is not surprising that there exist several uses and interpretations of the term.

In general, problems that deal with the stability of motion have concerned researchers for many years in many fields of engineering. Definitions for stability and for the related criteria and estimates of critical conditions as developed through the years are given by Stoker (1955). In particular, the contributions of Routh (1975) Thompson and Tait (1923) deserve particular attention. Some of these criteria find wide uses in problems of control theory (Lefschetz, 1965), of stability and control of aircraft (Seckel, 1964), and in other areas (Crocco and Cheng, 1956).

The class of problems falling in the category of parametric excitation, or parametric resonance, includes the best defined, conceived, and understood problems of dynamic stability. An excellent treatment and bibliography can be found in the book by Bolotin (1964). Another reference on the subject is Stoker's book (1950).

The problem of parametric excitation is best defined in terms of an example. Consider an Euler column, which is loaded at one end by a periodic axial force. The other end is immovable. It can be shown that, for certain relationships between the exciting frequency and the column natural frequency of transverse vibration, transverse vibrations occur with rapidly increasing amplitudes. This is called parametric resonance and the system is said to be dynamically unstable. Moreover, the loading is called parametric excitation.

Other examples of parametric resonance include (1) a thin flat plate parametrically loaded by inplane forces, which may cause transverse plate vibrations; (2) parametrically loaded shallow arches (symmetric loading), which under certain conditions vibrate asymmetrically with increasing amplitude; and (3) long cylindrical, thin shells (or thin rings) under uniform but periodically applied pressure, which can excite vibrations in an asymmetric mode. Thus, it is seen that, in parametric excitation, the loading is parametric with respect to certain deformation forms. This makes parametric resonance different from the usual forced vibration resonance. In addition, from these few examples of parametric excitation, one realizes that systems that exhibit bifurcational buckling under static conditions (regardless of whether the bifurcating static equilibrium branch is stable or unstable) are subject to parametric excitation.

Moreover, there exists a large class of problems for which the load is applied statically but the system is nonconservative. An elastic system is conservative when subjected to conservative loads; the reader is also referred to Ziegler's book (1968) for a classification of loads and reactions. An excellent review on the subject of stability of elastic systems under nonconservative forces is given by Herrmann (1967). He classifies all problems of nonconservative systems into three groups. The first group deals with follower-force problems, the second with problems of rotating shafts (whirling), and the third with aeroelasticity (fluid–solid interaction; flutter). All of these groups, justifiably or not, are called problems of dynamic stability. In the opinion of the author, justification is needed only for the first group. Ziegler (1956) has shown that critical conditions for this group of nonconservative systems can be obtained only through the use of the dynamic or kinetic approach to stability problems. The question of applicability of the particular approach was clearly presented by Herrmann and Bungay (1964) through a two-degree-of-freedom model. They showed that in some nonconservative systems, there exist two instability mechanisms, one of divergence (large deflections may occur) and one of flutter (oscillations of increasing amplitude). They further showed that the critical load for which the flutter type of instability occurs can be determined only through the kinetic approach, as illustrated in Chapter 11, while the divergence type of critical load can be determined by employing any one of the three approaches (classical, potential energy, or kinetic). It is understandable, then, why many authors refer to problems of follower-forced systems as dynamic stability problems. Furthermore, the problem of flow-induced vibrations in elastic pipes is another fluid–solid interaction problem that falls under the general heading of dynamic stability. The establishment of stability concepts, as well as of estimates for critical conditions, is an area of great practical importance. A few references (Au-Yang and Brown, 1977; Benjamin, 1961; Blevins, 1977; Chen, 1975a, b; Chen, 1978; Chen, 1981; Gregory, 1966; Hill and Swanson, 1970; Junger and Feit, 1972; King, 1977; Paidoussis, 1970; Paidoussis and Deksnis, 1970; Reusselet and Hermann, 1977; Scanlan and Simin, 1978) are provided for the interested reader. In addition, a few studies have been reported that deal with the phenomenon of parametric resonance in a fluid–structure interaction problem (Bohn and Hermann, 1974; Ginsberg, 1973; Paidoussis and Issid, 1976; Paidoussis and Sundararajan, 1975). For completeness, one should refer to a few studies of aeroelastic flutter (Dowell, 1969, 1970; Dugundji, 1972; Kornecki, 1970; Kuo et al., 1972).

Finally, a large class of structural problems that has received considerable attention and does qualify as a category of dynamic stability is that of impulsively loaded configurations and configurations that are suddenly loaded with loads of constant magnitude and infinite duration. These configurations under static loading are subject to either limit point instability or bifurcational instability with an unstable post-buckling branch (violent buckling). The two types of loads may be thought of as mathematical idealizations of blast loads of (1) large decay rates and small decay times and (2) small decay rates and large decay times, respectively. For these loads, the concept of dynamic stability is related to the observation that for sufficiently small values of the loading, the system simply oscillates about the near static equilibrium point and the corresponding amplitudes of oscillation are sufficiently small. If the loading is increased, some systems will experience large-amplitude oscillations or, in general, a divergent type of motion. For this phenomenon to happen, the configuration must possess two or more static equilibrium positions, and escaping motion occurs by having trajectories that can pass near an unstable static equilibrium point. Consequently, the methodologies developed by the various investigators are for structural configurations that exhibit snapthrough buckling when loaded quasi-statically.

Solutions to such problems started appearing in the open literature in the early 1950s. Hoff and Bruce (1954) considered the dynamic stability of a pinned half-sine arch under a half-sine distributed load. Budiansky and Roth (1962), in studying the axisymmetric behavior of a shallow spherical cap under suddenly applied loads, defined the load to be critical when the transient response increases suddenly with very little increase in the magnitude of the load. This concept was adopted by numerous investigators (Budiansky, 1967; Budiansky and Hutchinson, 1964; Hsu, 1967, 1968a, b, c; Simitses, 1974; Tamura and Babcock, 1975) in subsequent years because it is tractable to computer solutions. Finally, the concept was generalized in a paper by Budiansky (1967) in attempting to predict critical conditions for imperfection-sensitive structures under time-dependent loads.

Conceptually, one of the best efforts in the area of dynamic buckling under suddenly applied loads is the work of Hsu and his collaborators (Hsu, 1967, 1968a, b, c; Hsu et al., 1968). Hsu defined sufficiency conditions for stability and sufficiency conditions for instability, thus finding upper and lower bounds for the critical impulse or critical sudden load. Independently, Simitses (1965), in dealing with the dynamic buckling of shallow arches and spherical caps, termed the lower bound a minimum possible critical load (MPCL) and the upper bound a minimum guaranteed critical load (MGCL). Some interesting comments on dynamic stability are given by Hoff (1967). Finally, Thompson (1967) presented a criterion for estimating critical conditions for suddenly loaded systems.

The totality of concepts and methodologies used by the various investigators in estimating critical conditions for suddenly loaded elastic systems (of the last category) can be classified in the following three groups:

1. Equations of motion approach (Budiansky and Roth, 1962). The equations of motion are (numerically) solved for various values of the load parameter (ideal impulse, or sudden load), thus obtaining the system response. The load parameter at which there exists a large (finite) change in the response is called critical.
2. Total energy–phase plane approach (Hsu, 1967, 1968a, b, c; Hsu et al., 1968). Critical conditions are related to characteristics of the system's phase plane, and the emphasis is on establishing sufficient conditions for stability (lower bounds) and sufficient conditions for instability (upper bounds).
3. Total potential energy approach (Simitses, 1965). Critical conditions are related to characteristics of the system's total potential. Through this approach also, lower

and upper bounds of critical conditions are established. This last approach is applicable to conservative systems only. The concepts and procedure related to the last approach are explained next in some detail.

12.2   THE TOTAL POTENTIAL ENERGY APPROACH: CONCEPTS AND PROCEDURE

The concept of dynamic stability is best explained through a single-degree-of-freedom system. First, the case of ideal impulse is treated and then the case of constant load of infinite duration.

12.2.1   IDEAL IMPULSE

Consider a single-degree-of-freedom system for which the total potential (under zero load) curve is plotted versus the generalized coordinate (independent variable) (see Fig. 12.1). Clearly, points A, B, C denote static equilibrium points, and point B denotes the initial position ($\theta = 0$) of the system.

Since the system is conservative, the sum of the total potential $\overline{U}_T^o$ (under "zero" load) and the kinetic energy T^o is a constant C, or

$$\overline{U}_T^o + T^o = C \tag{1}$$

Moreover (see Fig. 12.1), since $\overline{U}_T^o$ is zero at the initial position ($\Theta = 0$), the constant C can be related to some initial kinetic energy T_i^o. Then

$$\overline{U}_T^o + T^o = T_i^o \tag{2}$$

Next, consider an ideal impulse applied to the system. Through the impulse–momentum theorem, the impulse is related to the initial kinetic energy T_i^o. Clearly, if T_i^o is equal to D (see Fig. 12.1) or $\overline{U}_T^o(\Theta_{II})$, the system will simply oscillate between Θ_I and Θ_{II}. On the other hand, if the initial kinetic energy T_i^o is equal to the value of the total potential at the unstable static equilibrium point C, $\overline{U}_T^o(C)$, then the system can reach point C with zero velocity ($T^o = 0$), and there exists a possibility of motion escaping (passing position C) or becoming unbounded. Such a motion is termed *buckled motion*. In the case for which motion is bounded and the path may include the initial point (B), the motion is termed *unbuckled motion*.

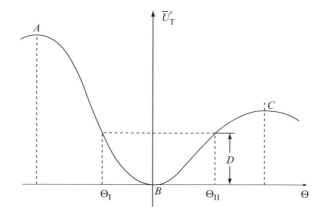

FIGURE 12.1   Total potential curve (zero load, one degree of freedom).

Through this, a concept of dynamic stability is presented and the necessary steps for estimating critical impulses are suggested. Note that once the unstable static equilibrium positions (points A and C) are established, the critical initial kinetic energy is estimated by

$$T_{\text{icr}}^{\text{o}} = \overline{U}_{\text{T}}^{\text{o}}(C) \tag{3}$$

Moreover, since T_i^{o} is related to the ideal impulse, the critical impulse is estimated through Eq. (3). Observe that an instability of this type can occur only when the system, under zero load, possesses unstable static equilibrium points. Furthermore, if position C corresponds to a very large and thus unacceptable position Θ (from physical considerations), one may still use this concept and estimate a maximum allowable (and therefore critical) ideal impulse. For instance, if one restricts motion to the region between Θ_{I} and Θ_{II}, the maximum allowable ideal impulse is obtained from Eq. (3), but with D or $\overline{U}_{\text{T}}^{\text{o}}(\Theta_{\text{II}})$ replacing $\overline{U}_{\text{T}}^{\text{o}}(C)$. Because of this, a critical or an allowable ideal impulse can be obtained for all systems (including those that are not subject to buckling under static conditons such as beams, shafts, etc.).

A similar situation exists for a two-degree-of-freedom system. Fig. 12.2 depicts curves of total potential lines corresponding to the value of the total potential at points A and A', in the space of $(\phi + \theta)$ and $(\phi - \theta)$. Note that ϕ and θ denote the two generalized coordinates. The total potential curves are typical for the mechanical model shown in Fig. 12.3. Further, note that $\phi = 0$ denotes symmetric behavior (see Simitses 1990). Returning to Fig. 12.2, points O, A, A', B, and C denote static equilibrium under zero load. The value of the total potential in the shaded area is smaller than the value of the total potential at points A and A'. Everywhere else the total potential is larger than that at points A and A'. Positions O and C correspond to stable static equilibrium positions, while positions A and A' (saddle points) and B correspond to unstable ones. Point O is the starting or natural unloaded position.

Here also, as in the case of the one-degree-of-freedom system, the work done by the ideal impulsive load is imparted into the system instantaneously as initial kinetic

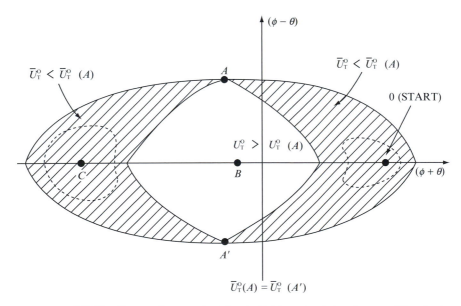

FIGURE 12.2   Curves of constant total potential (zero load, two degrees of freedom).

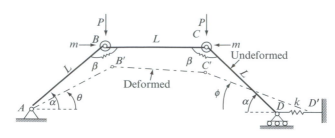

energy. Through the impulse–momentum theorem, the ideal impulse can be related to the initial kinetic energy. For small values of the initial kinetic energy, motion is confined to some small area surrounding the starting (stable) static point, O. In this case, there is no possibility of buckled motion. If the kinetic energy value is equal to the value of the total potential at point A (or A'), then there *exists a possibility* of the system reaching either position A or A' with zero kinetic energy and motion to take place toward point C (buckled motion). Note that if the value of the initial kinetic energy is equal to the value of the total potential at the unstable static position B (top of the hill), $U_T^o(B)$, then motion toward the far stable position C (buckled) is guaranteed. Thus, in the first case a possibility of escaping motion exists, while in the latter case escaping motion is guaranteed.

On the basis of the above explanations for both systems, the following definitions may now be given:

Possible locus or path. A possible locus or path on the total potential surface is one that corresponds at every point of the locus or path to a nonnegative kinetic energy.

Unbuckled motion. Unbuckled motion of the system is defined as any possible locus or path on the total potential surface that passes through or completely encloses only the near equilibrium point.

Note that for the case of the ideal impulse, the near static equilibrium point is also the starting point.

Buckled motion. If the possible locus or path passes through or encloses other static equilibrium points (stable or unstable), then the motion is defined as buckled.

Note that buckled motion may also be referred to as escaping motion, and the phenomenon as dynamic snap through buckling.

Minimum possible critical load (MPCL). The least upper bound of loads for which all possible loci correspond only to unbuckled motion. At the MPCL there exists at least one possible locus on the potential surface that the structure can follow to dynamically snap through.

Note that Hsu (1967, 1968a, b, c) and Hsu et al. (1968) refer to this bound as a sufficiency condition for stability (dynamic).

Minimum guaranteed critical load (MGCL). The greatest lower bound of loads for which no possible loci correspond to unbuckled motion.

Note that, in this case, dynamic snapthrough will definitely happen. This bound is termed by Hsu a sufficiency condition for instability (dynamic).

From the above definitions, it is evident that for one-degree-of-freedom systems, the two critical loads are coincident. On the other hand, for multi-degree-of-freedom systems, the critical load can only be bracketed between an upper bound (MGCL) and a lower bound (MPCL).

One final comment for the case of ideal impulse: Note from Fig. 12.1, in the absence of damping (as assumed), the direction of the ideal impulse is immaterial. If the system is loaded in one direction (say that the resulting motion corresponds to positive Θ), then a critical condition exists when the system reaches position C with zero kinetic energy. If the system is loaded in the opposite direction, then some negative position will be reached with zero kinetic energy; after that the direction of the motion will reverse, and finally the system will reach position C with zero kinetic energy. Both of these phenomena occur for the same value of the ideal impulse. This is also true for the two-degree-of-freedom system (see Fig. 12.2 and 12.3). The critical load (ideal impulse) is not affected by the direction of the loading. (Starting at point O, the system will initially move along the $\theta + \phi$ axis, to the right or to the left.)

12.2.2   CONSTANT LOAD OF INFINITE DURATION

Consider again a single-degree-of-freedom system. Total potential curves are plotted versus the generalized coordinate on Fig. 12.4. Note that the various curves correspond to different load values P_i. The index i varies from 1 to 5, and the magnitude of the load increases with increasing index value. These curves are typical of systems that, for each load value, contain at least two static equilibrium points A_i and B_i. This is the case when the system is subject to limit point instability and/or bifurcational buckling with unstable branching, under static application of the load (shallow arches and spherical caps, perfect or imperfect cylindrical and spherical shells, two-bar frames, etc.).

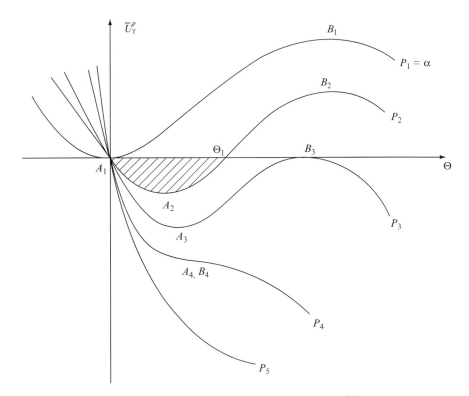

FIGURE 12.4   Total potential curves (one degree of freedom).

Given such a system, one applies a given load suddenly with constant magnitude and infinite duration. For a conservative system,

$$\overline{U}_T^P + T^P = C \tag{4}$$

The potential may be defined in such a way that it is zero at the initial position ($\Theta = 0$). In such a case, the constant is zero, or

$$\overline{U}_T^P + T^P = 0 \tag{5}$$

Since the kinetic energy is a positive definite function of the generalized velocity, motion is possible when the total potential is nonpositive (shaded area, in Fig. 12.4, for P_2). From this it is clear that for small values of the applied load, the system simply oscillates about the near static equilibrium position (point A_2). This is also an observed physical phenomenon. As the load increases, the total potential at the unstable point B_i decreases, it becomes zero (point B_3), and then it increases negatively until points A_i and $B_1(A_4, B_4)$ coincide (the corresponding load P_4 denotes the limit point under static loading). For loads higher than this (P_4), the stationary points (static equilibrium positions) disappear from the neighborhood. When the sudden load reaches the value corresponding to P_3, a critical condition exists, because the system can reach position B_3 with zero kinetic energy and then move toward larger Θ values (buckled motion can occur). Thus, P_3 is a measure of the critical condition. Note that the value of P_3 is smaller than the value of the limit point P_4. This implies that the critical load under sudden application (infinite duration) is smaller than the corresponding static critical load.

In this case, also, one may wish to limit the dynamic response of the system to a value smaller than B_3 (see Fig. 12.4), say Θ_1. Then in such a case, the maximum allowable and consequently the critical dynamic load is denoted by P_2.

As in the case of the ideal impulse, for systems of two or more degrees of freedom, upper and lower bounds of critical conditions can be established.

Figure 12.5 shows typical constant-potential (zero) lines in the space of the generalized coordinates $\phi + \theta$ and $\phi - \theta$ (see Fig. 12.3). In this case, the total potential is defined such that the constant in Eq. (4) is zero (in the absence of initial kinetic energy). Points O, A, A', B, and C denote the same static equilibrium positions as in Fig. 12.2. Note that in the shaded areas the total potential is nonpositive, and everywhere else the total potential is nonnegative. In Fig. 12.5a, the motion is confined to the shaded area enclosing point O (starting point) and the motion is unbuckled. As the value of the load increases, the value of the total potential at A and A' continuously decreases until it becomes zero at load P_{II} (Fig. 12.5b). At this load, the system can possibly snap through toward the far equilibrium position C and either oscillate about C or return to region (OAA') and oscillate about O, and so on. Load, P_{II} denotes the lower bound for a critical condition (MPLC) because there exists at least one path through which the system can reach position A or A' with zero kinetic energy and possibly escape toward (snap through to) position C. As the value of the applied load is further increased, buckled motion will always occur at some value P_{III} (larger than P_{II}). This load denotes the upper bound. The system shown on Fig. 12.3 is analyzed in detail in a later section. For this load case also, the concepts of dynamic stability are applied to continuous structural systems in later sections.

Moreover, these concepts and the related methodologies for estimating critical conditions are modified and applied to the case of suddenly loaded systems with constant loads of finite duration. These modifications are also presented in Simitses (1990). In addition, since most structural configurations in service are subject

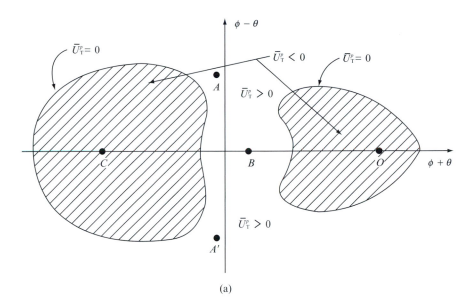

(a)

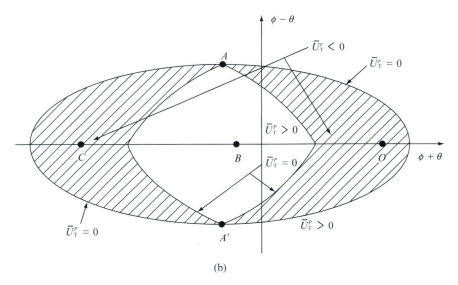

(b)

FIGURE 12.5 Curves of constant (zero) total potential (two degrees of freedom). (a) $|P_{\mathrm{I}}| < |P_{cr_\infty}|$ (MPCL). (b) $|P_{\mathrm{II}}| > |P_{\mathrm{I}}|$, $P_{\mathrm{II}} = P_{cr_\infty}$ (MPCL).

to static loads, the effect of static preloading on the critical dynamic conditions is presented by similar concepts (see Simitses 1990). An example of this would be a submarine resting at a depth of 1,000 feet (static preloading) and subjected to a blast loading (dynamic) of small decay rate and large decay time (sudden load of constant magnitude and infinite duration).

12.3 EXTENSION OF THE DYNAMIC STABILITY CONCEPT

The concept of dynamic stability, discussed in the previous section, is developed primarily for structural configurations that are subject to violent buckling under static loading. It is also observed that the concept can be extended, even for these systems, when one limits the maximum allowable deflection resulting from the sudden loads. This being the case, then, the extended and modified concept can be used for all structural configurations (at least in theory).

This is demonstrated in this section through a simple model. First, though, some clarifying remarks are in order.

All structural configurations, when acted on by quasi-static loads, respond in a manner described in one of Fig. 12.6 to 12.10. These figures characterize equilibrium positions (structural responses) as plots of a load parameter P versus some characteristic displacement θ. The solid curves denote the response of systems that are free of imperfections, and the dashed curves denote the response of the corresponding imperfect systems.

Figure 12.6 shows the response of such structural elements as columns, plates, and unbraced portal frames. The perfect configuration is subject to bifurcational buckling, while the imperfect configuration is characterized by stable equilibrium (unique), for elastic material behavior.

Figure 12.7 typifies the response of some simple trusses and two-bar frames. The perfect configuration is subject to bifurcational buckling, with a smooth (stable) branch in one direction of the response and a violent (unstable) branch in the other. Correspondingly, the response of the imperfect configuration is characterized by stable equilibrium (unique) for increasing load in one direction, while in the other, the system is subject to limit point instability.

Figure 12.8 typifies the response of troublesome structural configurations such as cylindrical shells (especially under uniform axial compression and of isotropic construction), pressure-loaded spherical shells, and some simple two-bar frames. These systems are imperfection-sensitive systems and are subject to violent buckling under static loading.

A large class of structural elements is subject to limit point instability. In some cases, unstable bifurcation is present in addition to the limit point. The response of

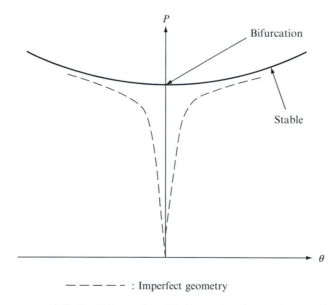

— — — — — : Imperfect geometry

FIGURE 12.6   Bifurcated equilibrium paths with stable branching.

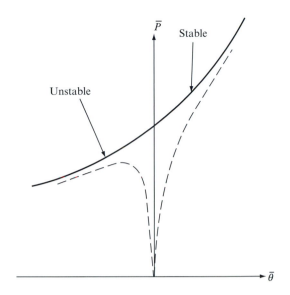

FIGURE 12.7 Bifurcated equilibrium paths with stable and unstable branches.

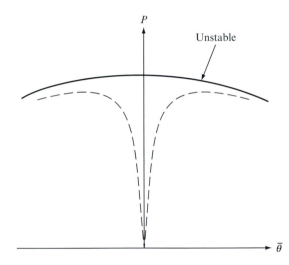

FIGURE 12.8 Bifurcated equilibrium paths with unstable branching.

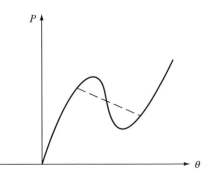

FIGURE 12.9 Snapthrough buckling paths (through limit point or unstable branching).

such systems is shown in Fig. 12.9. Two structural elements that behave in this manner are the shallow spherical cap and the low arch. Both elements have been used extensively in practice.

Finally, there is a very large class of structural elements that are always in stable equilibrium for elastic behavior and for all levels of the applied loads. These systems are not subject to instability under static conditions. Typical members of this class are beams and transversely loaded plates. For this class of structural elements, the load-displacement curve is unique and monotonically increasing (Fig. 12.10).

The concept of dynamic stability, as developed and discussed (see also Budiansky and Hutchinson, 1962; Hsu, 1967; Simitses, 1965), is always with reference to systems that are subject to violent buckling under static loading. This implies that dynamic buckling has been discussed for systems with static behavior as shown in Fig. 12.7 (to the left), 12.8, and 12.9.

In developing concepts and the related criteria and estimates for dynamic buckling, it is observed that, even for systems that are subject to violent (static) buckling, critical dynamic loads can be associated with *limitations* in deflectional response rather than escaping motion through a static unstable point. This is especially applicable to the design of structural members and configurations, which are deflection limited. From this point of view, then, the concept of dynamic stability can be extended to all structural systems.

The extended concepts are demonstrated through the simple mass-spring (linear) system shown in Fig. 12.11.

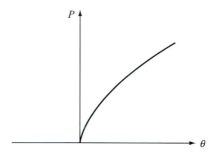

FIGURE 12.10 Unique stable equilibrium path.

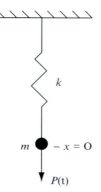

FIGURE 12.11 The mass-spring system.

12.3.1   THE MASS-SPRING SYSTEM

Consider the mass-spring (linear) system shown in Fig. 12.11. Consider a suddenly applied load $P(t)$ applied at $t = 0$. This load may, in general, include the weight (mg). In the case of finite duration, consider the weight to be negligible.

First, the problem of constant load suddenly applied with infinite duration is considered.

For this case, one may write the equation of motion and solve for the response by imposing the proper initial conditions:

$$\ddot{x} + \frac{k}{m}x = \frac{P}{m} \tag{6}$$

subject to

$$\dot{x}(0) = x(0) = 0 \tag{7}$$

where the dot denotes differentiation with respect to time.

By changing the dependent variable to

$$y = x + C \tag{8}$$

where C is a constant, the equation of motion and initial conditions become

$$\ddot{y} + \frac{k}{m}y = 0 \tag{9}$$

$$y(0) = -\frac{P}{k} \quad \text{and} \quad \dot{y}(0) = 0 \tag{10}$$

The solution is

$$y = \frac{P}{k}\cos\sqrt{\frac{k}{m}}t$$

and

$$x = \frac{P}{k}\left(1 - \cos\sqrt{\frac{k}{m}}t\right) \tag{11}$$

Note that

$$x_{\max} = \frac{2P}{k} \tag{12}$$

and it occurs at

$$\sqrt{\frac{k}{m}}t = \pi \quad \text{or at} \quad t = \pi\sqrt{\frac{m}{k}} = \frac{T_0}{2} \tag{13}$$

where T_0 is the period of vibration.

Note that if the load is applied quasi-statically, then

$$P_{\text{st}} = kx_{\text{st}} \tag{14}$$

From Eqs. (12) and (14), it is clear that if the maximum dynamic response $x_{\max}$ and maximum static deflection $x_{\text{st}_{\max}}$ are to be equal and no larger than a specified value X (deflection-limited response), then

$$P_{st} = 2P_{dyn} \qquad (15)$$

Because of this, many systems for which the design loads are dynamic in nature (suddenly applied of constant magnitude and infinite duration) are designed in terms of considerations but with design (static) loads twice as large as the dynamic loads, Eq. (15). Note that both loads (P_{st}, P_{dyn}) correspond to the same maximum (allowable) deflection X.

Next, the same problem is viewed in terms of energy considerations.

First, the total potential U_T^P for the system is given by

$$U_T^P = \frac{1}{2}kx^2 - Px \qquad (16)$$

and the kinetic energy T^P by

$$T^P = \frac{1}{2}m(\dot{x})^2 \qquad (17)$$

Note that the system is conservative, the kinetic energy is a positive definite function of the velocity (for all t), and $U_T^P = 0$ when $x = 0$. Then,

$$U_T^P + T = 0 \qquad (18)$$

and motion is possible only in the range of x values for which U_T is nonpositive (see shaded area of Fig. 12.12).

It is also seen from Eq. (16) that the maximum x value corresponds to $2P/k$.

Note that the static deflection is equal to P/k, Eq. (14) and point A on Fig. 12.12. Therefore, if the maximum dynamic response and maximum static deflection are to be equal to X, Eq. (15) must hold.

Now, one may develop a different viewpoint for the same problem. Suppose that a load P is to be applied suddenly to the mass-spring system with the condition that the maximum deflectional response cannot be larger than a specified value X. If the magnitude of the load is such that

$$\frac{2P}{k} < X \qquad (19)$$

we shall call the load dynamically subcritical.

When the inequality becomes an equality, we shall call the corresponding load dynamically critical. This implies that the system cannot withstand a dynamic load $P > kX/2$ without violating the kinematic constraint. Therefore,

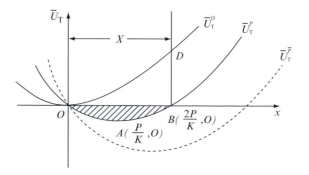

FIGURE 12.12   Total potential curves (suddenly loaded mass-spring system).

$$P_{\text{dyncr}} = \frac{kX}{2} \tag{20}$$

Moreover, on the basis of this concept, one may find a critical ideal impulse. The question, in this load case, is to find the ideal impulse such that the system response does not exceed a prescribed value X. From Fig. 12.12 and conservation of energy,

$$U_T^{\text{o}} + T^{\text{o}} = T_i^{\text{o}} \tag{21}$$

and T_i^o is critical if the system can reach position D with zero velocity (kinetic energy). Thus,

$$T_{i_{cr}}^{\text{o}} = U_T^{\text{o}}(D) = U_T^{\text{o}}(X) \tag{22}$$

From the impulse–momentum theorem, the ideal impulse, Imp, is related to the initial velocity and consequently to the initial kinetic energy:

$$\text{Imp} = \lim_{t \to 0} (Pt_0) = m\dot{x}_i \tag{23}$$

where $\dot{x}_i$ is the initial velocity magnitude (unidirectional case) and t_0 is the duration time of a square pulse.

From Eq. (23),

$$\dot{x} = \frac{\text{Imp}}{m} \tag{24}$$

and use of Eq. (17) yields

$$\dot{x}_i = \left(\frac{2T_i^o}{m}\right)^{\frac{1}{2}} \tag{25}$$

Since the critical initial kinetic energy is given by Eq. (22),

$$\text{Imp}_{cr} = (mk)^{\frac{1}{2}}X \tag{26}$$

Next, the following nondimensionalized parameters are introduced:

$$p = \frac{2P}{kX}, \quad \xi = \frac{x}{X}, \quad \tau = t\sqrt{\frac{k}{m}}$$
$$\overline{U}_T = \frac{2U_T}{kX^2}, \quad \overline{T} = \frac{2T}{kX^2}, \quad \overline{\text{Imp}} = \frac{2\,\text{Imp}}{X\sqrt{km}} \tag{27}$$

On the basis of this, Eq. (26) becomes

$$\overline{\text{Imp}}_{cr} = 2 \tag{28}$$

Similar examples, as well as this one, are treated in Simitses (1990) for the case of a step load of finite duration.

12.4   BEHAVIOR OF SUDDENLY LOADED SYSTEMS

So far, from the discussion of the subject of dynamic stability of suddenly loaded structural configurations, it is seen that the following phenomena are possible:

1. *Parametric resonance.* Systems that are subject to bifurcational buckling under static conditions are subject to parametric resonance if the loading is sudden (ideal impulse or constant load of infinite duration) and certain conditions are met.

   For example, a perfectly straight column (Wauer, 1980) loaded suddenly without eccentricity is subject to parametric resonance if the inplane motion is accounted for. What happens in this case, as shown by Wauer (1980), is that the inplane motion is periodic, which leads to a periodic axial force and yields a Mathieu–Hill type of equation for transverse motion and therefore a possibility of parametric resonance. (See also Simitses, 1990.)

   Similarly, a perfectly symmetric shallow arch loaded by a sudden symmetric loading may lead to parametric resonance in an antisymmetric mode because of the coupling between symmetric and antisymmetric modes.

   Parametric resonance for suddenly loaded systems is more of a theoretical possibility than an actual physical phenomenon, because neither the structural system nor the external (sudden) loading is free of imperfections.

   More details are presented in Appendix A of Simitses (1990). The material presented is by no means complete, but it suffices to make the discussion self-contained.

2. *Escaping motion type of instability.* Systems that are subject to the violent type of buckling under static loading can and do experience an escaping motion type of instability when suddenly loaded (Fig. 12.7 [to the left], 12.8, and 12.9). Examples of these include shallow arches, shallow spherical caps, certain two-bar frames, and imperfect cylindrical shells. The concepts discussed in Section 12.2 are for these systems. Moreover, the Budiansky–Roth criterion and the concepts developed by Hsu and his collaborators were developed for these systems. The physical phenomenon for this case is as follows: For small values of the load (suddenly applied) parameter, the system simply oscillates (linearly or nonlinearly) about the near static equilibrium point. As the load parameter is increased, a value is reached for which an escaping or large-amplitude motion is observed. This phenomenon is demonstrated through several examples in later sections and in Simitses (1990).

3. *Linear or nonlinear oscillatory motion.* Systems that under static loading are subject either to bifurcational (smooth) buckling with a stable postbuckling branch [Fig. 12.7 (to the right) and 12.6] or are not subject to buckling at all (Fig. 12.10) do not experience any type of dynamic instability. These systems, when suddenly loaded, simply oscillate about the stable static equilibrium position. Examples of these systems include the imperfect column, unbraced portal frames, and transversely loaded (suddenly) beams and thin plates.

   Note that, through the application of the extended concept of dynamic stability (Section 12.3), critical conditions (loads) may be found for these systems by imposing limitations on the dynamic response characteristics of the system (either a maximum allowable amplitude of vibrations, a maximum allowable inplane strain, or some other constraint).

12.5   SIMPLE MECHANICAL MODELS

Two single-degree-of-freedom mechanical models are employed in this section to demonstrate the concept of dynamic stability for the extreme cases of the ideal impulse and sudden constant load of infinite duration. These models are typical of imperfection-sensitive structural configurations. They are kept as simple as possible, so that the emphasis can easily be placed on the concepts rather than on complex mathematical theories. For each model, the static stability analysis, based on the total potential energy approach, is given in detail. In addition, the total energy phase plane approach is used for one model. For the same model, the equations of motion approach is also used, for demonstration and comparison purposes. The main emphasis, though, is placed on the total potential energy approach.

Finally, a few observations are presented that result from these simple studies but are general in applicability.

12.5.1   MODEL A: A GEOMETRICALLY IMPERFECT MODEL

Consider the model shown in Fig. 12.13. This model consists of two rigid bars of equal length L pinned together. The left bar is pinned on an immovable support A, while the right end of the second bar is pinned on a movable support C and loaded by a horizontal constant-directional force P. A linear spring of stiffness k connects the common pin B to an immovable support D, which is a distance L directly below support A. The initial geometric imperfection θ_o is an angle between the horizontal line joining supports A and C and bar AB (or BC). The deformed position is characterized by angle θ as shown (in its positive direction). For simplicity, the two rigid bars are assumed to be weightless, and the mass m of the system is concentrated at joint B.

12.5.1.1   Static Stability Analysis of Model A

The stability analysis of this model under quasi-static application of the load P is performed by employing the energy approach. Through this approach, equilibrium is characterized by

$$\frac{dU_T^p}{d\theta} = 0 \tag{29}$$

where U_T is the total potential, and the character of equilibrium (stable or unstable) by the sign of the second derivative.

The total potential is

$$\overline{U}_T^P = \frac{U_T^P}{kL^2} = \left(\sqrt{1+\sin\theta} - \sqrt{1+\sin\theta_o}\right)^2 - p(\cos\theta_o - \cos\theta) \tag{30}$$

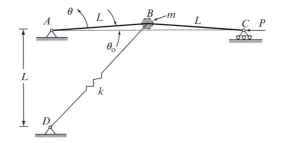

FIGURE 12.13   Geometry and sign convention for Model A.

where $p = 2P/kL$ and $\overline{U}_T^P$ denotes the nondimensionalized total potential. The superscript P implies "under load P."

The static equilibrium points are characterized by

$$p = \left(\sqrt{1 + \sin\theta} - \sqrt{1 + \sin\theta_o}\right) \frac{\cot\theta}{\sqrt{1 + \sin\theta}} \quad \text{for} \quad \theta_o \neq 0 \tag{31}$$

Note that, for $\theta_o = 0$ equilibrium is characterized by

$$\text{either } \theta = 0 \quad \text{or} \quad p = \cot\theta\left(\sqrt{1 + \sin\theta} - 1\right)/\sqrt{1 + \sin\theta} \tag{32}$$

Equilibrium positions are plotted in Fig. 12.14 as p versus $\theta - \theta_o$ for various values of the geometric imperfection θ_o. The stability test reveals that the dashed line positions are stable, while the solid line positions are unstable and snapping (violent buckling) takes place through the existence of a limit point. Also note that positions characterized by negative values for $\theta - \theta_o$ (not shown here) are stable and there is no possibility of buckling. Therefore, our interest lies in the area of $\theta_o > 0$ and $\theta > \theta_o$.

12.5.1.2 Dynamic Analysis: Ideal Impulse

Assume that the load P is suddenly applied with a very short duration time T_o and that the impulse (PT_o) is imparted instantaneously into the system as initial kinetic energy.

Through impulse–momentum, one obtains the following relation:

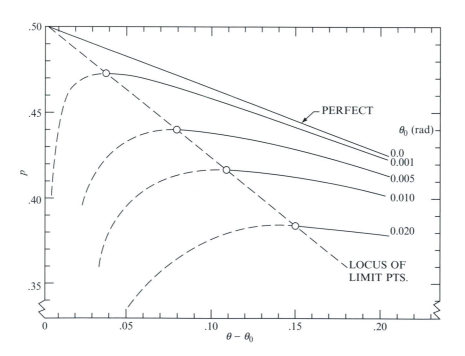

FIGURE 12.14   Load-displacement curves (Model A).

$$\lim_{T_o \to 0} (PT_o) = \frac{1}{2} \left(\frac{mL\dot{\theta}_o}{\sin \theta_o} \right) \tag{33}$$

where $\dot{\theta}_o = d\theta/dt$ at $\theta = \theta_o$.

Since the system is conservative,

$$U_T^o + T^o = \text{const.} = T_i^o \tag{34}$$

where U_T^o denotes the total potential "under zero load" and T^o is the kinetic energy, given by

$$T^o = \frac{1}{2}mL^2\dot{\theta}^2 \tag{35}$$

Note that T_i^o is the initial kinetic energy imparted instantaneously by the impulsive load.

The expression for $\overline{U}_T^o$ is given by

$$\overline{U}_T^o = \left(\sqrt{1 + \sin \theta} - \sqrt{1 + \sin \theta_o} \right)^2 \tag{36}$$

Figure 12.15 is a plot of $\overline{U}_T^o$ versus $\theta - \theta_o$.

According to Eq. (34), and since T^o is positive definite, motion is possible if and only if

$$T_i^o - U_T^o \geq 0 \tag{37}$$

This implies that, for a given initial kinetic energy, Eq. (35), and consequently a given impulse, say $T_i^o = D$ (see Fig. 12.15—total potential presented in nondimensionalized form), motion is confined in the region $\Theta_I < \theta - \theta_o < \Theta_{II}$. It is clearly seen then that, as long as $T_i^o = D < \overline{U}_T^o(C)$, the motion of the system is bounded, and it contains only the stable zero-load static equilibrium point B. Such a motion is termed

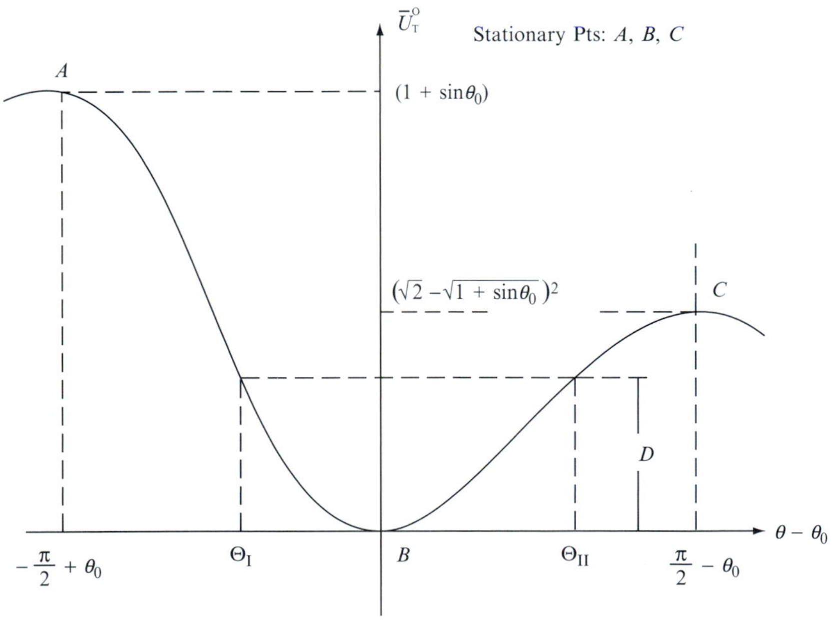

FIGURE 12.15   "Zero-load" total potential versus $\theta - \theta_o$ (Model A).

unbuckled. For the motion to cease to be unbuckled, that is, to become unbounded, and cease to include only the initial stable static equilibrium point B,D must be at least equal to the value of $\overline{U}_T^o$ at the unstable static point C. Then that point C can be reached with zero velocity, and the motion can become unbounded. Clearly, if D is even slightly higher than the $\overline{U}_T^o$ value at point C, the motion does become unbounded, and it can contain other static equilibrium points, such as point C. Such a motion is called *buckled*, and a critical condition exists when the impulse is large enough to satisfy the relation

$$T_{i_{cr}}^o = U_T^o(C) \tag{38}$$

Introducing nondimensionalized time and load parameters

$$\tau = t(2k/m)^{\frac{1}{2}}, \quad \tau_o = T_o(2k/m)^{\frac{1}{2}}, \quad p = 2P/kL \tag{39}$$

then

$$\overline{T}_i^o = \frac{T^o}{kL^2} = \frac{1}{2}\left(\frac{m}{k}\right)\dot{\theta}_0^2 = \left(\frac{d\theta}{d\tau}\right)^2_{\theta_o} \tag{40}$$

where $\dot{\theta}_0$ is the initial angular speed.

From Eq. (33) one obtains

$$\lim_{T \to 0}(p\tau_o) = \frac{2}{\sin\theta_o}\left(\frac{d\theta}{d\tau}\right)_{\theta_o} \tag{41}$$

The zero-load static equilibrium positions are obtained by requiring the total potential, Eq. (36), to have a stationary value, or

$$\frac{d\overline{U}_T^o}{d\theta} = 0 = \left(\sqrt{1 + \sin\theta} - \sqrt{1 + \sin\theta_o}\right)\cot\theta/\sqrt{1 + \sin\theta} \quad \text{for } \theta_o \neq 0 \tag{42}$$

This requirement yields

$$\theta = \theta_o \quad \text{or} \quad \theta = \pm\pi/2 \tag{43}$$

Through the second derivative (variation) of $\overline{U}_T^o$ with respect to θ, it can easily be shown that position $\theta = \theta_o$ is a stable one (point B on Fig. 12.15) whereas positions $\theta = \pm\pi/2$ (points C and A on Fig. 12.15) are unstable.

A critical condition exists when $\overline{T}_i^o = \overline{U}_T^o\ (\theta = \pi/2)$, or

$$\overline{T}_{i_{cr}}^o = \left(\sqrt{2} - \sqrt{1 + \sin\theta_o}\right)^2 \tag{44}$$

From Eqs. (12) and (13),

$$(p\tau_o)_{cr} = \frac{2}{\sin\theta_o}\overline{T}_{i_{cr}}^{\frac{1}{2}} = 2\left(\sqrt{2} - \sqrt{1 + \sin\theta_o}\right)/\sin\theta_o \tag{45}$$

Two observations are worth mentioning at this point: (1) Because this is a one-degree-of-freedom model, the critical impulse $(p\tau_o)_{cr}$ given by Eq. (45) represents both the minimum possible (MPCL) and minimum guaranteed (MGCL) critical load. Although the concept presented so far is clear, and it leads to a criterion and estimate of the critical condition, it might be impractical when applied to real structures. In the particular example shown so far, it is clear that, according to the presented concept of dynamic instability, buckled motion is possible if the system is allowed to reach the position $\theta = \pi/2$. In many cases such positions may be considered excessive, especially in deflection-limited designs. In such cases, if θ cannot be larger than a specified value, then the allowable impulse is smaller and its value can be found from Eq. (35), if C is replaced by the maximum allowable value of θ, say Θ_L. In this case,

$$(p\tau_o)_{cr} = 2\left(\sqrt{1 + \sin \Theta_L} - \sqrt{1 + \sin \theta_o} \right) / \sin \theta_o \qquad (46)$$

Related to this discussion is the broad definition of stability proposed by Hoff (1967): "A structure is in a stable state if admissible finite disturbances of its initial state of static or dynamic equilibrium are followed by displacements whose magnitudes remain within allowable bounds during the required lifetime of the structure." (2) Finally, as already mentioned in this chapter (Section 12.1.2), the sense of the impulsive load (in the absence of damping) has no effect on the critical condition.

If the load is applied (extremely short duration) to the right instead of the left (see Fig. 12.13) then the system tends to move with negative values for $\theta - \theta_0$ (see Fig. 12.15; the system would move toward Θ_1). The critical value for the initial kinetic energy is still given by Eq. (44), because the system would reach position E (see Fig. 12.15) with zero velocity, reverse its motion, pass through the stable static equilibrium position B, and then reach the unstable static equilibrium point C with zero kinetic energy (buckled motion, thus, is possible).

12.5.1.3   Dynamic Analysis: Sudden Constant Load of Infinite Duration

For this case, the sum of the total potential and kinetic energy is zero:

$$\overline{U}_T^P + \overline{T} = 0 \qquad (47)$$

Figure 12.16 shows plots of $\overline{U}_T^P$ versus $\theta - \theta_o$ (in radians) for various values of the applied load p. It is seen from this figure that for $p < 0.432$ motion is confined between the origin and $\theta - \theta_o < A$, or the motion is unbuckled. A critical condition exists when the motion can become unbounded by including position A'' (buckled motion).

Thus, the critical load is found by requiring [see Eq. (47)] that $\overline{U}_T^P$ be zero at the unstable static equilibrium position, A'' (see Fig. 12.16; the curves on this figure correspond to $\theta_o = 0.005$).

Numerically, the critical dynamic load is found by solving the following two equations simultaneously:

$$\left(\sqrt{1 + \sin \theta} - \sqrt{1 + \sin \theta_o} \right)^2 - p(\cos \theta_o - \cos \theta) = 0 \qquad (48)$$

$$p = \left(\sqrt{1 + \sin \theta} - \sqrt{1 + \sin \theta_o} \right) \cot \theta / \sqrt{1 + \sin \theta} \qquad (49)$$

subject to the condition $d^2\overline{U}_T^P/d\theta^2 < 0$, at the solution of Eqs. (48) and (49).

The inequality condition ensures that $\overline{U}_T^P = 0$, Eq. (48), at an unstable static equilibrium position. The simultaneous solution of Eqs. (48) and (49) (two equations in the unknowns θ and p) yields the dynamic critical load and the corresponding position of the unstable static equilibrium point A'' (see Fig. 12.16).

Values of critical dynamic loads for the case of suddenly applied constant loads of infinite duration are shown graphically in Fig. 12.17 for various small imperfection angles θ_o, and they are compared to the corresponding static critical loads (see Fig. 12.14).

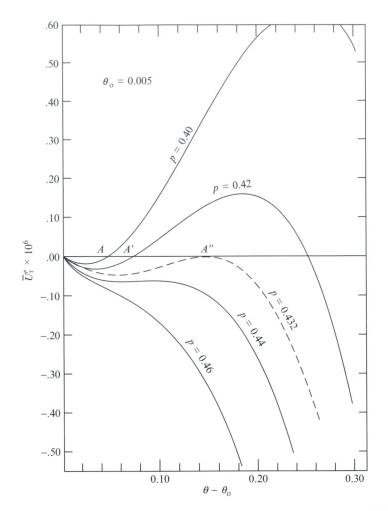

FIGURE 12.16 Total potential versus displacement for various loads (Model A).

For this load case also, since the system is a single-degree-of-freedom system, the minimum possible (MPCL) and minimum guaranteed (MGCL) critical loads are one and the same. Furthermore, if the value of θ is limited by other considerations (say the maximum allowable θ value is such that $\theta - \theta_0$ is equal to the value denoted by A' on Fig. 12.16), then there is no escaping motion type of instability, but the value $p = 0.42$ (see Fig. 12.16) would be a measure of the maximum allowable sudden (dynamic) load and therefore critical (in the sense that the kinematic constraint is not violated for loads smaller than $p = 0.42$).

12.4.2 MODEL B: A LOAD IMPERFECTION MODEL

Model B, shown in Fig. 12.18, is representative of eccentrically loaded structural systems, exhibiting limit point instability. The bar is rigid and of length L, the spring is linear of stiffness k, and the load eccentricity is denoted by e. The bar is assumed to be weightless, and the mass m of the system is concentrated on the top of the rod, point B.

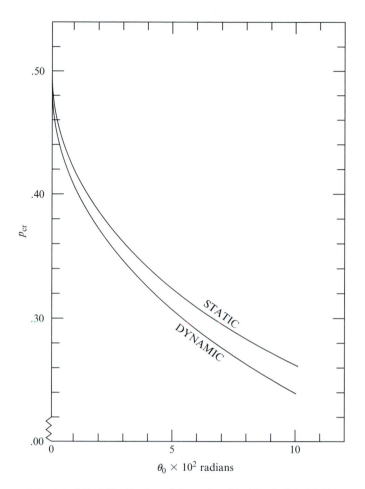

FIGURE 12.17   Static and dynamic critical loads (Model A).

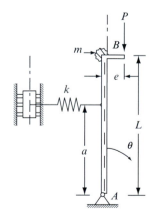

FIGURE 12.18   Geometry and sign convention for Model B.

Static Stability Analysis of Model B

For this model also, the energy approach is employed in the static stability analysis.

The total potential is given by

$$U_T^P = \frac{1}{2}ka^2 \sin^2\theta - PL\left(1 - \cos\theta + \frac{e}{L}\sin\theta\right) \tag{50}$$

First, nondimensionalized parameters are introduced:

$$\bar{e} = \frac{e}{L}, \quad p = \frac{PL}{ka^2}, \quad \overline{U}_T^P = \frac{2U_T^P}{ka^2} \tag{51}$$

With the aid of Eqs. (51), the expression for the total potential becomes

$$\overline{U}_T^P = \sin^2\theta - 2p(1 - \cos\theta + \bar{e}\sin\theta) \tag{52}$$

For equilibrium,

$$\frac{d\overline{U}_T^P}{d\theta} = 0 = 2\sin\theta\cos\theta - 2p(\sin\theta + \bar{e}\cos\theta) \tag{53}$$

From this equation, one obtains all of the static equilibrium positions. These are plotted in Fig. 12.19 for both positive eccentricity (as shown in Fig. 12.18) and negative eccentricity. The positions (of static equilibrium) corresponding to zero eccentricity are also shown.

If $\bar{e} = 0$, the static equilibrium positions are characterized by (see Eq. 53)

$$\sin\theta = 0 \rightarrow \theta = 0 \quad \text{and} \quad p = \cos\theta \tag{54}$$

On the other hand, if $\bar{e} \neq 0$, the static equilibrium positions are characterized by

$$p = \sin\theta/(\tan\theta + \bar{e}) \tag{55}$$

Note that if $\bar{e}$ is replaced by $-\bar{e}$ and θ by $-\theta$, the load deflection relation, Eq. (55), does not change. This is reflected in Fig. 12.19 by the two curves, one corresponding to $\bar{e} = A^2$ and the other to $\bar{e} = -A^2$.

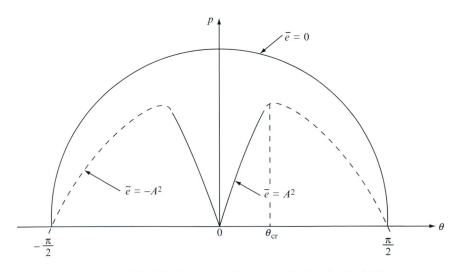

FIGURE 12.19   Positions of static equilibrium for Model B.

The stability or instability of the static equilibrium positions, Eq. (55), is next established through use of the second variation. If we restrict the range of θ values to $0 < \theta < \pi/2$, we study the sign of the second derivative of the total potential evaluated at static equilibrium points:

$$\frac{d^2 \overline{U}_T^P}{d\theta^2} = \cos 2\theta - p(\cos \theta - \bar{e} \sin \theta) \tag{56}$$

and

$$\left.\frac{d^2 \overline{U}_T^P}{d\theta^2}\right|_{\text{equil.pts.}} = \frac{\cos^2 \theta}{\tan \theta + \bar{e}}\left(\bar{e} - \tan^3 \theta\right) \tag{57}$$

It is clearly seen from Eq. (57) that the sign of the second derivative (evaluated at static equilibrium points) depends on the sign of $\bar{e} - \tan^3 \theta$. Thus

if $\tan^3 \theta < \bar{e}$, we have stability,

and if $\tan^3 \theta > \bar{e}$, we have instability $\tag{58}$

The stable positions correspond to the solid lines of the $\bar{e} = \pm A^2$ curves of Fig. 12.19, while the dashed lines characterize unstable static positions. When $\tan^3 \theta = \bar{e}$, $p = p_{\text{cr}}$ and substitution for this θ into Eq. (55) yields

$$p_{\text{cr}} = \left(1 + \bar{e}^{\frac{2}{3}}\right)^{-\frac{3}{2}} \tag{59}$$

Similar arguments can be used for $-\pi/2 < \theta < 0$ and the results are the same.

Dynamic Analysis: Ideal Impulse

Following the same procedure as for Model A, one can easily establish critical conditions for this load case.

The expressions for the zero-load total potential and kinetic energy are given by

$$\overline{U}_T^o = \sin^2 \theta \tag{60}$$

$$\overline{T}^o = \frac{2T^o}{ka^2} = \frac{2I}{2ka^2}\left(\frac{d\theta}{dt}\right)^2 \tag{61}$$

where I is the moment of inertia of the mass of the system about the hinge A (see Fig. 12.18)

Introducing a nondimensionalized time parameter τ, where

$$\tau = t\left(ka^2/I\right)^{\frac{1}{2}} \tag{62}$$

the expression for the kinetic energy becomes

$$\overline{T}^o = \left(\frac{d\theta}{d\tau}\right)^2 \tag{63}$$

Use of the angular impulse momentum theorem yields

$$\left[\lim_{T_o \to 0} (PT_o)\right]\bar{e} = I\left(\frac{d\theta}{dt}\right)_i \tag{64}$$

and in terms of nondimensionalized parameters where T_o and τ_o are taken as small as one wishes and the limit sign is thus dropped (but implied),

$$(p\tau_o) = \frac{1}{\bar{e}}\left(\frac{d\theta}{d\tau}\right)_i \tag{65}$$

where i implies *initial* velocity and/or kinetic energy.

Note from Eq. (60) that there exist three static equilibrium points under zero load. These correspond to $\theta = 0$ and $\theta = \pm\pi/2$ (see Fig. 12.19). Thus, a critical condition exists if the ideal impulse is instantaneously imparted into the system as initial kinetic energy, $\overline{T}_i^o$, of sufficient magnitude for the system to reach the unstable static points ($\pm\pi/2$; see Fig. 12.19 for stability or instability) with zero kinetic energy. In such a case, buckled motion is possible, and

$$\overline{T}_{i_{cr}}^o = \sin^2(\pm\pi/2) = 1 \tag{66}$$

From Eqs. (63) and (65),

$$(p\tau_o)_{cr} = 1/\bar{e} \tag{67}$$

It is observed for this model also that the sense of the impulsive load does not affect the critical condition. The only difference is that dynamic instability can take place by escaping motion either though unstable position $\theta = \pi/2$ or $\theta = -\pi/2$.

Moreover, in deflection-limited situations, say $|\theta| < |\Theta_L|$ where $|\Theta_L| < \pi/2$, the maximum allowable (and therefore critical) impulse is given by

$$(p\tau_o)_{cr} = \frac{\sin\Theta_L}{\bar{e}} \tag{68}$$

12.4.2.3 Dynamic Analysis: Sudden Constant Load of Infinite Duration

In a similar manner as for Model A, a critical condition exists if buckled motion can take place. This is possible if the total potential is zero at an unstable static equilibrium position. The critical condition is obtained through the simultaneous solution of the following two equations (in two unknowns p and θ):

$$\overline{U}_T^P = 0 = \sin^2\theta - 2p(1 - \cos\theta + \bar{e}\sin\theta) \tag{69}$$

and

$$p = \sin\theta/(\tan\theta + \bar{e}) \tag{70}$$

subject to the condition

$$\left.\frac{d^2\overline{U}_T^P}{d\theta^2}\right|_{\text{equil.pt.}} < 0 \tag{71}$$

Note that Eq. (70) characterizes static equilibrium positions.

Results are presented graphically in Fig. 12.20 for several values of the load eccentricity, and they are compared to the static (critical) conditions. Note that the static curve represents a plot of Eq. (59).

It is important to note that for both Models A and B the total potential, $\overline{U}_T^P$, is defined in such a way that it is zero at the initial (unloaded) position. Thus, in the absence of initial kinetic energy, the energy balance for both models is given by Eq. (5) of Section 12.1, or

$$\overline{U}_T^P + \overline{T}^P = 0 \tag{72}$$

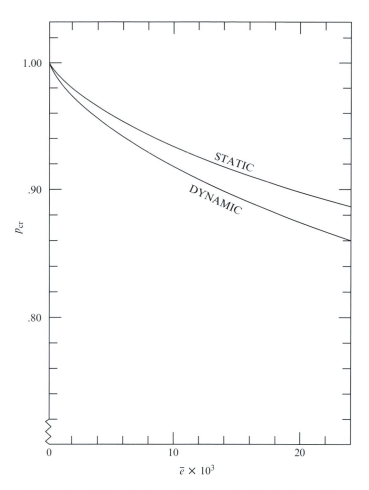

FIGURE 12.20   Static and dynamic critical loads (Model B).

Model B is used to demonstrate the other two approaches that are discussed in previous sections.

12.5.2.4   The Total Energy–Phase Plane Approach

First the case of the ideal impulse is treated. This approach is also based on the total energy balance (conservation of energy), but instead of associating critical conditions with characteristics of the total potential surface (under "zero load" for the ideal impulse), the critical condition is associated with characteristics of the system phase plane.

Conservation of energy requires

$$\overline{U}_{\mathrm{T}}^{\mathrm{o}} + \overline{T}^{\mathrm{o}} = \overline{T}_{\mathrm{i}}^{\mathrm{o}} \tag{73}$$

Use of Eqs. (60), (63), and (65) yields

$$\sin^2 \theta + \left(\frac{d\theta}{d\tau}\right)^2 = [(p\tau_{\mathrm{o}})\bar{e}]^2 \tag{74}$$

This equation is plotted on the phase plane ($\dot{\theta} = d\theta/d\tau$ versus θ curves) for various values of the right-hand side (Fig. 12.21).

Clearly, if

$$[(p\tau_o)\bar{e}]^2 < 1 \tag{75}$$

Eq. (74) denotes a closed curve about the null position ($\dot{\theta} = \theta = 0$) in the phase plane. In this case the motion is called unbuckled (see Fig. 12.21). When $[p\tau_o\bar{e}]^2 = 1$, Eq. (74) denotes a curve that can escape the closed loop and thus the motion becomes buckled. Therefore,

$$(p\tau_o)_{\text{cr}} = \frac{L}{e} = \frac{1}{\bar{e}} \tag{76}$$

Clearly, the result is the same as before, Eq. (67).

Next, the case of a sudden constant load of infinite duration is considered.

Use of Eqs. (69) and (63) yields the following expression for the total energy:

$$\dot{\theta}^2 + \sin^2\theta - 2p(1 - \cos\theta + \bar{e}\sin\theta) = 0 \tag{77}$$

This equation is shown qualitatively on Fig. 12.22b for different values of the sudden load. Fig. 12.22a shows total potential curves for various values of the applied load starting from zero. The two figures are shown together and clearly demonstrate the applicability of both concepts in establishing critical conditions. The symbol p_{cr} is used on Fig. 12.22 to denote the critical load for the case of suddenly applied loads with infinite duration.

Note on Fig. 12.22 that for sudden loads smaller than the critical, the system simply oscillates about the near static equilibrium position A_i. At the critical load, escaping (buckled) motion is possible through the unstable static equilibrium position B_3.

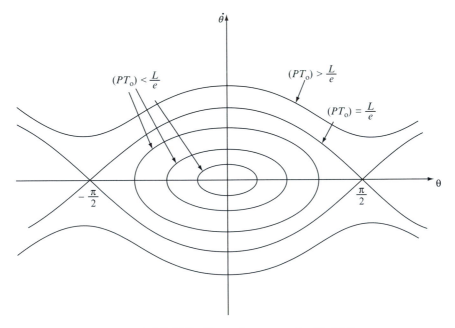

FIGURE 12.21   Phase plane curves for Model B.

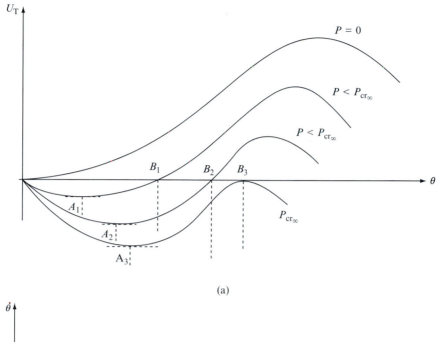

(a)

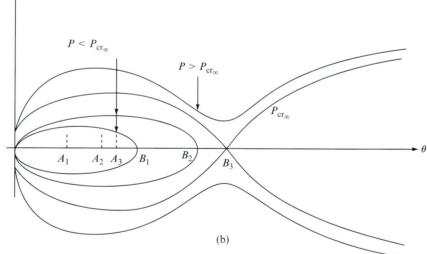

(b)

FIGURE 12.22 Critical conditions for Model B: constant load of infinite duration. (a) Total potential curves. (b) The phase plane.

For two-degree-of-freedom systems, dealing with phase plane curves is considerably more complex. As the number of degrees of freedom increases, the complexity increases exponentially to the point of intractability. As far as the continuum is concerned, this approach can be used only by reducing the phase space to a finite-dimensional space by constraining the motion. This means that the deformation of the continuum is represented by a finite number of degrees of freedom (Ritz, Galerkin, finite-element and finite-difference methods).

12.5.2.5 The Equations of Motion Approach

As stated in Section 12.1 this approach was first applied by Budiansky and Roth (1962) in finding critical conditions for a pressure-loaded, clamped, shallow, thin, spherical shell. The pressure was assumed to be applied suddenly with constant magnitude and infinite duration. The equations of motion are solved for several values of the load parameter, starting from a small value and incrementing it. At low values of the load parameter, the system experiences small oscillations. The maximum response amplitude, w_{max}, increases smoothly with λ. Fig. 12.23 gives a qualitative description of the phenomenon. At some level of λ, the maximum response amplitude experienced a large jump. The λ value at which this jump takes place is identified as the critical dynamic load, λ_{cr}.

For Model B (see Fig. 12.18) the equation of motion is given by

$$I\frac{d^2\theta}{dt^2} + M(\theta) = 0 \tag{78}$$

where $M(\theta)$ is the restoring moment, which can be expressed in terms of the contributions of the spring force and the externally applied force, or

$$M(\theta) = ka^2 \sin\theta\cos\theta - PL\left(\sin\theta + \frac{e}{L}\cos\theta\right) \tag{79}$$

Note that θ is a function of time, and the sudden force P is a step function of time.

Substitution of the expression for $M(\theta)$, Eq. (74), into the equation of motion, Eq. (75), and use of the nondimensionalized parameters, Eqs. (51) and (62), yields

$$\frac{d^2\theta}{d\tau^2} + \sin\theta\cos\theta - p(\sin\theta + \bar{e}\cos\theta) = 0 \tag{80}$$

This equation is solved numerically (using a finite-difference scheme), for $\bar{e} = 0.02$ and several values of the load parameter p: $p = 0.1, 0.5, 0.8, 0.85$, and 0.9. The results are presented graphically on Fig. 12.24 and 12.25. Fig. 12.25 shows plots of $\theta(\tau)$ versus time τ for various values of the load parameter. Note that for $p < p_{cr_\infty}$ (see

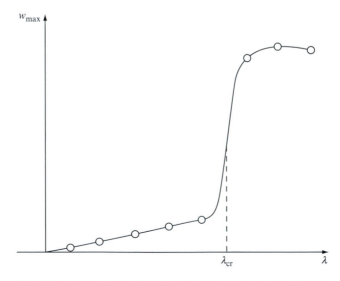

FIGURE 12.23   Description of the Budiansky–Roth criterion of dynamic stability.

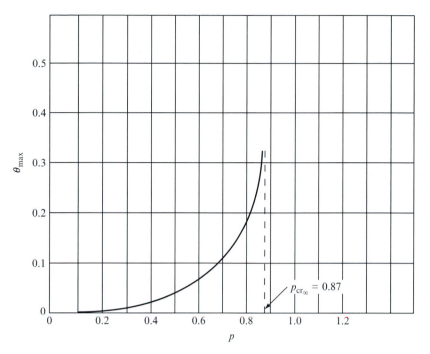

FIGURE 12.24 A plot of θ_{max} versus p (Model B).

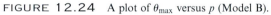

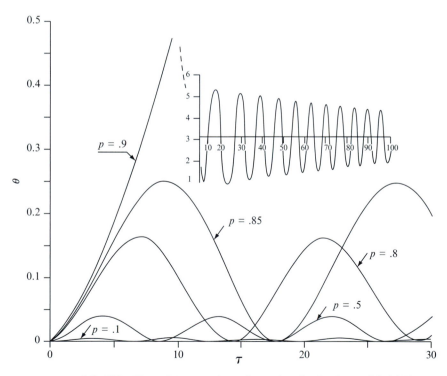

FIGURE 12.25 Plots of θ versus time τ for various load values p (Model B).

Fig. 12.8), the motion is simply oscillatory. The oscillations are between zero and a maximum amplitude that is much smaller than $\pi/2$. They seem to take place about the near static stable equilibrium position. For $p > p_{cr_\infty}$ ($p = 0.9$, Fig. 12.25), the motion has a very large amplitude ($\theta_{max} > \pi/2$), and it appears to be oscillating about the value of π. The important observation here is that if the range of allowable θ values is extended beyond $\pm\pi/2$, then $\theta = \pm\pi$ is a stable static equilibrium position, and the system tends to oscillate about this far ($\theta = \pi$) static position. Moreover, the amplitude decreases with time, because the force p yields restoring moments about the hinge, and the position $\theta = \pi$ is an asymptotically stable position (even in the absence of damping). Asymptotically stable means that for $t \to \infty$, the system will come to rest at this position. Fig. 12.25 shows a plot of the maximum response amplitude versus the load parameter. Clearly, there is a large jump in the maximum amplitude of oscillation, θ_{max}, between $p = 0.85$ and $p = 0.90$. According to the Budiansky–Roth criterion, p_{cr} is estimated to be 0.87, which is in excellent agreement (as expected for a single-degree-of-freedom system) with the value obtained from energy considerations.

PROBLEMS

1. Consider the model shown on Fig. P12.1. The bars are rigid and hinged. For simplicity, assume the mass of the system to be concentrated at the hinge between the two bars. Find (numerically) for the entire range of α values ($0° \le \alpha < 90°$)
   a. Values for the critical ideal impulse.
   b. Values for the critical dynamic load for the case of infinite duration. Compare these values to the static critical load values.

   For this problem use (i) the total potential energy approach and (ii) the total energy-phase plane approach.

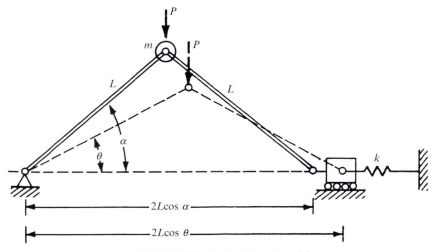

FIGURE P12.1   Problem 1 model.

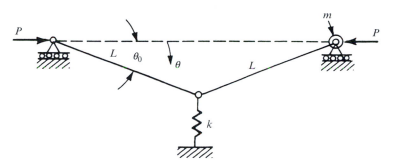

FIGURE P12.2 Problem 2 model.

2. Consider the mechanical model shown in Fig. P12.2. The two bars are rigid and the spring stiffness is constant, k. Assume that the mass of the system is concentrated at the right-end hinge.
   a. Give a complete static analysis.
   b. Find (numerically), for $\theta_o = 0.001$, $\theta_o = 0.05$, $\theta_o = 0.10$, (i) values for the critical ideal impulse and (ii) values for the critical dynamic load (constant magnitude of infinite duration). Use the potential energy approach.

3. Consider the mechanical model shown in Fig. P12.3. The two bars are rigid and the rotational spring stiffness is constant, β. The constant k may be determined from geometric considerations and θ_o is some small geometric imperfection. Assume that the system mass is concentrated at the right-end hinge.
   a. Give a complete static analysis.
   b. Find (numerically), for $\theta_o = 0.01$, 0.05, 0.10, and 0.20 (i) values for the critical ideal impulse and (ii) values for the critical dynamic load (constant load of infinite duration). Use the potential energy approach.

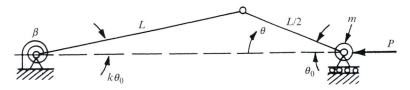

FIGURE P12.3 Problem 3 model.

REFERENCES

Au-Yang, M. K. and Brown, S. J. Jr. (editors). (1977). *Fluid Structure Interaction Phenomena in Pressure Vessel and Piping Systems*, PVP-PB-026. ASME, New York.

Benjamin, T. B. (1961). Dynamics of a system of articulated pipes conveying fluid. *Proc. R. Soc. London Ser. A*, 26, pp. 452–486.

Blevins, R. D. (1977). *Flow-Induced Vibration*. Van Nostrand-Reinhold, New York.

Bohn, M. P. and Herrmann, G. (1974). The dynamic behavior of articulated pipes conveying fluid with periodic flow rate. *J. Appl. Mech.*, 41, 1, pp. 55–62.

Bolotin, V. V. (1964). *The Dynamic Stability of Elastic Systems* (translated by V. I. Weingarten et al.). Holden-Day, San Francisco.

Budiansky, B. (1967). Dynamic buckling of elastic structures: Criteria and estimates. *Dynamic Stability of Structures* (edited by G. Herrmann). Pergamon, New York.

Budiansky, B. and Hutchinson, J. W. (1964). Dynamic buckling of imperfection-sensitive structures. *Proc. XI International Congress of Applied Mechanics*, Munich.

Budiansky, B. and Roth, R. S. (1962). Axisymmetric dynamic buckling of clamped shallow spherical shells. Collected Papers on Instability of Shell Structures. NASA TN D-1510.

Chen, S. S. (1975a). Vibration of a row of circular cylinders in a liquid. *J. Eng. Ind., Trans. ASME.* 91, 4, pp. 1212–1218.

Chen, S. S. (1975b). Vibration of nuclear fuel bundles. *Nucl. Eng. Des.*, 35, 3, pp. 399–422.

Chen, S. S. (1978). Crossflow-induced vibrations of heat exchanger tube banks. *Nucl. Eng. Des.*, 47, 1, pp. 67–86.

Chen, S. S. (1981). Fluid damping for circular cylindrical structures. *Nucl. Eng. Des.*, 63, 1, pp. 81–100.

Crocco, L. and Cheng, S. I. (1956). *Theory of Combustion Instability in Liquid Propellant Rocket Motors*, AGARD Monograph No. 8. Butterworths, London.

Dowell, E. (March 1969). Nonlinear flutter of curved plates. *AIAA J.*, 7, 3, pp. 424–431.

Dowell, E. (March 1970). Panel flutter, a review of the aeroelastic stability of plates and shells. *AIAA J.*, 8, 3, pp. 385–399.

Ginsberg, J. H. (1973). The dynamic stability of a pipe conveying a pulsatile flow. *Int. J. Eng. Sci.*, 11, pp. 1013–1024.

Gregory, R. W. and Paidoussis, M. P. (1966). Unstable oscillation of tubular cantilevers conveying fluid. *Prod. R. Soc. London Ser. A*, 293, pp. 512–527.

Herrmann, G. (Feb. 1967). Stability of equilibrium of elastic systems subjected to nonconservative forces. *Appl. Mech. Rev.*, 20, 2, pp. 103–108.

Herrmann, G. and Bungay, R. W. (1964). On the stability of elastic systems subjected to nonconservative forces. *J. Appl. Mech.*, 31, 3, pp. 435–440.

Hill, J. L. and Swanson, C. P. (1970). Effects of lumped masses on the stability of fluid conveying tubes. *J. Appl. Mech.*, 37, 2, pp. 494–497.

Hoff, N. J. (1967). Dynamic stability of structures. *Dynamic Stability of Structures* (edited by G. Herrmann). Pergamon, New York.

Hoff, N. J. and Bruce, V. C. (1954). Dynamic analysis of the buckling of laterally loaded flat arches. *Q. Math. Phys.*, 32, pp. 276–288.

Hsu, C. S. (1967). The effects of various parameters on the dynamic stability of a shallow arch. *J. Appl. Mech.*, 34, 2, pp. 349–356.

Hsu, C. S. (June 1968a). Equilibrium configurations of a shallow arch of arbitrary shape and their dynamic stability character. *Int. J. Nonlinear Mech.*, 3, pp. 113–136.

Hsu, C. S. (1968b). On dynamic stability of elastic bodies with prescribed initial conditions. *Int. J. Nonlinear Mech.*, 4, 1, pp. 1–21.

Hsu, C. S. (1968c). Stability of shallow arches against snap-through under timewise step loads. *J. Appl. Mech.*, 35, 1, pp. 31–39.

Hsu, C. S., Kuo, C. T. and Lee, S. S. (1968). On the final states of shallow arches on elastic foundations subjected to dynamic loads. *J. Appl. Mech.*, 35, 4, pp. 713–723.

Junger, M. and Feit, D. (1972). *Sound, Structures and Their Interaction*. MIT Press, Cambridge, Mass.

King, R. 1977. A review of vortex shedding research and its applications. *Ocean Eng.*, 4, pp. 141–171.

Koning, C. and Taub, J. (1933). Impact buckling of thin bars in the elastic range hinged at both ends. *Luftfahrtforschung*, 10, 2, pp. 55–64 (translated as NACA TM 748 in 1934).

Kornecki, A. (July 1970). Traveling wave-type flutter of infinite elastic plates. *AIAA J.*, 8, 7, pp. 1342–1344.

Kuo, G. C., Morino, L. and Dugundji, J. (Nov. 1972). Perturbation and harmonic balance methods of nonlinear panel flutter. *AIAA J.*, 10, 11, pp. 1479–1484.

Lefschetz, S. (1965). *Stability of Nonlinear Control of Airplanes and Helicopters*. Academic Press, New York.

Morino, L. (March 1969). Perturbation method for treating nonlinear panel flutter problems. *AIAA J.*, 7, 3, pp. 405–411.

Paidoussis, M. P. (1970). Dynamics of tubular cantilevers conveying fluid. *J. Mech. Sci.*, 12, 2, pp. 85–103.

Paidoussis, M. P. and Deksnis, B. E. (1970). Articulated models of cantilevers conveying fluid: The study of a paradox. *J. Mech. Eng. Sci.*, 42, 4, pp. 288–300.

Paidoussis, M. P. and Issid, N. T. (1976). Experiments on parametric resonance of pipes containing pulsatile flow. *J. Appl. Mech., Trans. ASME*, 98, pp. 198–202.

Paidoussis, M. P. and Sundararajan, C. (1975). Parametric and combination resonances of a pipe conveying pulsating fluid. *J. Appl. Mech.*, 42, 4, pp. 780–784.

Reusselet, J. and Herrmann, G. (1977). Flutter of articulated pipes at finite amplitude. *Trans. ASME*, 99, 1, pp. 154–158.

Routh, E. J. (1975). *Stability of Motion* (edited by A. T. Fuller). Taylor & Francis, Halsted Press, New York. (Originally it appeared in 1877.)

Scanlan, R. H. and Simin, E. (1978). *Wind Effects on Structures: An Introduction to Wind Engineering.* Wiley, New York.

Seckel, E. (1964). *Stability and Control of Airplanes and Helicopters.* Academic Press, New York.

Simitses, G. J. (1965). Dynamic snap-through buckling of low arches and shallow spherical caps. June Ph.D. Dissertation, Department of Aeronautics and Astronautics, Stanford University.

Simitses, G. J. (1974). On the dynamic buckling of shallow spherical caps. *J. Appl. Mech.*, 41, 1, 1974, pp. 299–300.

Simitses, G. J. (1989). *Dynamic Stability of Suddenly Loaded Structures*, Springer-Verlag, New York.

Stoker, J. J. (1950). *Non-Linear Vibrations in Mechanical and Electrical Systems.* vol. II. Interscience, London.

Stoker, J. J. (1955). On the stability of mechanical systems. *Commun. Pure Appl. Math., VIII*, pp. 133–142.

Tamura, Y. S. and Babcock, C. D. (1975). Dynamic stability of cylindrical shells under step loading. *J. Appl. Mech.*, 42, 1, pp. 190–194.

Thompson, J. M. T. (1967). Dynamic buckling under step loading. *Dynamic Stability of Structures* (edited by G. Herrmann). Pergamon, New York.

Thompson, W. and Tait, P. G. (1923). *Treatise on Natural Philosophy*, part I. Cambridge University Press, Cambridge, England. (It was first published in 1867.)

Wauer, J. (1980). Uber Kinetische Verweigungs Probleme Elasticher Strukturen unter Stosseblastung. *Ingenieur-Arch.*, 49, pp. 227–233.

Ziegler, H. (1956). On the concept of elastic stability. *Advances in Applied Mechanics*, vol. 4, pp. 351–403. Academic Press, New York.

Ziegler, H. (1968). *Principles of Structural Stability.* Blaisdell, Waltham, Mass.

---

* From G.J. Simitses, *Dynamic Stability of Suddenly Loaded Structures* (Chapter 1 and part of Chapter 2), Springer-Verlag, 1989. Reprinted with permission.

APPENDIX

Work- and Energy-Related Principles and Theorems

This appendix will summarize the work and energy principles (and theorems derived from these) that have been used in stability analysis and are directly referred to in this text. In addition, some explanations and definitions will be given to facilitate understanding and application of these concepts. Because of this, only the principles and derived theorems associated with virtual work will be treated. Complementary energy and complementary virtual work concepts, principles, and theorems are not included. The student interested in an extensive and thorough treatise of all the work and energy principles is referred to the texts of Argyris (1960), Dym and Shames (1973), Fung (1965), and Langhaar (1962). One of the first and best-written texts on the subject (with numerous applications on structural problems) is the book by Hoff (1956).

A.1   STRAIN ENERGY

A deformable body is said to be perfectly elastic if the state of stress and the corresponding state of strain are the same for the same level of the external forces regardless of the order of application of the loads and of whether this level is during loading or unloading of some or all of the loads. This statement is clearly understandable when related to the simple tensile test. If the stress-strain relation for such a test is the same during the loading and unloading processes, the behavior is called elastic and the specimen is called a perfectly elastic body.

If a perfectly elastic body is under the action of external loads (distributed and concentrated forces, distributed and concentrated moments), the body deforms and work is done by these external loads. This work, in the absence of kinetic energy (quasistatic application of the loads), is stored in the system. Because of the assumption that the material is perfectly elastic, the work done by the loads can be regained if the loads are quasistatically decreased to zero. The energy stored in the system is known as the strain energy.

If we consider a deformable body at state I and apply a set of loads that strain the body to state II, and if we use a cartesian reference frame, x, y, z, the work done by these forces, W_e, is equal to the strain energy, U_i, and it is given by (for small strains)

$$W_e = U_i = \int_V \left[\int_I^{II} \left(\sigma_{xx} d\varepsilon_{xx} + \sigma_{yy} d\varepsilon_{yy} + \sigma_{zz} d\varepsilon_{zz} + \sigma_{xy} d\gamma_{xy} + \sigma_{yz} d\gamma_{yz} + \sigma_{zx} d\gamma_{zx} \right) \right] dV \quad (1a)$$

or

$$U_i = \int_V \overline{U}_i dV \quad (1b)$$

where $\overline{U}_i$ is defined as the strain-energy density (strain energy per unit volume).

The existence of the strain-energy density function and the energy balance expressed by Eqs. (1) is in agreement with the first and second laws of thermodynamics for isentropic processes. In this case, the energy stored in the system is called internal energy. In addition, if the process is a reversible isothermal one, then the stored energy is often called the free energy (see Dym and Shames, 1973 and Fung, 1965). In effect, the strain-energy density represents the energy that can be converted to mechanical work in a reversible adiabatic or isothermal process.

Since the strain-energy density at a point depends on the state of strain, the incremental strain-energy density, which is a perfect differential for perfectly elastic behavior, $d\overline{U}_i$, is given by

$$d\overline{U}_i = \frac{\partial \overline{U}_i}{\partial \varepsilon_{xx}} d\varepsilon_{xx} + \frac{\partial \overline{U}_i}{\partial \varepsilon_{yy}} d\varepsilon_{yy} + \frac{\partial \overline{U}_i}{\partial \varepsilon_{zz}} d\varepsilon_{zz} + \frac{\partial \overline{U}_i}{\partial \gamma_{xy}} d\gamma_{xy} + \frac{\partial \overline{U}_i}{\partial \gamma_{yz}} d\gamma_{yz} + \frac{\partial \overline{U}_i}{\partial \gamma_{zx}} d\gamma_{zx} \quad (2)$$

From Eqs. (1), it can be seen that

$$\sigma_{xx} = \frac{\partial \overline{U}_i}{\partial \varepsilon_{xx}}, \quad \sigma_{yy} = \frac{\partial \overline{U}_i}{\partial \varepsilon_{yy}}, \quad \sigma_{zz} = \frac{\partial \overline{U}_i}{\partial \varepsilon_{zz}}$$

$$\sigma_{xy} = \frac{\partial \overline{U}_i}{\partial \gamma_{xy}}, \quad \sigma_{yz} = \frac{\partial \overline{U}_i}{\partial \gamma_{yz}}, \quad \sigma_{zx} = \frac{\partial \overline{U}_i}{\partial \gamma_{zx}} \quad (3)$$

When the material follows Hooke's law (linearly elastic behavior), then

$$\overline{U}_i = \frac{1}{2} \left(\sigma_{xx} \varepsilon_{xx} + \sigma_{yy} \varepsilon_{yy} + \sigma_{zz} \varepsilon_{zz} + \sigma_{xy} \gamma_{xy} + \sigma_{yz} \gamma_{yz} + \sigma_{zx} \gamma_{zx} \right) \quad (4)$$

If the linear stress-strain relations are used in Eq. (4) in terms of Poisson's ratio, ν, and Young's modulus of elasticity, E, the strain-energy density can be expressed solely either in terms of strains or in terms of stresses.

THREE-DIMENSIONAL CASE

$$\begin{aligned} \overline{U}_i = {} & \frac{E}{2(1+\nu)(1-2\nu)} \left[(1-\nu)\left(\varepsilon_{xx}^2 + \varepsilon_{yy}^2 + \varepsilon_{zz}^2\right) \right. \\ & \left. + 2\nu\left(\varepsilon_{xx}\varepsilon_{yy} + \varepsilon_{yy}\varepsilon_{zz} + \varepsilon_{zz}\varepsilon_{xx}\right) + \frac{1-2\nu}{2}\left(\gamma_{xy}^2 + \gamma_{yz}^2 + \gamma_{zx}^2\right) \right] \end{aligned} \quad (5a)$$

$$\begin{aligned} \overline{U}_i = {} & \frac{1}{2E} \left[\left(\sigma_{xx}^2 + \sigma_{yy}^2 + \sigma_{zz}^2\right) - 2\nu\left(\sigma_{xx}\sigma_{yy} + \sigma_{yy}\sigma_{zz} + \sigma_{yy}\sigma_{xx}\right) \right. \\ & \left. + 2(1+\nu)\left(\sigma_{xy}^2 + \sigma_{yz}^2 + \sigma_{zx}^2\right) \right] \end{aligned} \quad (5b)$$

TWO-DIMENSIONAL CASE

1. Plane stress (x-y plane):

$$\overline{U}_i = \frac{1}{2(1-\nu^2)}\left[\varepsilon_{xx}^2 + \varepsilon_{yy}^2 + 2\nu\varepsilon_{xx}\varepsilon_{yy} + \frac{1-\nu}{2}\gamma_{xy}^2\right] \tag{6a}$$

$$\overline{U}_i = \frac{1}{2E}\left[\sigma_{xx}^2 + \sigma_{yy}^2 - 2\nu\sigma_{xx}\sigma_{yy} + 2(1+\nu)\sigma_{xy}^2\right] \tag{6b}$$

2. Plane strain (x-y plane):

$$\overline{U}_i = \frac{E}{2(1+\nu)(1-2\nu)}\left[(1-\nu)\left(\varepsilon_{xx}^2 + \varepsilon_{yy}^2\right) + 2\nu\varepsilon_{xx}\varepsilon_{yy} + \frac{1-2\nu}{2}\gamma_{xy}^2\right] \tag{7a}$$

$$\overline{U}_i = \frac{1}{2E}\left[(1-\nu^2)\left(\sigma_{xx}^2 + \sigma_{yy}^2\right) - 2\nu(1+\nu)\sigma_{xx}\sigma_{yy} + 2(1+\nu)\sigma_{xy}^2\right] \tag{7b}$$

ONE-DIMENSIONAL CASE

For this case, let us consider an Euler-Bernoulli beam with the x-z plane as a plane of structural symmetry. Then $\sigma_{xy} = \sigma_{xz} = \sigma_{yy} = \sigma_{yz} = \sigma_{zz} = 0$ and,

$$\overline{U}_i = \frac{E}{2}\varepsilon_{xx}^2 \tag{8a}$$

$$\overline{U}_i = \frac{1}{2E}\sigma_{xx}^2 \tag{8b}$$

A.2 THE PRINCIPLE OF VIRTUAL DISPLACEMENT OR VIRTUAL WORK

Before we state this principle, we must first clearly explain what is meant by "virtual displacement." A virtual displacement is a hypothetical displacement which must be compatible with the constraints for a given problem. If we deal with a particle, a virtual displacement is a single vector without any limitations on magnitude and direction. If we deal with a *rigid* body, a virtual displacement is a displacement field $\overline{u}(x, y, z)$ which must be compatible with the requirement that the body be rigid (see Fig. A.1). Note that in Fig. A.1a, in addition to the rotation θ, the system may be translated, and the combination comprises a virtual displacement. Lastly, when we deal with a cohesive deformable continuum, the virtual displacement must be compatible with (1) the constitution of the medium, and (2) the associated method of analysis. The latter statement means that the virtual displacements must be consistent with the theory and its related kinematic assumptions that lead to the field equations that govern the response of the system to any set of external causes.

First, what we mean by compatible with the constitution of the medium is that, since we deal with a cohesive continuum, the virtual displacement components must be single-valued continuous functions of positions (material points coordinates) with continuous derivatives. Second, compatible with the associated method of analysis implies the following: (1) Since we are interested in deformations in the analysis of

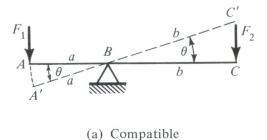

(a) Compatible

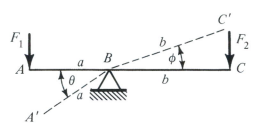

(b) Incompatible

FIGURE A.1   Compatible and incompatible virtual displacements for a rigid bar.

deformable bodies, then the virtual displacement is a deformation field $\bar{u}(x, y, z)$ which is consistent with the kinematic constraints on the bounding surface. (2) Since there are different approximating theories describing the kinematics of the problem, as for example the theory of small deformation gradient, the virtual deformation field must be consistent with these approximations. Therefore, a deformation field which has these properties is referred to as a kinematically admissible field, and thus any kinematically admissible field can be used as a virtual displacement. Finally, the reason the virtual displacement is called hypothetical is that during virtual displacement the forces, internal and external, are kept constant, which is not compatible with the behavioral response of systems, in general.

The principle of virtual displacements or virtual work may be stated as follows:

> A body is in equilibrium, under a given system of loads, if and only if for any virtual displacement the work done by the external forces is equal to the strain energy.

Note that:
1. A principle in mechanics is like an axiom in mathematics. There is no proof of a principle, although one may show its equivalence to another principle or law.
2. If we realize that a virtual displacement is kinematically admissible and that the forces are kept constant during virtual displacements, the principle holds for deformable bodies as well as rigid bodies and particles. In the case of rigid bodies and particles, the strain energy is zero.

3. The mathematical expression for the principle is

$$\delta_\varepsilon W = \delta_\varepsilon U_i \tag{9}$$

for deformable bodies, and

$$\delta_\varepsilon W = 0 \tag{10}$$

for rigid bodies and particles. In Eqs. (9) and (10), $\delta_\varepsilon W$ represents the work done by the external forces and $\delta_\varepsilon U_i$ is the strain energy during a virtual displacement denoted by the subscript ε.

A few simple applications of the principle are given below.

1. A Particle under N Forces.   Given a particle under the application of N forces $\mathbf{F}_i$, according to the principle, this particle is in equilibrium if and only if

$$\delta_\varepsilon W = 0$$

Let $\bar{u}$ be a virtual displacement. Then by the principle

$$\mathbf{F}_1 \cdot \mathbf{u} + \mathbf{F}_2 \cdot \mathbf{u} + \cdots + \mathbf{F}_N \mathbf{u} = 0$$

or

$$(\mathbf{F}_1 + \mathbf{F}_2 + \mathbf{F}_3 + \cdots + \mathbf{F}_N) \cdot \mathbf{u} = 0$$

$$\left(\sum_{i=1}^{N} \mathbf{F}_i \right) \cdot \mathbf{u} = 0 \tag{11}$$

For this to be true, either $\sum_{i=1}^{N} \mathbf{F}_i$ is normal to $\mathbf{u}$ or zero. But since $\mathbf{u}$ is any displacement vector, then $\sum_{i=1}^{N} \mathbf{F}_i$ must be zero. This is in complete agreement with the necessary and sufficient conditions for equilibrium of a particle under static loads which are derived from Newton's second law.

2. The Fulcrum Problem.   Consider the rigid bar of Fig. A.2.

The virtual displacement consists of a translation in the positive y-direction and a rotation θ as shown. (Note that the rigid bar ACB which is originally straight remains straight as $A'C'B'$ during the virtual displacement (compatible with the fact that the bar is rigid).

The work done by the forces during the virtual displacement is zero

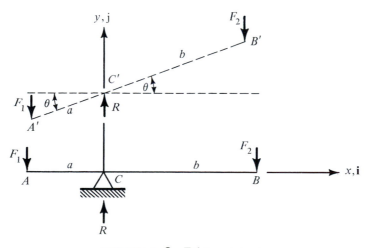

FIGURE A.2   Fulcrum geometry.

$$(-F_1\,\mathbf{j})\cdot\mathbf{AA'} + (R\mathbf{j})\cdot\mathbf{CC'} + (-F_2\mathbf{j})\cdot\mathbf{BB'} = 0 \qquad (12)$$

where $\mathbf{AA'}$, $\mathbf{CC'}$, and $\mathbf{BB'}$ are position vectors from A to A', C to C', and B to B', respectively

$$\mathbf{AA} = a(1 - \cos\theta)\mathbf{i} + (d - a\sin\theta)\mathbf{j}$$
$$\mathbf{CC'} = d\mathbf{j} \qquad (13)$$
$$\mathbf{BB'} = -b(1 - \cos\theta)\mathbf{i} + (d + b\sin\theta)\mathbf{j}$$

Substitution of Eqs. (13) into Eq. (12) yields

$$-F_1(d - a\sin\theta) + Rd - F_2(d + b\sin\theta) = 0$$

or

$$(-F_1 + R - F_2)d + (F_1 a - F_2 b)\sin\theta = 0 \qquad (14)$$

Since d and θ are independent (we can have $d \neq 0$ and $\theta \equiv 0$ or $\theta \neq 0$ and $d \equiv 0$), then

$$F_1 + F_2 = R \quad \text{and} \quad F_1 a = F_2 b \qquad (15)$$

These equations are in complete agreement with the necessary and sufficient conditions for equilibrium of a rigid body (sum of forces equals zero, and sum of moments about C equals zero, respectively).

3. Extension of a Bar. Consider the straight bar shown in Fig. A.3. Making the usual linear theory assumptions (small deformation gradients and linearly elastic behavior) and reducing the problem to a one-dimensional one, we may write

$$\varepsilon_{xx} = \frac{du}{dx}$$
$$\sigma_{xx} = E\varepsilon_{xx} \qquad (16)$$

where u is a function of x only. If we allow $\bar{u}$ to denote a virtual displacement, then $\bar{u}$ must be kinematically admissible and $\bar{u}(0) = 0$. On this basis, the corresponding virtual strain is given by

$$\bar{\varepsilon}_{xx} = \frac{d\bar{u}}{dx} \qquad (17)$$

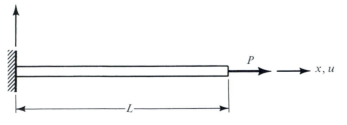

FIGURE A.3   Bar geometry.

The virtual work and corresponding strain energy are given by

$$\delta_\varepsilon W = P\bar{u}(L) \tag{18a}$$

and

$$\delta_\varepsilon U_i = \int_V \sigma_{xx}\bar{\varepsilon}_{xx}dV$$
$$= \int_0^L \sigma_{xx}A\frac{d\bar{u}}{dx}dx \tag{18b}$$

Integration by parts yields

$$\delta_\varepsilon \overline{U}_i = (\sigma_{xx}A\bar{u})\Big|_0^L - \int_0^L \frac{d}{dx}(\sigma_{xx}A)\bar{u}\,dx \tag{19}$$

Since $\bar{u}(0) = 0$, by the principle of virtual work we obtain

$$[(P - A\sigma_{xx})\bar{u}]_{x=L} - \int_0^L \frac{d}{dx}(\sigma_{xx}A)\bar{u}\,dx = 0 \tag{20}$$

Because $\bar{u}$ is a virtual displacement, then

$$\frac{d}{dx}(\sigma_{xx}A) = 0 \longrightarrow \sigma_{xx}A = \text{constant}$$

and $P = A\sigma_{xx}$ since $\bar{u}(L)$ is arbitrary. The system is in equilibrium if and only if $\sigma_{xx} = P/A = \text{constant}$.

A.3 DERIVATIVES OF THE PRINCIPLE OF VIRTUAL WORK

A number of principles, theorems, and methods may be considered as direct derivatives from the principle of virtual work. The most pertinent of these derivatives are listed in this section.

A.3.1 THE PRINCIPLE OF THE STATIONARY VALUE OF THE TOTAL POTENTIAL

If the system is conservative, then the work done by the loads from the zero deformation state (strain-free position) to a final state is equal to the negative change in the total potential of the external forces, U_p. If this potential is defined such that it is zero at the zero deformation state, then

$$-W = U_p \tag{21}$$

Next, the variation in the work done during virtual displacements is related to the variation in the potential of the external forces by

$$\delta_\varepsilon W = -\delta_\varepsilon U_p \tag{22}$$

Substitution of Eq. (22) into Eq. (9) yields

$$\delta_\varepsilon U_i + \delta_\varepsilon U_p = 0$$

or

$$\delta_\varepsilon\left(U_i + U_p\right) = \delta_\varepsilon U_T = 0 \tag{23}$$

where U_T is called the total potential (energy) of the system. This equation implies that:

An elastic deformable system is in equilibrium (static) if and only if the first variation of the total potential vanishes for every virtual displacement.

Note that virtual displacements must be kinematically admissible and that the loads and stresses remain constant during such deformations.

If we consider that the total potential is a function of N deformation parameters q_i, and that a virtual displacement can be taken to be any one of these q_i's, then by Eq. (23)

$$\delta_\varepsilon U_T = \frac{\partial U_T}{\partial q_i}\delta q_i \quad i = 1, 2, \ldots, N \tag{24}$$

Since δq_i is arbitrary, this equation implies that

$$\frac{\partial U_T}{\partial q_i} = 0 \quad i = 1, 2, \ldots, N \tag{25}$$

for static equilibrium.

By Eq. (25) the total differential of U_T must be zero, or

$$dU_T = \sum_{i=1}^{N} \frac{\partial U_T}{\partial q_i} dq_i = 0 \tag{26}$$

This argument may be extended to a deformable system, and we conclude that the vanishing of $\delta_\varepsilon U_T$ implies the vanishing of dU_T. Next, since dU_T vanishes at stationary points (relative minima, maxima, or saddle points), and U_T is said to have a stationary value at such points, then Eq. (23) may be interpreted as the mathematical expression of the following principle.

An elastic deformable system is in equilibrium (static) if and only if the total potential has a stationary value.

An equivalent statement to the above is:

Of all possible kinematically admissible deformation fields in an elastic conservative system, for a specified level of the external loads and the corresponding internal loads, only those corresponding to equilibrium (static) make the total potential assume a stationary value.

This statement is known as the *principle of the stationary value of the total potential*. In reality, it is a theorem because it is derived from and proven by a basic principle, the *principle of virtual work*.

A.3.2   THE PRINCIPLE OF THE MINIMUM TOTAL POTENTIAL

The above theorem is easily extended to an equivalent of the Lagrange-Dirichlet theorem (see Chapter 1) for an elastic conservative system by requiring the stationary value to be a relative minimum. If this happens, the equilibrium is stable. Often, this theorem is referred to as the *principle of the minimum total potential*, and it is given below.

Of all possible kinematically admissible deformation fields in an elastic conservative system, for a specified level of the external loads and the corresponding internal loads, only those that make the total potential assume a minimum value correspond to a stable equilibrium.

A.3.3   CASTIGLIANO'S FIRST THEOREM (PART I)

Another derivative of the principle of virtual work is Castigliano's first theorem, part I.

Consider an elastic system under the action of N concentrated loads, P_j, (forces and moments). Let y_j denote the components of deformation (or rotations) at the points of applications of the forces (or moments) and in the directions of these loads. If δy_j denote virtual displacements, then the virtual work is given by

$$\delta W = \sum_{j=1}^{N} P_j \delta y_j \tag{27}$$

If we can express the deformation components of the material points on the body in terms of the y_j components, the stresses, strains, and consequently the strain energy of the elastic system become functions of the y_j components, the structural geometry, and the elastic behavior (stress-strain law which need not necessarily be linearly elastic). If we now give each component y_j a small variation δy_j (virtual displacement), then

$$\delta U_i = \sum_{j=1}^{N} \frac{\partial U_i}{\partial y_j} \delta y_j \tag{28}$$

By the principle of virtual work

$$\sum_{j=1}^{N} \left(\frac{\partial U_i}{\partial y_j} - P_j \right) \delta y_j = 0 \tag{29}$$

Therefore, since the virtual displacements are independent, we have the mathematical expression of Castigliano's first theorem

$$\frac{\partial U_i}{\partial y_j} = P_j \tag{30}$$

Note that this theorem applies to elastic systems regardless of the behavior (nonlinear elastic behavior as well). For applications, see Przemieniecki (1968) and Oden (1967).

One important application of the theorem is in finding reaction forces for structural systems. For example, if the deformation component is known to be zero at some point and for a given direction, first we let y_r exist; then we express U_i in terms of y_r. Finally, the sought reaction is equal to $(\partial U_i/\delta y_r)$, evaluated at $y_r = 0$, according to Eq. (30).

A.3.4   THE UNIT-DISPLACEMENT THEOREM

Another important derivative of the principle of virtual work is the unit-displacement theorem. This theorem is used to determine the load P_r (force or moment) necessary to maintain equilibrium in an elastic system when the distribution of true stresses is known. Let the true stresses be given by $(\sigma_{xx}, \sigma_{yy}, \ldots, \sigma_{zx})$. Consider a virtual displacement δy_r at the point of application and in the direction of P_r.

This virtual displacement produces virtual strains $\delta\varepsilon_{ij}^r$ and according to the principle of virtual work

$$P_r\delta y_r = \int_v \left(\sigma_{xx}\delta\varepsilon_{xx}^r + \sigma_{yy}\delta\varepsilon_{yy}^r + \cdots + \sigma_{zx}\delta\gamma_{zx}^r\right)dV \tag{31}$$

In a linearly elastic system, the virtual strains $\delta\varepsilon_{ij}^r$ are proportional to y_r:

$$\delta\varepsilon^r = \varepsilon^r\delta y_r \tag{32}$$

where ε^r represents compatible strains due to a unit virtual displacement ($\delta y_r = 1$). Assuming, therefore, that $\delta y_r = 1$, Eq. (30) becomes

$$P_r = \int_v \left(\sigma_{xx}\varepsilon_{xx}^r + \sigma_{yy}\varepsilon_{yy}^r + \cdots + \sigma_{zx}\gamma_{zx}^r\right)dV \tag{33}$$

This equation is the mathematical expression of the unit-displacement theorem, which is stated below:

> The force necessary to maintain equilibrium under a specified stress distribution (which is derived from a specified deformation state) is given by the integral over the volume of true stresses τ_{ij} multiplied by strains ε_{ij}^r compatible with a unit displacement at the point and in the direction of the required force.

This theorem, because of Eq. (32), is restricted to a system with linearly elastic behavior. For a more extensive treatment and applications, see Hoff (1956), Oden (1967), Pestel and Leckie (1963), and Przemieniecki (1968).

Some authors refer to the above as the unit-dummy-displacement method (not a theorem). This method or theorem may be used very effectively for the calculation of stiffness properties of structural elements employed in matrix methods of structural analysis (Pestel, 1963 and Przemieniecki, 1968).

A.3.5 THE RAYLEIGH-RITZ METHOD

A variational formulation of a boundary-value problem is very useful for the approximate computation of the solution. One of the most widely used approximate methods is the Rayleigh-Ritz or simply Ritz method. This method was first employed by Lord Rayleigh (1945) in studies of vibrations and by Timoshenko (1961) in buckling problems. The method was refined and extended by Ritz (1909), and since then it has been applied to numerous problems in applied mechanics including deformation analyses, stability, and vibrations of complex systems. Although the method is based on the variational formulation of a specific problem, it may be considered as a derivative of the principle of the stationary value of the total potential when applied to elastic systems under quasistatic loads.

The basic ideas of the method are outlined by using as an example the deformation analysis of a general three-dimensional elastic system under the application of quasi-static loads (stable equilibrium). For a more rigorous treatment of the method from a mathematical (variational) point of view, refer to the texts of Courant and Hilbert (1953), Gelfand and Fomin (1963), and Kantorovich and Krylov (1958).

An elastic system consists of infinitely many material points; consequently, it has infinitely many degrees of freedom. By making certain assumptions about the nature of the deformations, we can reduce the elastic system to one with finite degrees of freedom. For instance, the deformation components u, v, and w may be represented by a finite series of kinematically admissible functions multiplied by undetermined constants

$$u(x, y, z) = \sum_{i=1}^{N} \overline{u}_i(x, y, z) = \sum_{i=1}^{N} a_i f_i(x, y, z)$$

$$v(x, y, z) = \sum_{i=1}^{N} \overline{v}_i(x, y, z) = \sum_{i=1}^{N} b_i g_i(x, y, z) \qquad (34)$$

$$w(x, y, z) = \sum_{i=1}^{N} \overline{w}_i(x, y, z) = \sum_{i=1}^{N} c_i h_i(x, y, z)$$

Note that, if we use small-deformation gradient theory, what is meant by kinematic admissibility is that the functions f_i, g_i, and h_i must be single-valued, continuous, differentiable, and must satisfy the kinematic boundary conditions. Then with Eqs. (34) the total potential, which is a functional, becomes a function of the 3-N undetermined constants a_i, b_i, and c_i or

$$U_T[u, v, w] = U_T(a_i, b_i, c_i) \qquad (35)$$

Now, since the functions f_i, g_i, and h_i are kinematically admissible, the virtual displacements can be taken as

$$\delta u = \delta a_i f_i, \quad \delta v = \delta b_i g_i, \quad \delta w = \delta c_i h_i \qquad (36)$$

and the variation in the total potential is given by

$$\delta U_T = \sum_{i=1}^{N} \left(\frac{\partial U_T}{\partial a_i} \delta a_i + \frac{\partial U_T}{\partial b_i} \delta b_i + \frac{\partial U_T}{\partial c_i} \delta c_i \right) \qquad (37)$$

Therefore, the elastic system is in equilibrium if

$$\frac{\partial U_T}{\partial a_i} = 0, \quad \frac{\partial U_T}{\partial b_i} = 0, \quad \frac{\partial U_T}{\partial c_i} = 0 \quad i = 1, 2, \ldots, N \qquad (38)$$

Equations (A-37) represent a system of $3N$ linearly independent algebraic equations in the $3N$ undetermined constants a_i, b_i, and c_i. The solution of this system yields the values for these constants, and substitution into Eqs. (34) leads to the approximate expressions for the deformation components u, v, and w. Once these are known, we can evaluate the strains from the kinematic relations, and consequently the stresses from the constitutive relations. Thus the analysis is complete because we know the state of deformation and the stress and strain at every material point.

A number of questions arise, as far as the method is concerned, regarding the choice of the functions f_i, g_i, and h_i and the accuracy of the solution (convergence). These questions are discussed rigorously and in detail in Berg (1962), Courant and Hilbert (1953), Gelfand and Fomin (1963), and Kantorovich and Krylov (1958). In summary, some of the important conclusions, in answer to these questions, are:

1. The Rayleigh-Ritz method is applicable to variational problems which satisfy the sufficiency conditions for a minimum (maximum) of a functional. The central idea is that of a minimizing (maximizing) sequence. A sequence $\overline{u}_1, \overline{u}_2, \ldots, \overline{u}_N$ (and consequently $\overline{v}_1, \overline{v}_2, \ldots, \overline{v}_N$, and $\overline{w}_1, \overline{w}_2, \ldots, \overline{w}_N$) of kinematically admissible functions is called a minimizing (maximizing) sequence if

$$U_T \left(\sum_{i=1}^{N} \bar{u}_i, \ \sum_{i=1}^{N} \bar{v}_i, \ \sum_{i=1}^{N} \bar{w}_i \right)$$

converges to the minimum (maximum) of $U_T \, [u, v, w]$ as N increases.

2. A minimizing (maximizing) sequence converges to a minimizing (maximizing) function, if constructed properly, for all one-variable problems (beams, columns) and all two- and three-variable problems (plates and shells) in which the order of the Euler differential equation is at least four.

3. A properly constructed minimizing (maximizing) sequence must be complete. This means that we select a set $\bar{u}_1, \bar{u}_2, \ldots, \bar{u}_N$ of admissible functions such that any admissible function, including the minimizing (maximizing) function, and its derivatives can be approximated arbitrarily closely by a suitable linear combination

$$a_1 \bar{u}_1 + a_2 \bar{u}_2 + \cdots + a_N \bar{u}_N \tag{39}$$

For example, in one-dimensional problems, $(L - x)x^{n+1} \, (n = 0, 1, 2, \ldots)$ is a complete set vanishing on the boundary of the interval $0 \leq x \leq L$. Similarly, $\sin(n\pi x/L) \, (n = 1, 2, \ldots)$ vanishes for the same case. Again in one-dimensional problems, $x^2(L - x)^{n+2} \, (n = 0, 1, 2, \ldots)$ is a complete set vanishing on the boundary, along with its first derivative, of the interval $0 \leq x \leq L$. Similarly, $\cos(n\pi x/L) - \cos[(n + 2)\pi x/L] \, (n = 0, 1, 2, \ldots)$ is a complete set for this latter case.

4. When the Rayleigh-Ritz method is used for beam, plate, and shell problems, it leads to fairly accurate expressions for the deformations. If one is interested in rotations, moments, and transverse shears, the accuracy decreases, respectively, because these quantities expressed in terms of deformations require higher derivatives of the deformations, and the derivatives are less accurate approximations than the functions themselves.

5. Also, because of the reasons given in item 4, equilibrium at a point is not satisfied exactly. Stresses computed through approximate deformations do not, in general, satisfy equilibrium equations.

The application of the Rayleigh-Ritz method to stability problems is presented in Chapter 5 of this text. As an application of the method for stable equilibrium, consider the beam shown in Fig. A.4. Using pure bending theory, the total potential is given by

$$U_T = \frac{EI}{2} \int_0^L (w'')^2 dx - Pw\left(\frac{L}{2}\right) \tag{40}$$

Since $\sin(m\pi x/L) \, (m = 1, 2, \ldots, N)$ satisfy the kinematic boundary conditions, let

$$w = \sum_{m=1}^{N} a_m \sin \frac{m\pi x}{L} \tag{41}$$

Substitution of Eq. (41) into Eq. (40) yields

$$U_T = \frac{EIL}{4} \sum_{m=1}^{N} a_m^2 \left(\frac{m\pi}{L}\right)^4 - P \sum_{m=1}^{N} a_m \sin \frac{m\pi}{2} \tag{42}$$

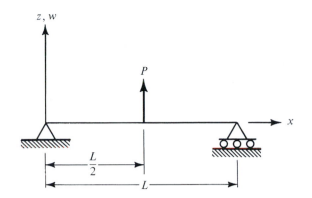

FIGURE A.4 Beam geometry.

By the principle of the minimum of the total potential,

$$\frac{EIL}{2}a_m\left(\frac{m\pi}{L}\right)^4 - P\sin\left(\frac{m\pi}{2}\right) = 0 \quad m = 1, 2, \ldots, N \tag{43}$$

Equations (43) represent a decoupled system of N linear algebraic equations in $a_m(m = 1, 2, \ldots, N)$. The solution is

$$a_m = \frac{2P}{LEI}\left(\frac{L}{m\pi}\right)^4\sin\frac{m\pi}{2}$$

and

$$w = \sum_{n=1}^{N}\frac{2P}{LEI}\left(\frac{L}{m\pi}\right)^4\sin\frac{m\pi}{2}\sin\frac{m\pi x}{L} \tag{44}$$

From Table A.1, we see that the convergence is very rapid. Although the convergence for the deformation is very rapid, this is not so for the moment and shear. For example, the shear, $V(x)$, is given by

$$V(x) = -EIw''' = \frac{2P}{\pi}\sum_{m=1}^{N}\frac{1}{m}\sin\frac{m\pi}{2}\cos\frac{m\pi x}{L} \tag{45}$$

In addition, because the minimizing sequence is orthogonal, N can be taken as infinity, in which case

TABLE A.1 Comparison with the exact solution

| Solution | $\frac{EI}{PL^3}w$ at | | | |
	L/8	L/4	3L/8	L/2
Exact	0.00765	0.01432	0.01904	0.02083
One-term	0.00806	0.01452	0.01891	0.02053
Two-term	0.00786	0.01432	0.01906	0.02081
Three-term	0.00765	0.01432	0.01904	0.02083

TABLE A.2   Closed form of $\sum\limits_{m=1}^{\infty} f(m)$

$f(m)$	m	Closed Form	
$1/m^2$	all	$\pi^2/6$	
$1/m^2$	odd	$\pi^2/8$	
$(-1)^{m+1}/m^2$	all	$\pi^2/12$	
$1/m^4$	all	$\pi^4/90$	
$1/m^4$	odd	$\pi^4/96$	
$1/(m^2 + a^2)$	all	$\frac{1}{2}[(\pi/a)\coth\pi a - 1/a^2]$	$a \neq 0$
$\sin mx/m$	all	$\pi - x/2$	$0 \leq x \leq 2\pi$
$\sin mx/m$	odd	$\pi/4$	$0 \leq x \leq \pi$
$\cos mx/m^2$	even	$x^2/4 - \pi x/4 + \pi^2/24$	$0 \leq x < \pi$
$\cos mx/m^2$	odd	$-\pi x/4 + \pi^2/8$	$0 \leq x < \pi$
$(-1)^{m+1}\sin mx/m^3$	all	$(x/12)(\pi^2 - x^2)$	$-\pi \leq x \leq \pi$
$\dfrac{2m\sin 2mx}{(2m-1)(2m+1)}$	all	$(\pi/4)\cos x$	$0 \leq x < \pi$
$\cos mx/4$	odd	$\pi^4/96 - (\pi^2/16)x^2 + (\pi/24)x^3$	$0 \leq x < \pi$
$\sin mx/m^5$	odd	$(\pi^4/96)x - (\pi^2/48)x^3 + (\pi/96)x^4$	$0 \leq x \leq \pi$

$$w(x) = \frac{2PL^3}{EI\pi^4} \sum_{m=1}^{\infty} \frac{1}{m^4} \sin\frac{m\pi}{2} \sin\frac{m\pi x}{L} \tag{46}$$

In such cases, if the series can be closed or if all of the terms are considered, the Rayleigh-Ritz method gives exact results. The series may be closed when evaluated at a point or in general (for any x) through different mathematical operations such as the calculus of residues (Phillips, 1954), integration of series (Carslaw, 1960 and Franklin, 1960), and others. (See Table A.2 for typical examples.)

REFERENCES

Argyris, J. H. and Kelsey, S. (1960). *Energy Theorems and Structural Analysis*. Butterworth Scientific Publications, London.

Berg, P. W. (1962). "Calculus of Variations." Ch. 16. *Handbook of Engineering Mechanics* (edited by W. Flügge), McGraw-Hill Book Co., New York.

Carslaw, H. S. (1930). *Introduction to the Theory of Fourier's Series and Integrals*. Dover Publications, Inc., New York.

Courant, R. and Hilbert, D. (1953). *Methods of Mathematical Physics*. Vol. 1. Interscience Publishers, New York.

Dym, C. L. and Shame, I. H. (1973). *Solid Mechanics: A Variational Approach*. McGraw-Hill Book Co., New York.

Franklin, P. (1960). "Basic Mathematical Formulas." in *Fundamental Formulas of Physics* (edited by D. H. Menzel), Dover Publications, Inc., New York.

Fung, Y. C. (1965). *Foundations of Solid Mechanics*. Prentice-Hall, Inc., Englewood Cliffs, N. J.

Gelfand, I. M. and Fomin, S. V. (1963). *Calculus of Variations*. (Translated from the Russian and edited by R. A. Silverman) Prentice-Hall, Inc. Englewood Cliffs, N. J.

Hoff, N. J. (19560. *The Analysis of Structures*. John Wiley & Sons, Inc., New York.

Kantorovich, L. V. and Krylov, V. I. (1958). *Approximate Methods of Higher Analysis*, 4th ed. (translated from the Russian by C. D. Benster), Interscience Publishers, New York.

Langhaar, H. L. (1962). *Energy Methods in Applied Mechanics*. John Wiley & Sons, Inc., New York.

Oden, J. T. (1967). *Mechanics of Elastic Structures*. McGraw-Hill Book Co., New York.

Pestel, E. C. and Leckie, F. A. (1963). *Matrix Methods in Elastomechanics*. McGraw-Hill Book Co., New York.

Phillips, E. G. (1954). *Functions of a Complex Variable with Applications*. Oliver and Boyd, London.

Przemieniecki, J. S. (1968). *Theory of Matrix Structural Analysis*. McGraw-Hill Book Co., New York.

Rayleigh, J. W. S. (1945). *Theory of Sound*. Dover Publications, Inc., New York.

Ritz, W. (1909). "Ueber eine neue Methode zur Lösung gewisser Variationsprobleme der mathematischen Physic." *J. Reine Angew. Math*. Vol. 135, pp. 1–61.

Timoshenko, S. P. and Gere, J. M., (1961). *Theory of Elastic Stability*. McGraw-Hill Book Co., New York.

AUTHOR INDEX

Subject Index